C0-CCM-107

The Social Origins of Democratic Socialism in Jamaica

NELSON W. KEITH

NOVELLA Z. KEITH

The Social Origins of Democratic Socialism in Jamaica

TEMPLE UNIVERSITY PRESS Philadelphia

Temple University Press, Philadelphia 19122

Published 1992
Printed in the United States of America

The paper used in this publication meets the minimum requirements of American National Standard for Information Sciences—Permanence of Paper for Printed Library Materials, ANSI Z 39.48-1984 ♾

Library of Congress Cataloging-in-Publication Data

Keith, Nelson W.
The social origins of democratic socialism in Jamaica / Nelson W. Keith and Novella Z. Keith.
p. cm.
Includes bibliographical references and index.
ISBN 0-87722-906-6 (cloth)
1. Jamaica—Economic conditions. 2. Jamaica—Economic policy. 3. Jamaica—Politics and government—1962– 4. Social classes—Jamaica. 5. Socialism—Jamaica. 6. Capitalism—Jamaica. I. Keith, Novella Zett. II. Title.
HC154.K45 1992
338.97292—dc20 91-21325

To Cordelia Evangeline (Eva) Keith
and Philip Timothy Keith (deceased)
and to Carla Zett
and Alessandro ("The Grape") Zett

It must be noted that the PNP was not against private enterprise but was concerned with monitoring the capitalist mode of production.

Richard L. Bernal

If the purpose of politics is the distribution of favours, one had better make sure that there are enough favours to go round or those who are not favoured will rise up in due course and smite those who are.

Michael Manley

C O N T E N T S

List of Tables and Figures *xi*

Abbreviations *xiii*

Preface *xv*

Introduction *xix*

P A R T **I**
Jamaica in Crisis

1 The Setting *3*

2 Democratic Socialism versus National Popularism *17*

P A R T **II**
The PNP and Radicalism

3 Impulses for Change: The Push from Below *53*

4 The Politics of the PNP: The Emergence of a Rational Practice *83*

P A R T **III**
The Progressive Accumulation of Capital

5 The Capitalists *113*

6 The Strategic Middle Class *139*

7 The Subordinate Classes *155*

8 The Rise of a New Politics *177*

PART **IV**
Charting National Popularism

9 National Popularism and the State *209*

10 Safeguarding Class Alliances: Resource Distribution under National Popularism *237*

11 State-Directed Change: Some Questions *267*

Notes *279*

Bibliography *293*

Index *311*

LIST OF TABLES AND FIGURES

Tables

1 Selected Indicators of the Importance of Bauxite to Jamaica, 1953–1972 *11*

2 Composition of the Gross Domestic Product (GDP) for Selected Years, 1950–1985 *41*

3 Composition of the National Income, 1930 *126*

4 Contribution of Economic Sectors to the Gross Domestic Product, 1950–1970 *133*

5 Diversification of Exports and Imports, 1970 and 1980 *212*

6 Real Growth Rates of Gross Domestic Product for Selected Industrial Sectors, 1975–1978 *227*

7 Foreign Exchange Drain, 1971–1973 *240*

Figures

1 Major Class Expressions Common to Democratic Socialism and National Popularism and Their Alternative Interpretations *32*

2 The Capitalists *42*

3 Intermediate Classes *44*

4 Subordinate Classes *46*

ABBREVIATIONS

AMC	Agriculture Marketing Corporation
API	Agency for Public Information
BITU	Bustamante Industrial Trade Union
CCPP	Central Committee of Primary Producers
CDF	Capital Development Fund (later incorporated into the Consolidated Development Fund)
CEO	Community Enterprise Organization
CMEA	Council for Mutual Economic Assistance
FF	Farmers' Federation
FP	Farmers' Party
IBA	International Bauxite Association
IBRD	International Bank for Reconstruction and Development
IDB	Inter-American Development Bank
IDT	Industrial Disputes Tribunal
IMF	International Monetary Fund
ITAC	Independent Trade Union Action Council
JADC	Jamaica Agricultural Development Corporation
JAS	Jamaica Agricultural Society
JEA	Jamaica Exporters' Association
JEF	Jamaica Employers' Federation
JIA	Jamaica Imperial Association
JIDC	Jamaica Industrial Development Corporation

JLP	Jamaica Labour Party
JMA	Jamaica Manufacturers' Association
JNEC	Jamaica National Exports Corporation
JPL	Jamaica Progressive League
JWTU	Jamaica Workers' and Tradesmen's Union
KFCA	Kingston Federation of Citizens' Association
KSAC	Kingston and St. Andrew Corporation
LRIDA	Labour Relations and Industrial Disputes Act
NEC	National Executive Council (PNP)
NFM	National Freedom Movement
NHT	National Housing Trust
NIS	National Insurance Scheme
NLP	National Labour Party
NRA	National Reform Association
NWU	National Workers' Union
OPEC	Organization of Petroleum Exporting Countries
PMILSLA	Poor Man's Improvement Land Settlement and Labour Association
PNP	People's National Party
PNPYO	People's National Party Youth Organization
PPP	People's Political Party (Garveyite, 1920s)
PPP	People's Progressive Party (Millard Johnson, 1962)
PSOJ	Private Sector Organisation of Jamaica
SEP	Special Employment Programme
STC	State Trading Company
SWCC	Sugar Workers' Cooperative Council
TUC	Trades Union Congress
UAWU	University and Allied Workers' Union
UTASP	Union of Technical, Administrative and Supervisory Personnel
WLL	Workers' Liberation League
WPJ	Workers' Party of Jamaica
YSL	Young Socialist League

PREFACE

▶ Sources of the Data

At the best of times, data gathering in Third World countries presents difficulties of all kinds. The present case was no exception. Often it was not possible to isolate certain statistics, which made us more dependent on qualitative data. In other cases, important statistics either were not gathered, were partial, or were slanted to meet partisan exigencies: statistics were found to be manipulated for specific ends.

Data gathering took place primarily during the course of five field trips to Jamaica between 1976 and 1980. These were supplemented by a greater number of trips of shorter duration that were used to gather data or conduct interviews.

The data for this study were obtained mainly from the following sources: (1) published information, including government documents, publications, and newspaper articles; (2) writings of non-governmental entities: the Institute of Social and Economic Research (ISER) at the University of the West Indies, Mona, Jamaica; the Private Sector Organization of Jamaica (PSOJ); the Jamaica Employers' Federation (JEF); the Jamaica Chamber of Commerce; the Jamaica Manufacturers' Association (JMA); the International Bauxite Association (IBA), and the Institute of Jamaica; (3) unpublished manuscripts (doctoral and master's theses, ongoing research papers, etc.); (4) secondary sources; (5) private collections (E.E.A. Campbell's at the Tom Redcam Library, Cross Roads, Kingston); and (6) in-depth interviews of two types. The first type of interview was with officials and influential members of key organizations, including political parties (the PNP, JLP, and WPJ); government ministries and government and quasi-government agencies such as the Jamaica Industrial Development Corporation, Jamaica Development Bank, JAMAL (a literacy program), Jamaica National Export Corporation, Small Businessmen's Association, Jamaica Bauxite Institute, National Planning Agency; labor unions (the NWU, BITU, UAWU, United Portworkers' Union); and organizations representing business interests (such as the Jamaica Manufacturers' Association, Jamaica Chamber of Commerce, Private Sector Organization of Jamaica, Jamaica Exporters' Association, and Jamaica Employers' Federation). The second type of interview was held with members of the middle class straddling the professions, government, and business. Our sampling was designed to identify the contours of an emerging fraction of the middle class that we have termed the bureaucratic/entrepreneurial fraction. In all, some 150 of these interviews were held, supplemented by extensive discussions with members of the class. Attempts were made to include members from all areas of the political, professional, and entrepreneurial spectrum. Anonymity was either specifically requested or implied as a condition of freely exchanging information.

We are all too aware of the dangers inherent in researching emerging phenomena, with few signposts to guide analysis. In order to enhance the reliability of information obtained from our interviewees, we have taken care to check and double check our data through additional interviews and/or the use of quantitative data.

▶ Acknowledgments

This book has benefited from the involvement of many people. We would especially like to acknowledge the critical input of Aggrey Brown of the University of the West Indies, Mona, Jamaica, and Jay Mandle of Colgate University, Hamilton, New York. The work is much the better for their suggestions, issue takings, and incisive editorial criticisms. Our next quota of thanks goes to Carolyn Adams of Temple University, Philadelphia, for her reading of the initial copy of the manuscript and her continuing encouragement. We wish also to thank Russ Keith, our man in Jamaica, who helped so vitally in arranging interviews with key people and in tracking down key documents. Special thanks are fully deserved by Doris Braendel of Temple University Press, whose diligent professionalism served in no small way to get us to this moment. And finally, we would like to acknowledge the financial support of the Rutgers University Research Council and Stockton State College Research and Professional

Development Grants, which partially defrayed expenses incurred on research in Jamaica.

Our parents, to whom this book is dedicated, know fully well our reasons for the bestowal of this honor.

INTRODUCTION

Jamaica is an island nation state situated ninety miles to the southwest of the southernmost tip of Cuba. A former British sugar colony, it has a population of about 2.3 million people. Like most other plantation economies, Jamaica has suffered from underdevelopment, which has led to discontent and unrest.

Perhaps the most famous demonstration of this discontent in recent times is the labor rebellion of 1938. This rebellion, which was driven by economic deprivation and political powerlessness, forced into place new economic and political forms. Democracy and political power improved, and there was a steady institutionalization of liberal democracy as the formal requirements for a competitive political system with free and fair elections and related institutions designed to maintain public liberties were set in place. After a process of gradual decolonization that began in 1944, Jamaica secured its independence on August 6, 1962.

A measure of economic diversification was introduced when traditional agricultural exports were joined by non-traditional exports, tourism, and, later, manufacturing. The discovery of bauxite in 1953 contributed mining and first-stage processing industry to the economy. Nonetheless, the economic models based on industrialization have not lived up to expectations, especially those of the popular classes. While the capitalists and certain sections of the middle class have prospered, the popular classes have been constantly waging a losing battle with depressed (and depressing) social conditions. By the late 1960s the economy was in the throes of a crisis, and the system could not be sustained by traditional formulas. The crisis required a restructuring of institutions and class relations. The needs were quantitative and qualitative.

In 1974, after the economy suffered another series of setbacks aggravated by a downturn in the world economy, a government led by Michael Manley and the People's National Party (PNP) declared Jamaica a democratic socialist state. By 1976 the project was in deep trouble; four years later this regime was replaced by a government led by the Jamaica Labour Party (JLP), a staunchly conservative, pro-capitalist party.

The argument presented in this book is that the socio-economic conditions on which the PNP fashioned its agenda of democratic socialism were really more fertile ground for an agenda for change that we term "national popularism." This is one of a number of transitional forms of peripheral capitalism that occurs when the essentially capitalist framework of the economy is being altered to make way for new patterns of capital accumulation. National popularism intervenes in the accumulation process by attempting to mediate unstable and potentially destructive class relations in the interest of an ongoing capitalist agenda. Because the levels of resource distribution required to maintain the alliance are not usually met, this form of capitalist mediation is predictably fragile at its very core. Its class alliances are brittle and unstable.

We face a difficulty in that the distinction between national popularism and democratic socialism is not immediately apparent. Democratic socialism also rests upon a foundation of brittle and unstable class relations. These result mainly from the tensions generated from burdening a weak productive base with non-productive policies. One example occurs when the manufacturer accuses the state of diverting too much revenue to 'socialist' or equity-related programs. Yet another difficulty is that both frameworks generate class expressions (ideological forms and political action) that are superficially alike.

A sensitive class analysis is the way out of this dilemma. This book undertakes this task and shows that national popularism more suitably explains the emergence and failure of the PNP's project. We do not discuss prescriptions that would have made the project work. Nor do we dispute the commonly held viewpoint—correctly held we are sure—that Michael Manley attempted, at crucial points, to install a democratic socialist state. The evidence is simply overwhelming, as the text will support.

At the same time, there is much to suggest that the regime deserted the common theoretical perspicacity that a project of this kind demands. It is evident that such erroneous interpretation was the result of at least two major factors. One was a misconception of what the state could singularly accomplish. It is, for example, dangerous to think that the state could easily impose socialism from above without the necessary popular class support. Many of the regime's policies appear to have been introduced in this way. The other factor involves the leadership of Michael Manley. A long tradition of paternalist politics, one that effectively sets political leadership apart from the electorate, has inspired this leader to acts bearing the stamp of the "Great Man with a Vision" single-handedly making history. One is simultaneously struck by a sense of admiration and incredulity as one is terrified by the obvious mistakes made when this David, armed by what he knows "is right," takes on Goliath.

To test the validity of these models we employ a limited number of variables. In combination, they provide support for our thesis that the PNP project could not go beyond the modification of the capitalist economy because the structural support for the systemic transformation envisioned by democratic socialism was simply not present. In the end, the basic capitalist foundations of the Jamaican economy remained intact. Consequently, our analysis casts doubt on the very popular view that democratic socialism was denied a fighting chance, and that it might have succeeded but for the outside interference of the United States. One can hardly deny the acts of destabilization initiated by the United States; they are sufficiently well known. The charge that the regime was a servant of the capitalists is at bottom also true. But the service was achieved in a far more complex manner than the straightforward "lackeyism" that some critics have directed at Manley and his regime. The regime was likewise guilty of waste and gross mismanagement, but these were not enough to topple it. And those who charge that the regime was about the business of promoting the welfare of the middle class and the subordinate classes are also correct, but that partial truth must be extracted from very complex processes that often appear to be contradictory.

What the book reveals are formative class relations promoted by new patterns of capital accumulation taking place within an economy that remained fundamentally capitalist. Most of the significant activity took place within the capitalist class, the middle class, and certain segments of the subordinate classes, including the Rastafarians. At the center of such activity were attempts to shift and modify class configurations to meet the requirements of the economy—generating and redistributing resources in keeping with the needs of the class structure.

Although a class analysis is at the center of our work, we part company with orthodoxy in several important ways. We have, for example, utilized the concept of interpellations in our analysis of the role of Rastafarianism in the consciousness of various classes (Chapter 7). The state is also given more latitude than many scholars of underdevelopment would accord it. However, we do not find persuasive the new formulations (by Kostas Vergopoulos, for example) that in an undeveloped economy the state is the most important social relation, more important than the mode of production itself. The state was not as powerful as Michael Manley thought, as demonstrated by the fact that dominating capitalist relations managed to hold it in check.

Our class analysis is historical, seeking to establish the patterns of class evolution and social relations that correspond to the realities of Jamaican society. We have interpreted events, actions, and opinions about policies and other issues to gain insight into the activities and expressions upon which the structure and meaning of social class depend. The book relies greatly on the identification of class trends and tendencies and probes their meaning by assessing their impact and influence on outcomes in social, economic, political, and ideological terms. These terms are viewed only in relation to processes bearing directly or causally on those inequalities creating class conflict.

The book is divided into four parts. Part I consists of two chapters: Chapter 1 deals with the main contours of the crisis in broad economic and political terms. Chapter 2 is the main theoretical chapter. It discusses the essential features of democratic socialism and national popularism and formulates criteria for testing the validity of these competing models. These analyses are followed by the charting of important changes in class relations and in the state, as the economy propelled itself through qualitatively different sub-periods from 1938 to 1980.

Part II consists of two historical chapters (Chapters 3 and 4) that discuss the organic links between the PNP and radicalism. Most of the major analyses of democratic socialism either gloss over or completely ignore the crucial implications of these links. We argue that a special relationship exists that can be traced to the formative years of the PNP's founding in 1938. That relationship is anchored to basic liberal democratic principles often expressed in militant and politically disruptive forms capable of being mistaken for socialism. The crisis we are studying, occurring in the late 1960s and early 1970s, reveals substantial continuities with this liberal democratic tradition. It is defined by ideological and political forms that are almost identical to those that gave the labor rebellion of 1938 its particular character. At the time of the crisis this special relationship between the PNP and radicalism was working itself out in a fashion not unlike the political drama of the labor rebellion period. Essentially, the middle and popular classes employed militancy, political disruption, intimidation, and the like—all against a liberal democratic backdrop—to dramatize their disappointment in the PNP's performance.

In a survey conducted in December 1981, only a year after the defeat of the Manley regime, we find that given a choice between indicators supporting socialism and those aligned to liberal democracy (for instance, government control of the economy versus reforms such as a minimum wage), respondents appeared to favor those policies introduced by the Manley regime that reflected liberal democracy (Stone, 1986, p. 69). Also, influential middle class supporters of the PNP, who were expected to be in the vanguard of democratic socialism, quickly showed themselves to have supported the regime mainly to enhance their fortunes in a liberal democratic setting. These chapters provide the historical bases for the existence of the ongoing relationship and serve to integrate it within the democratic socialist debate.

Part III contains four chapters. The first three trace the accumulation process with regard to the social classes most crucial to the crisis. We dwell in particular on certain class fractions whose practices set the stage for the eventual entry of democratic socialism. Fractions are intraclass divisions based on economic practice (for instance, agricultural, commercial, and industrial capitalists). These divergent economic practices in turn spawn distinctive politico-economic interests that the fractions work to project onto the state. In stable circumstances, intracapitalist rivalry notwithstanding, one fraction of the capitalist class is dominant and able to organize state action around its interests. Crises, on the other hand, are frequently marked by shifts in dominance from one fraction to another—shifts that may spell the weakness and occasional displacement of the class from power, as may occur through socialist revolutions. Far from revealing weaknesses that would allow the emergence of socialism, our analyses indicate classes caught in the swell of capitalist relations in modification. Chapter 5 analyzes the capitalists, Chapter 6 looks closely at the movements of strategic elements of the middle class (the bureaucratic/entrepreneurial fraction), and Chapter 7 analyzes the changing relations within the subordinate classes. More politically oriented, Chapter 8 analyzes the new politics created by the twists, shifts, and turns in class relations occurring on the eve of the crisis. We find that although the activities of the popular classes were brimming with militancy and indignant energy, their politics and ideology had little to do with socialism. Indeed there did not exist a critical mass of socialist consciousness or activity to institute the "socialist" aspect of the democratic socialist agenda. Capitalist forces and relations were preemptive. For instance, the capitalists were caught in intraclass rivalries but not of the sort that foreshadows the decay of the socio-economic order—a condition precedent of socialism, of even the hybrid version that the regime outlined. The PNP's principal theoretical document, *Principles and Objectives*, speaks of the calculated destruction of capitalism and infers the taming of the capitalist class.

The last section, Part IV, charts the path of national popularism through the policies, activities, and politics of the PNP regime between 1974 and 1980. In Chapter 9 we analyze the state's project with regard to the basic requirements needed by the state to carry out the project: the politico-economic principles on which the agenda depended and the capability and level of commitment within state apparatuses whose cooperation (or, at least, neutrality) was required for implementation. In Chapter 10, we analyze the regime's policies in the context of the class mediations of the state. Our contention is that the main purpose of the policies was to shore up capitalism. Chapter 11 is a brief conclusion.

Jamaica in Crisis

P A R T

I

The Setting

C H A P T E R

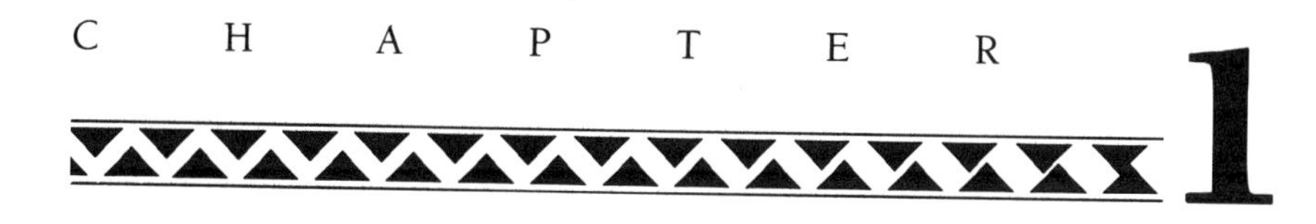

In March 1972 the People's National Party (PNP) defeated the Jamaica Labour Party (JLP) in national elections. This rotation, which had occurred regularly after each party had completed two terms in office, assumed a new twist two years into the PNP's new term, when, in 1974, the party abruptly announced the introduction of democratic socialism. This doctrine was never free from profound ambiguities. Its most clearly understood meaning, however, was that it would combine capitalist and socialist socio-economic principles and that parliamentary politics would continue to play a direct role in the ongoing political process. At the outset, these policies received appreciable support from all social classes. The new regime also benefited from fractures and dislocations in the international economic system that weakened the metropolitan economies in their relations with the Third World.

Democratic socialism emerged in the teeth of crisis. In 1972 unemployment stood at 23.6 percent. A world economy in recession and the dramatic rise in the price of oil dictated by OPEC deepened local distress. Between 1972 and 1974, imports of fruits, vegetables, and other necessities increased by nearly 71 percent, while the energy bill increased from $50 million to $180 million. Foreign investment, on which the economy depended excessively, began to decline (see Palmer, 1979). At the same time, inflation rates rose steeply to more than double the world average. On the whole, an unusually heavy responsibility was placed on the state, which, expanding its role in demand management beginning with 1972, created most of the new employment.

▶ Background of the Crisis

These economic and socio-political events were manifestations of a crisis rooted in the political economy of Jamaica. During the 1950s and the 1960s Jamaica experienced a profound transformation, as a fundamentally agricultural, labor surplus economy was gradually replaced by modern industries, bauxite mining and exports, tourism, commerce, and services. The change was government-sponsored and designed to meet the needs of a rapidly increasing labor force. The governmental blueprint was a program of "industrialization by invitation," which structured inducements to foreign capital to channel productive ventures in directions likely to meet local needs. Labor-intensive ventures were to be an important component of the program, which featured incentives for both import substitution and export promotion. American and Canadian aluminum companies entered the economy to develop the capital-intensive bauxite industry (Girvan, 1971). Many of the most profitable investments also came under the control of foreign capital (Jefferson, 1972).

The program had uneven results. On the positive side, the economy grew at an annual average rate of 6.7 percent between 1950 and 1968. There was noticeable improvement in the standard of living, resulting in part from upward movement in a changing class structure. Nonetheless, self-sustaining growth and socio-economic development proved more elusive than Gross Domestic Product (GDP) measurements might suggest. In fact, by the late 1960s it was evident that Jamaica had reached an impasse and needed new developmental formulas. Rather than develop our own analysis of the root causes of the crisis, we focus below on local interpretations, which have a bearing on the policies put in place to counteract its impact.

The Anger toward Foreign Capital

The disappointing economic performance was enough to stoke a growing anger with foreign capital especially toward the late 1960s. The developmental formulas applied by Jamaican planners were of metropolitan provenance. The somewhat simplistic argument was made that the crisis was due to faulty conceptualization and advice from sources such as the World Bank, rather than to local ineptitude or mismanagement. According to foreign advisers, if specific developmental formulas were followed and certain economic targets were achieved, the economy would "take off" into growth and development.[1] While the targets were amply met, socio-economic development was elusive. The statistical story is told by a comparison of the World Bank indices of development and Jamaica's performance from 1950 to 1968 (Adams, 1969, p. 5; Girvan, 1971, p. 237; and Jefferson, 1972, pp. 53, 65, 185, 188):

World Bank Targets	Jamaica's Performance
Growth of GDP: 5 percent per annum	1950–68: 6.7 percent per annum
Gross Domestic Savings Rate: 10 percent of GDP	1950–67: 21 percent of GDP
Export Earnings as percent of Total Income: 10 percent	1950–68: 34.5 percent of GDP, 42.5 percent of Total National Income
Growth of Exports (value): 4 percent per annum	1953–68: 8.7 percent per annum (goods); 12.8 percent per annum (services)
Capital/Output Ratio: 3:1	1950–68: 3.1:1

As most local economists, saw it, the World Bank recipes had merely made the local economy dependent. Between 1959 and 1969, the income of foreign investors far exceeded the economy's receipts in capital inflows, thereby making Jamaica a net exporter of capital (Jefferson, 1972, p. 285). Heavy doses of dependency theory infused critiques that revolved around three main issues of foreign derivation:[2] leakages of valuable foreign exchange earned by the new industries, which depended heavily on imports; the creation of economic enclaves, integrated vertically into the networks of multinational corporations rather than into a national economy; and the siphoning off of profits by these same transnational combines, which controlled prices, markets, and information. For good measure, local factors were also blamed such as the excessive profiteering of local merchants, whose mark-ups were legendary, and the excessive consumerism of a growing middle class.

In tourism, for instance, one of the leading export industries during the period, leakage may well have approached some 50 percent of the industry's gross intake (Jefferson, 1972, pp. 175–179). For manufacturing as a whole, the import content was nearly always upwards of 40 percent (Ayub, 1981, pp. 60–61). This partly explained the inability of industrialization to create employment for an expanding labor force. In fact, by the end of the 1960s, unemployment was approaching the levels of the 1940s—that is, the period immediately following the 1938 labor rebellion that prompted these new economic policies. The growing volume of foreign capital merely entrenched dependence. The import orientation, one of its major forms, was linked to continued foreign investment and foreign loans to maintain a favorable balance in the foreign reserves account. Nonetheless, the trade balance remained uniformly negative, except for the years 1962 and 1963; by 1970 the trade deficit was $153 million, fully 35 percent of the value of imports. In 1972, visible exports rose by $29.1 million, but imports of consumer goods accounted for 80 percent of this figure. ("$" always refers to Jamaican dollars. Other currency will be specified.) The debt service connected with this disappointing economic performance was dismal (*Economic and Social Survey, Jamaica,* 1972). It is this lopsided economic "development" that angered the PNP under Michael Manley and helped to promote economic nationalism and, later, the democratic socialist project.

But this pattern of economic activity did not benefit only the foreign investors. Among the capitalists, the export agriculturalists were eclipsed by the merchant fraction of the class, which profited immensely from distribution. Members were not above excessive profiteering (Jefferson, 1972) and posed a real threat to progressive socio-economic development, as an influential development expert of the period warned the Jamaican government (Balogh, 1970, pp. 311, 315). Other beneficiaries whose stock was promoted at the expense of the subordinate classes were from the middle class. Their advance was directly in phase with the consumerism created by the government's development policies. In numerical terms, the data bore out expected discrepancies. In 1959, a mere 10 percent of the population accounted for almost 50 percent of the national income; in 1971, the poorest 60 percent of the population received 19 percent of the national income while 31 percent of it went to the wealthiest 5 percent. These discrepancies created tensions between and within social classes.

Dependency Theory and the Search for Solutions

It would be correct to assume that these new perspectives were reinforced by a new wave of theory and practice. In the 1960s and 1970s the PNP began to embrace the increasingly influential dependency theory, abandoning neo-classical economics, which was increasingly suspect. It was repeatedly advanced that Jamaica's socio-economic salvation lay in breaking the bonds of dependence on the metropolitan powers. Self-reliance and, when necessary, "negotiated" dependence were the preferred options. Anger toward foreign capital, supported by the new theoretical strains, fueled a call for economic nationalism, which the PNP embraced in the late 1960s. In the words of Michael Manley, who assumed leadership of the PNP in 1968 (1970, p. 105),

> Jamaica has fallen into the same trap as many other developing countries by thinking that the indiscriminate granting of tax incentives to foreign capital—regardless of the contribution which the particular capital can make to development, or of the posture of that capital in the society—will necessarily contribute to progress.

The major assault on foreign capital was led by a politically active segment of young intellectuals, many of whom assisted in giving shape to the radicalism in the air. Their contributions ranged from economic criticism to recommendations on a wide range of social, political, and economic issues.

The New World Group spearheaded the main attack upon the U.S.- and Canadian-owned bauxite industry. A political action group whose leadership was heavily represented by graduates from the University of the West Indies at Mona, the New World Group steered common debate along radical paths. Indeed, the antecedents for reorganizing the bauxite industry—which the PNP began doing in 1974—can be traced directly to Norman Girvan, one of its prominent members. On the question of bauxite revenues, for example, he showed that while the flow of taxes and royalties had been constant, these were at the mercy of the industry's oligopolistic structure, that is, its near total control of the market. The manner in which revenues were procured was devised in 1953, revised in 1957 and again in 1966, but the modifications were still based on an inherently exploitative master formula. As Girvan (1971, pp. 63–64) explained it, such agreements would favor the government if costs were rising or the value of the product was falling. However, when the value of the product rose relative to costs, the government lost a significant portion of potential revenue. He estimated that these losses amounted to BWI$25 million between 1957 and 1962.

These disclosures contributed a great deal to the case for Jamaicanizing the industry. Additionally, the "saga of Kitimat" (we have so labeled the refrain because of its frequent use by Jamaican technocrats) added its own urgency. The saga is tersely summarized by Isaiah A. Litvak and Christopher J. Maule (1975, p. 46):

> In 1962, Alcan operations in Canada were dependent upon a reliable supply of raw material from Guyana, with Guyanese bauxite

> accounting for about 50 per cent of Alcan's aluminum production in Canada; its main smelter in Arvida, Quebec, was almost totally dependent upon Guyanese bauxite, while the smelter at Kitimat, British Columbia, received Guyanese bauxite during the Canadian winter. In Canada, the company supported two industrial communities . . . and was the major industry in four Quebec towns. Alcan employed 17,000 in Canada, which accounted for $95 million in wages, and paid government taxes amounting to $14.3 million. Exporting products valued over $227 million, Alcan ranked as Canada's sixth largest exporter. [Note: These are Canadian dollars.]

The case for economic nationalism was made even more compelling by disclosures that a ton of bauxite processed into semi-fabricated aluminum increased in value from between BWI$14 and BWI$28 to BWI$350 (Jefferson and Girvan, 1971, p. 220).

Norman Girvan's work is one example of many other efforts influenced by dependency theory. Economists Havelock Brewster and Clive Thomas (1967) brought some of these insights to the subject of regional economic development. Part of this plan focused on the ailing sugar industry—a plan that not only challenged the economics of the Sugar Manufacturers' Association (SMA) but pitted the old ruling class (the planters) against the new breed of university-educated professionals (meritocracy). In debate, Sir Robert Kirkwood, overlord of the SMA and manager of sugar estates linked to the powerful British corporation Tate and Lyle, crossed swords with these educated sons of ex-slaves; in the prickly exchange, the young economists did not display any inclination for genuflecting—a sure sign that Quashie's new day was breaking![3] Other dependency-influenced works focused on the banana industry (Beckford, 1967) and politics (Millette, 1968). This is by no means a complete list.

These new perspectives were not restricted to rarefied intellectual circles. The mass of the people was heavily courted. In this respect, *Abeng*, a weekly anti-establishment tabloid founded by many of these intellectuals, holds an important place. While it was erratic and intemperate at times, the tabloid addressed a catholic range of issues in a deliberate attempt at political education. Bridging the communication gap between the urban and the rural areas was a top priority. So, too, was its policy to provide the general reading public, particularly the subordinate classes, a radical alternative to the pro-establishment, conservative newspaper, the *Daily Gleaner*. *Abeng* developed a keen sense for identifying and articulating the prominent issues of the moment: race, bauxite, tourism, education, cultural issues, and so on. At nearly all times, it embraced an ideological point of view obviously informed by the radicalism then blowing in the wind: the Black Power Movement, Marxism-Leninism, and dependency theory. It is small wonder that it was put out of commission, yet not before the tabloid served to agitate the backwaters of public opinion.[4]

The New World Group, for its part, engaged in an extensive research and information dissemination campaign. The now defunct *New World Quarterly* testified to this fact. But its members were also involved in active grass roots political organizing. In the tradition of action team politics, these activists went into the field to "parley with the people." Their presence was marked in the rural parishes where Group members and proselytes would hold forums. In order of need, there would be the words of encouragement and solidarity for the die-hard supporter and political fieldworker; gentle but insistent cajoling of the waverer; and one-on-one sessions by the star performer (usually from Kingston) with the recalcitrant opposition. This brand of politics, which we term "politics by informed messenger," first made its appearance during this period. In the past, politicians usually left the "education" of their constituencies to caretakers and the *Daily Gleaner*; the main role of the traditional caretaker, in turn, was keeping the politician informed. Likewise, the educated had nearly always treated politics as a realm of utter defilement.

▶ Third Worldism and the Global Crisis

It should be evident that Jamaicans were growing increasingly discontented with the prevailing socio-economic formulas—a fact which fueled support for democratic socialism. At the same time, developments abroad assisted in providing the climate for the policies of the PNP. Fractures and dislocations within world capitalism were accompanied by a corresponding rise of a Third World–oriented ideology and practice ("Third Worldism") that embraced the broadly held Third World notion that the traditional control by, and dependence on, metropolitan countries must be replaced by a renegotiated posture of less dependence and greater self-reliance (Chaliand, 1977; pp. 17–24). The epochal weakness of world capitalism was linked to the challenge posed by Western Europe and Japan to U.S. leadership, the monetary instability that eventually would lead to the abandonment of the gold standard, and the resulting intra-capitalist rivalry. Into this breach stepped the Organization of Petroleum Exporting Countries (OPEC), which led the way for a push toward the cartelization of strategic raw materials. OPEC's quadrupling of the price of crude oil in 1973 set the stage for Jamaica's own initiatives in the mining industry (confirmed by interview with Michael Manley)[5] and greatly boosted the confidence and ideology of the Third World bloc.

▶ The PNP Policies of the 1970s

This coexistence of domestic economic pressure, a favorable conjuncture in world capitalist relations, and the consolidation of Third Worldism provides the orbit for the PNP policies of the 1970s. With the announcement of democratic socialism, the PNP unveiled a plan for national transformation. On the economic front, the plan called for direct government intervention in the economy to open up opportunities in the industrial sector and buttress the food-producing capacities of agriculture. At its foundations were drastic changes in the way Jamaica managed bauxite, potentially one of its chief resources.

The Bauxite Policies

The PNP's election manifesto of 1972 foreshadowed the policies of 1974: the party would create the National Bauxite Commission to assist in reformulating development strategy. Its task would be to devise ways to utilize bauxite as the engine for national development.

Jamaica figured prominently in the production of bauxite and alumina (the product of first-stage processing), which provide a crucial strategic input of the large industrial economies. At the end of World War II, average annual world output of bauxite stood at 755,000 tons; by 1973, production had increased tenfold, reaching 75 million tons (Manley, 1987, chap. 1). Of the 87 percent of U.S. bauxite requirements obtained from foreign sources in 1974, 60 percent was supplied by Jamaica (*International Bauxite Association Digest,* March 1977, appendix).

Since the post-war period, the bauxite industry has occupied a prominent role in Jamaica's economy. For the period of 1953–1973, it accounted for 10.5 percent of the local GDP, 29 percent of the total tax on corporate profits, 33 percent of the national income, and 65 percent of the value of all exports (Reid, 1977, p. 18). The foreign exchange receipts from bauxite were crucial to the Jamaican economy, which was heavily dependent on imports. In 1976 these earnings represented 46 percent of the nation's total intake (*Economic and Social Survey, Jamaica,* 1976). Table 1 highlights the importance of the industry to the Jamaican economy on the eve of democratic socialism.

TABLE 1
Selected Indicators of the Importance of Bauxite to Jamaica, 1953–1972

Years	Foreign investment as % of total investment (averages)	Direct foreign investment in industry (millions of $)	Industry exports as % of all exports (averages)
1950–57	33.0*	101.0	24.7*
1958–60	24.0	9.0	47.0
1961–65	12.2	56.0	46.4
1966–69	28.5	380.0	49.9
1970–72	38.7	117.0	64.0
	Total investment	663.0	

* For 1953–57.

Source: National Planning Agency, Kingston, Government of Jamaica.

The bauxite policies were multifaceted, aiming to wrest control from the multinationals (or at least gain a significant share of it) and make of the industry the engine of national growth. At the center were a producers' association and government-sponsored joint ventures involving the multinationals themselves as well as regional partners.

The International Bauxite Association (IBA)

The new bauxite policies started auspiciously in March 1974 with the formation of the International Bauxite Association (IBA). The headquarters were located in Kingston, a testimony to the leading role Jamaica played in creating the new organization. Australia, Guinea, Guyana, Jamaica, Sierra Leone, Surinam, and Yugoslavia—the original members—accounted for 65 percent of all the bauxite mined in 1973.

IBA's charge was to redress the inequitable balance between the producers and the multinationals—to be an antidote to dependence. The bauxite-producing nations as a whole were plagued by oligopolistic practices, in particular, monopoly pricing and control over distribution. The IBA would be a cohesive mechanism ensuring uniform prices and the necessary solidarity in negotiation. With access to valuable intra-industry information, its members would now get at the source of under- and over-invoicing, manipulated interest charges, and spurious royalties for which the industry was known. Serious thought was also devoted to a vertically integrated industry: technical exchanges took place between the IBA and

OPEC; the erection of processing plants and smelters and the ownership of a shipping fleet were explored.

Joint Ventures and Country/Country Agreements[6]

Joint ventures and country/country agreements were logical complements to the IBA's function. To begin, Jamaica established a new relationship with the mining companies.

Under the terms of the new agreements, Jamaica would participate in bauxite mining and alumina operations, aiming for 51 percent of ownership; bauxite and alumina produced by these operations would be available to build a Jamaican vertically integrated industry. The Jamaican government would purchase all lands held by the mining companies, with provisions allowing for future mining rights. Finally, a production levy—initially 7.5 percent of the price of aluminum ingot—was imposed on the portion of operations that remained under foreign ownership. The levy would mean an eightfold increase in government revenue from bauxite—from $25 million to $200 million.

Country/country agreements involving Jamaica, Mexico, Venezuela, Trinidad and Tobago, and Guyana were the second prong of the restructuring. In principle, these agreements were a storybook case of regional cooperation, enabling countries to make maximum use of their natural resources by participating in a regionally integrated aluminum industry. Trinidad and Tobago, Mexico, Venezuela and Guyana, with ample energy reserves, would provide the energy needed for smelting and aluminum fabrication (Guyana would also provide alumina); Jamaica would supply the bauxite and alumina obtained in her contracts with the mining companies. Mexico and Venezuela were also to undertake other ventures in Jamaica, including a cement plant to be erected by Venezuela.

Industrial and Agricultural Policies

Industrial and agricultural policies were directly dependent on the success of the bauxite policies. The bauxite levies, imposed on May 15, 1974, were designed for developmental purposes. Section 12 of the Bauxite (Production Levy) Act of 1974 established a Capital Development Fund that would provide for government investment in bauxite and alumina production, finance projects geared to internal industrialization, finance the transformation of subsistence agriculture into larger and more productive entities including sugar cooperatives and collective farming, and provide funds for a limited number of non-recurrent undertakings (the Urban Development Corporation, for example).

To further bolster its "socialist" image, the state used the Capital Development Fund to gain substantial ownership of certain public utilities. Within days of the first bauxite levy, the government bought the Jamaica Omnibus Company; by the end of 1975, it owned a controlling interest in the Jamaica Public Service Company and the Jamaica Telephone Company.

The *Green Paper on Industrial Development Programme* (GPID) outlined industrial policies that were in line with common Third World aims to improve the bloc's market share in manufactures from 7 percent to 25 percent by the year 2000 (UNIDO, 1975). This blueprint for industrialization in light manufactures projected an investment pool of $520 million to be deployed from bauxite sources. Of this sum, approximately $260 million was earmarked for state investment, either wholly or in partnership with foreign capital (GPID, n.d., p. 18). The state would be involved in eight projects, five of them joint ventures (GPID, n.d., pp. 3–4). Light manufacturing activities such as toolmaking were favored, as were ventures in agribusiness. The stress on food production served the two processes of developing sectoral linkages to stem the outflow of foreign exchange for food purchases and to increase employment (GPID, n.d., pp. 66–75). Overall, it was estimated that the program would provide at least 40,000 new jobs in the early 1980s. Doubtless in keeping with the regime's socialist posture, the new ventures would be decentralized, both administratively and geographically.

At the center of the agricultural policies was Project Land Lease, a land reform program that involved lands owned, bought, or leased by the government. These lands were to be distributed in three phases based on differing criteria as to size of plot, categories of farmers, and possession. The state itself operated model farms as demonstration sites that it was hoped would attract potential farmers from the urban areas. Tanzania's communitarian model was influential here; Michael Manley is a close friend of Julius Nyerere, then the Tanzanian president, who visited Jamaica in 1974 during the course of these developments.

These policies leave little doubt that a profound transformation was envisioned, an impression that is confirmed when we contemplate the many other domestic policies unveiled at the same time. We have listed them chronologically. A more complete list can be obtained from Norman Girvan, Richard Bernal, and W. Hughes (1980), from which this version is adapted.

PNP Initiatives, 1972–1977

1972
- Literacy program (JAMAL)
- Public housing
- Youth training
- Community health services
- Special employment program
- Lowering of voting age to 18

1973
- Free education (secondary and university)
- Free uniforms (primary school)
- Food subsidies
- Improved cultural training facilities
- National Youth Service

1974
- National minimum wage
- Loans for farm development
- Increase in National Insurance Scheme pensions
- New facilities for small business
- Sugar cooperatives
- Food outlets under Agricultural Marketing Corporation

1975
- Worker participation
- Compulsory recognition of trade unions

1976
- National Housing Trust

1977
- State Trading Company (STC)
- National Commercial Bank
- Small Business Development Company

Foreign Policy

The PNP had clearly taken a pro-Third World stance since its progressive adoption of economic nationalism in the mid-1960s (see Chapter 8). Much had been occurring on the international scene to support such a stance. Since the early 1970s, Third World members of the IMF's Committee of Twenty had become highly visible and effective.[7] They were active members of the Development Committee, a joint committee of the IMF and the World Bank, on which they represented the Third World's views on monetary matters. They were also effective members of the International Development Association, which provided "soft" loans for the World Bank (Helleiner, 1974, p. 347; Howe, 1975, p. 115). Indeed they were generally seen to be "very active and [to] take their job very, very seriously" (Schweitzer, 1976, p. 216). Of course, OPEC's hiking the price of crude in 1973 formed the pièce de resistance. Third Worldism was on the upswing.

The PNP's Third World stance was struck by Michael Manley, who had for some time been emerging as a prominent Third World leader. He came to local politics with impressive credentials, being the younger son of Norman Washington Manley, a national hero, the principal founder of the PNP in 1938, and perhaps the chief architect of Jamaica's entry into modern politics. A foreign policy rooted in political non-alignment and directed toward Third World solidarity really began with Michael Manley's trip to Africa in 1969. In his subsequent address to the PNP conference, he dwelled on the importance of that continent for Third World development and solidarity (cited in Hearne, 1976, p. 180):

> I intend to lay the foundations from now for this idea of co-operation among the under-developed nations because I believe if we in the under-developed parts of the world had only the diplomatic vision to see that we could forge links of steel around the world that would give us for the first time an opportunity to throw the begging pans away.

In 1970, writing in the *Foreign Affairs Quarterly* (Manley, 1970), he exhorted the Third World to develop what he termed "a common economic diplomacy." In 1973, he seized the opportunity at the Non-Aligned Conference in Algiers to stress solidarity in practice as well as in ideology:

> We suggest . . . that the capacity for action, depends upon the construction of a methodology appropriate to each situation and its needs. We are not certain that the Movement is giving enough attention to the search for methods. The danger which can threaten everything that is promised by the Non-Aligned Movement, is that it should lose its credibility by the over-articulation of its promise, while under-delivering in its purpose. (Hearne, 1976, pp. 202–203)

That posture translated concretely into the opening of full diplomatic relations with Cuba and the People's Republic of China. Indeed Michael Manley's growing ties with Cuba formed one of the main bases for his difficulties with the United States. In 1975, for example, he visited Cuba and signed accords that ensured Cuba's involvement in the regime's developmental agenda. On the economic side, it embraced the New International Economic Order (NIEO), a blueprint for international development with stress on greater autonomy for the Third World, the local ramifications of which can be clearly seen in the PNP's development policies. Jamaica's involvement encompassed not only the rhetorical and ideological support of PNP government ministers, but the presentation of technical papers as well.[8]

Yet, in spite of this vast network of seemingly well laid plans, the PNP's transformative agenda failed. Democratic socialism enjoyed uneven success until mid-1976, when it went into irreversible decline. The sources of its failure were many: managerial and political ineptitude, destabilization by the United States, massive internal contradictions to the theory and practice of democratic socialism, and the downturn in the international economy. Most importantly, however, structural conditions, rooted primarily in the dynamics of local classes and politics, did not favor democratic socialism but rather followed the contours of an alternative analytical framework—national popularism.

Democratic Socialism versus National Popularism

C H A P T E R **2**

In 1974, when the PNP announced that democratic socialism would henceforth provide the guiding principles for Jamaican society and economy, both the concept and its implications were only vaguely understood. In fact, the regime's formulations went through three distinct definitional phases, in an evolutionary process that had little to do with a quest for clarity.

The very elusive and ambiguous nature of democratic socialism as concept and practice makes any attempt to define it problematic. To avoid that attempt, however, would be to risk teleology and fail to accord proper weight to the inevitable nuances and situation-specific aspects of democratic socialism. We do not, naturally, accept the regime's definitions as given. The political arena, perhaps of necessity, abounds with vague and self-serving pronouncements—a weakness to which the PNP regime was not immune. Our tack, then, is to give due weight to the regime's formulations, while at the same time inserting them within the context of broadly accepted theoretical principles.

▶ Democratic Socialism

The PNP Defines Its Agenda

The regime's initial formulations of democratic socialism are captured by Michael Manley's statements that "socialism is love" and "the philosophy that best gives expression to the Christian ideal of equality of all God's children" (PNP, 1974: 17). Both definitions were aimed at specific targets:

the reference to love was directed at Rastafarianism, which was then enjoying a dramatic rise in political importance (Chapter 8). The other definition assured the electorate of the PNP's Christian groundings. Since its adoption of socialism in 1940, the PNP was always suspected of being anti-Christian. In Jamaica then and now, communism is considered anti-Christian and Jamaicans remain a very religious people.

A third agenda is served by definitions of this sort. Brotherhood, equality, voluntarism (that is, unwarranted belief in the power of the leader's vision and exhortations) denial (or reluctant admission) of the existence of social classes—all these themes that pour forth from the writings and utterances of the regime's spokespersons—are mere affirmations of the foundation principles of the party. These principles we have termed "Manleyism," to dramatize the philosophical basis hammered out largely by Norman Manley between 1938 and 1952 (Chapters 3 and 4).

Manleyism embodies notions that were modified to meet the new conditions facing the younger Manley. These notions were translated into strategies, cast largely in the image of the deep-seated desires and ideologies of the Jamaican people, which, however, served to restrain potentially disruptive radical impulses. In the turbulent days of the late 1930s, Manleyism kept at bay the volatile remnants of Bedwardism, Garveyism, the ascendant thrusts of Rastafarianism, and Marxism-Leninism. The process of containment that the older Manley employed involved using the intrinsic democratic and capitalist impulses that underlay these displays of radicalism to mediate the class contradictions of those times. The younger Manley engaged in a project not too dissimilar in crucial respects.

While the PNP's definition of democratic socialism is vague, eclectic, and often determined in an ad hoc and ex post facto manner, its main principles shunt between African Socialism and Marxism-Leninism. The former is separated from Marxism-Leninism by the class struggle. Michael Manley found the notion of the natural harmony between Africans agreeable. As Issa G. Shivji (1976, p. 4) states it, advocates posit a peculiarly African principle of "communalism." There is no parallel of class divisions and class conflict; "the class problem in Africa is, therefore, one of prevention" (Sigmund, 1967). In Nyerere's words, traditional African society had "hardly any room [for the kind of] parasitism" that creates such divisions.[1] In fact, Manley (1974, p. 26) cites "the Nyerere model" as one without parallel in "marry[ing] the ancient and eternal ideal of democracy in the sense of government by the people with the 'natural tendency' of the people." If the natural tendency of Africans manifests itself in communalism and one-party states, that of Jamaicans comes in the form of the "ability to accept that the vote is the natural end product of a dispute and that a majority decision is conclusive of an issue" (Manley, 1974, p. 30). According to Michael Manley, Jamaicans are formally democratic by nature.

Do social classes play a part in democratic socialism? At the earliest stage of theory development (between 1969 and 1974), Manley speaks only of "hierarchies which reflect economic necessity." Jamaica must avoid the development of classes, he warns, by fostering a "family model in which there are no classes of children, but only children of different skills enjoying different rewards, but with equal claims upon parental love and concern" (Manley, 1974, p. 41). So, as in the Kenyan case, the class problem is "largely one of prevention."

Those who know Jamaica as a former slave colony transformed into a plantation economy and ushered into the modern period by a labor rebellion in 1938, should be taken aback by this characterization. It turns out that the PNP under Michael Manley is an extension of the PNP founded by his father, Norman Manley, in 1938. The notions of the common good undergirding the old Manleyism tried to deny class divisions and class conflict (Chapter 3); those related to the new Manleyism would finally concede class divisions but not class conflict.

The realities of political power and political struggle contributed to a significant conceptual and definitional metamorphosis of democratic socialism. The second phase begins with the PNP's victory in the 1972 elections. The final chapter of *The Politics of Change*, which was written in May of 1973, revealed beginning difficulties with the notion of the "natural tendency" around which the methodology of democratic socialism was to be built. There were four areas of clear confrontation between the PNP regime and the more privileged members of the society. These were taxation, national service, state intervention in the economy, and foreign policy. The substantial increases in property taxes introduced by the regime greatly offended the wealthy landowners. The new prime minister expressed confidence that the storm would blow over: "it is inevitable that those who have more will object to being taxed more. This is a completely predictable and normal reaction which clearly will not last" (Manley, 1974, p. 212). "Upper class unease" and "resistance" resulted from the regime's plan to introduce national service. The plan was designed to "transcend the divisiveness that is inherent in the differences of personal attainment" by exposing the underprivileged to opportunities on a par with "the elite" (Manley, 1974, p. 209).

By 1974, when democratic socialism was formally unveiled, the regime had retreated from the "family model" advanced by Manley. Belief in the voluntarism of the electorate had also begun to fade. As the practical side to the Jamaican's supposed respect for the majority vote began to surface, such terms as "upper class" and "privileged members of society" started to creep into the political lexicon. Michael Manley informs us that the emergence of these sectional interests (class interests) resulted from "inchoate fear" that would be dispelled by "reassurance: how to make the better-off members of society realize that the progress of the poor is not at their expense but, rather, a condition to be desired and the goal that all must seek if social stability is to be preserved" (Manley, 1974, p. 203). But it was evident that his notion of the "natural tendency" was being dislodged by the pressures within the class structure. The new formulations now reflected terms and concepts one usually associates with Marxist-Leninist socialism, namely, "the social and economic exploitation of the poor and the middle classes" and "the elimination of exploitation through creating equality for all" (People's National Party, 1974). This would be achieved through socialism, with a "mixed economy within the context of the socialist organisation planned for Jamaica" (Manley, statement in the House of Representatives, 1974, in Hearne, 1976, p. 166). A mixed economy meant a combination of capitalist and socialist policies. Still, these would be undergirded by Jamaicans' "natural tendency" derived from their heritage and traditions.

The joining of socialist rhetoric and appeals to enlightened reason and voluntarism continued even as class conflict was growing in the months prior to the next elections (December 1976). Speaking in Montego Bay in April 1976, Manley stated:

> To help us make [democratic socialism] work, we need lawyers, nurses, everybody, doctors. My appeal to you is that you can make the difference. Socialism cannot turn back. That is here and the alternative to democratic socialism, you had better understand is anarchist violence and maybe even bloody revolution. I believe that this country can avoid a bloody revolution if its people . . . tap the roots of their patriotism and their christianity, and make a commitment in a brotherhood (*sic*) to say let us work for democratic socialism (Hearne, 1976, p. 300).

The third phase in the development of democratic socialism involved the regime's relationship with Marxism-Leninism. The relationship was unavoidable for a variety of reasons: First, the PNP has seen itself as a socialist party since 1940, though Michael Manley took pains to distance it from communism.[2] Nonetheless, the basic liberal democratic doctrine of the party had much in common with Marxism-Leninism, allowing the PNP to comfortably use rhetoric that would appeal to potential communists. As in its early days, the PNP was the only party with which they had some ideological compatibility. And the party, still functioning as the armature for radical and unconventional political ideas, even attracted within its fold avowed Marxists like D. K. Duncan, who would become minister of national mobilization and general secretary of the PNP.

Second, the orbit of radicalism into which democratic socialism was inserted was an amalgam of Black Power, dependency theory, and Marxism-Leninism, and rhetorical forms were used interchangeably and at times quite indistinguishably. The growing influence of these radical agendas compelled the PNP to take stock of Marxism-Leninism in a climate that in theory could find this agenda agreeable.

Third, new patterns of interclass resource distribution gave ample notice that old forms of political mediation were inadequate. Heretofore patronage had been the monopoly of the state, distributed largely through political parties; now it was aggressively fanning out into the private sector. Members of the construction industry, for example, were expected to pad their payroll with ghost workers. "They would come with a note saying that Mr. X [often a politician or a henchman] sent them," as one interviewee explained. The ritual, we understand, was accompanied by an unmistakeable threat. This form of intimidation also now began to undermine the old paternalism that existed between capitalists and members of the subordinate classes. The traditional "tenk yuh, Buckie Massa" affectations were giving ground to truculence (Chapter 10).

Fourth, the political script called for containment of the kind the party had effected in the period after the 1938 labor rebellion (Chapter 4). Once again, as it did in the years 1938 to 1952, the party would have to face the challenge of Marxism-Leninism, not so much to openly destroy it—this would reduce its effective political rhetoric—as to contain that agenda on the strength of political definitions "from below."

In the militant political atmosphere of the late 1960s and the early 1970s—the time of the crisis—the Marxists-Leninists attacked democratic socialism and the regime at every turn in their newly created journal, *Socialism!* Marxism-Leninism was not substantively incorporated into democratic socialism, but it obviously maintained a place in the PNP's reckoning, as the regime's occasional references to class divisions and class conflict admit, at least by implication.

In 1976 democratic socialism took on a new form. The PNP's victory in the elections of that year was impressive; it captured forty-seven of the sixty seats in Parliament. Nonetheless, the economic downturn was now posing enormous problems for state mediation of the growing conflicts between and within the classes (Chapter 10). As Carl Stone's polls revealed, the alliance between the regime, the poor, and the middle class was shifting. Middle-class support declined by half (81.4 percent to 40.2 percent), little less dramatic than we decline in support of the upper classes (75 percent to 27 percent). The support of the subordinate classes, on the other hand, increased from 43.3 percent to 71.6 percent (*Daily Gleaner*, August 6, 1980, p. 7). The shift in support, as we shall see, had little to do with a project driven by a progressive, transformative class consciousness. Mostly, it was based on the narrowly self-interested, material expectations of the bulk of all social classes (Chapters 8, 9, and 10).

The crucial importance of the subordinate classes to the PNP brought Marxism-Leninism again into serious reckoning. Communists and their supporters were relatively insignificant numerically—about 3 percent of the electorate at that time—but they were known to be well organized and active mainly in the rural parishes. Marxism-Leninism was also kept close to the PNP because of D. K. Duncan, the party's chief ideologue, its most effective organizer, and perhaps the party's most popular figure between 1976 and 1978. Drawn to Duncan and the party was the PNP Youth Organization (PNPYO), an influential group of activists cut from the cloth of Marxism-Leninism. The PNPYO was to cause the PNP a lot of criticism from within and from the JLP, as its unyielding orthodox line would at times be proffered on the party's behalf or ascribed to it by virtue of their association. Another communist cell was the Brigade League, formed toward the end of 1977. The Brigade was the outgrowth of Jamaican workers and technicians trained in Cuba as a result of the Jamaica/Cuba accords of 1975.[3]

Some degree of strategic distancing was required. The PNP needed to shore up its non-communist image. On the other hand, some contact with, or recognition of, communism was imperative to pander to the orthodox Marxism-Leninism of D. K. Duncan and his followers. Thus, a contradiction emerged: while the PNP was distancing itself from the PNPYO and the Brigade League, it was being pursued by the Workers' Liberation League (WLL), another group of communists that formed the nucleus of the Workers' Party of Jamaica (WPJ). Just months before the elections of 1976, the tone of the WLL changed dramatically from caustic criticism to support of the PNP. In a flush of self-criticism, the WLL conceded its failure to appreciate that "liberal opposition to imperialism" (that is, democratic socialism under the PNP) "can, at the present stage of the anti-imperialist and democratic movement . . . , have the effect of moving forward the movement as a whole" (*Socialism!* 3, October 10, 1976, p. 11). The League then urged support for the PNP because the party was committed to promoting "the national movement" and "because the revolutionary-democratic forces [in Jamaica] are insufficiently united and organised to provide an alternative independent of the present PNP" (*Socialism!* 3, October 10, 1976, p. 11). Communist cadres worked alongside PNP candidates in that election. On October 31, 1976, just prior to that event, the general secretary of the WLL, in a speech on the Jamaica Broadcasting Corporation (JBC), exhorted the electorate to vote for the PNP. In a section of the broadcast, he stated that while the PNP's policies were 'halfwayism,' they were moving toward socio-economic emancipation (*Socialism!* 3, November 11, 1976, p. 9)

This did not mean, however, that Marxism-Leninism was now a part of the democratic socialist equation. Over the same period, the PNP Executive Committee had been hammering out a definitive document on democratic socialism. This document, *Principles and Objectives,* published between late 1978 and early 1979, attempted to resolve the dilemma in a way that proved quite confusing. The PNP conceded class conflict and the class struggle, asserting that "the struggle between classes can only be resolved through the building of socialism" (p. 13). In a near retraction, however, the document went on to state that the "natural alliance" promoting the socialist transformation of Jamaica would include the working class, the small farmers, the "middle strata," "reclaimed elements from the lumpen proletariat, . . . and capitalists who are willing and able to contribute to the building of a just society." In actuality, then, the class struggle was recognized only to be quickly superseded by a union of all these elements "around clear objectives in a politics of purpose" (p. 14). The language may have alluded to "scientific socialism," but in the end the consensus formula prevailed.

The PNP was engaged in a balancing act. On the one hand, the formula outlined above was supposed to assure the capitalists of a positive role in the new order. On the other hand, this statement clarified the issue of communism, which had been raised by one of the party's departing souls, Allan Isaacs, minister of mining, who in December 1975 leaked important documents to the JLP and accused the PNP of going over to communism. Third, the party's concession to "scientific socialism" was directed to forces within the PNP itself. Dr. D. K. Duncan, the general secretary of the PNP from June 1974, had in part commanded this concession. On the other hand, the PNP has been anchored to a political philosophy—Manleyism—which has historically been open to radical politics. But while the cat and mouse game between the PNP and Marxism-Leninism fits squarely within this tradition, these same traditions also alert us to the difficulties lying in the path of any meaningful integration.

As our brief review demonstrates, democratic socialism in Jamaica took a circuitous route. Initially, the foundations on which it was to be built highlighted the democratic political process, the Christian principles of brotherhood and equality, the ideals of equal opportunity and equal rights, and a determination to prevent the exploitation of Jamaicans (Hearne, 1976, p. 158). The theory informing appropriate policies tried to reconcile contradictory strains with regard to class conflict and the class struggle, although positions that considered them endemic in Jamaican capitalism were, in the end, more prevalent than those embracing the family model of which Manley spoke. Other particularities of the theory were that the elimination of bourgeois domination involves strategies and policies based on a class alliance that includes some of the capitalists, that this alliance is made possible by the penchant for democratic compromise rooted in the traditions of the Jamaican people, and that complementing these characteristics are those of enlightened reason and voluntarism, which place the necessity of natural justice for all over the divisiveness of narrow interests.

Criticism of the PNP's Democratic Socialism

The ethical aspects of any brand of socialism remain largely uncontended, as socialists far and wide declare some variant of natural justice, freedom, and equality as the foundation of their future society. When Leopold Senghor declared Senegalese socialism "existential and lyrical," our apt response should be a combination of a chuckle and an admission. We admit that a socialist society, being a society as yet unrealized, possesses a utopian dimension, but we chuckle when we ponder the practical and theoretical details associated with founding it on Senghor's vision. We are free to speculate on the ends, but the means for installing a socialist society depend almost totally on the social, political, economic, and ideological dynamics of the society being, or to be, so transformed. Quite clearly, normative injunctions based on freedom, justice, equality of distribution, and the like speak to changes in social and economic arrangements anterior to and determinative of them. It is doubtful that these social and economic redresses are identical to moral injunctions such as the "hunger for spiritual nourishment" around which Senegalese socialism was theoretically organized (Sigmund, 1967, p. 245).

The notion of the "natural tendency" of the Jamaican people is a troublesome one. If democratic compromise is common to the Jamaican's heritage and traditions, so, too, is class conflict. Logically, both must be included in the mix of strategies and policies. If we agree that material life conditions social life, civilization and its many expressions (see, for instance, Amin, 1976, pp. 25–26), the evaluation process seems a simple one. Classes and class conflict are the cause and the effect of structures and processes creating, eliminating, and recreating the material conditions of life. How a people make decisions about matters related to these conditions is clearly secondary to the processes through which the conditions manifest themselves.

We recognize that, at this level, we are only dealing with definitional issues and that the test of the regime's agenda necessarily requires a much more encompassing approach, a discussion of which we undertake below. Nonetheless, we are troubled, at the outset, by the continuously contradictory position of the PNP on the role of the class struggle.

To us, the differences between capitalism and democratic socialism are differences in class dynamics and class power. It is quite true that democratic socialism is a transitional stage in which a mix of capitalist and socialist structures and practices must necessarily coexist. It is equally true that strategies for building democratic socialism are far from settled. However, this does not mean that we must accept initiatives and policies as democratic socialist simply because they are so defined. How to tell whether a project so labeled is actually headed in that direction becomes a matter of weighing policies and political practices and evaluating their contributions, real and potential, to future outcomes.

Given the nature of democratic socialism, which we conceive more as a process than as an end state, selecting appropriate outcomes may present problems. The answer is not, we suggest, in a formulaic comparison that matches given policies or initiatives instituted at any given point with those of an ideal state (socialism) toward which one is moving. Policies do not have a priori, intrinsic class positions. For instance, in the early 1970s the Pakistani state purchased all local commercial banks and made investments in public industry that were four times greater than aggregate investment in private industry (Ahmad, 1978). These initiatives were related to the profound crisis that followed the separation of Bangladesh and were removed from socialism. Similarly, marked participation in the economy by many of the Arab states, particularly in the oil producing countries, has had little to do with socialism, occasional socialist labels notwithstanding. We must recognize, therefore, that the fact that the Jamaican state embarked on a mixed economy and other typically socialist projects (state food farms, for instance) does not constitute sufficient evidence for a democratic socialist path.

Socialism is about a shift in relative class power from the dominant (capitalist) classes to the subordinate classes. The state is an entity through which such a shift can be effected, but it must be remembered that it itself is or, through the development process, is destined to become, a creature of class relations. The fact that the state in many ex-colonies developed considerable power and, occasionally, seeming autonomy from the class structure, does not mean that the state is free to impose socialism as state policy, without regard for the class structure, or, indeed, that the state is above classes. In all such cases, the real test must be the extent to which such policies and initiatives lead to a shifting of relative class power. This is the important criterion that enables one to determine objectively whether or not a country is engaged on a democratic socialist path. Rather than searching for policies and events that in themselves can be taken as indicators of democratic socialism, the evaluation process should proceed on the basis of whether these policies and events amount to some critical mass that initiated a perceptible shift in class relations that increased the power of the subordinate classes. Without such a bias, even at the level of definition, policies and their evaluations could become self-serving indeed.

Thus it is entirely appropriate, as Michael Manley and other proponents of "many socialisms" claim, to fashion a transformative path that is historically and situationally specific, but the specificity must be grounded in a study of the evolution and organization of social classes, one that goes beyond the voluntarism and philosophical pronouncements of the PNP leadership. We do not reject outright the possibility of a strong state instituting socialism "from above." However, such a project can be sustained only by an appropriate configuration of class forces in and around the state (Saul, 1986, p. 225). As our work demonstrates, this was not the case in Jamaica.

In fact, our work flows from the basic contention that democratic socialism, regardless of the sincere intentions of many within and outside the PNP, was never a possibility. We contend that the class basis for it was absent, that the state from which the agenda emerged gave too much credence to its own autonomy and ability to sound the clarion call and unite all classes. This type of class alliance is far better explained by employing the analytical model offered by national popularism.

▶ National Popularism

Social change in the Third World—change directed against or in support of colonialism or imperialism—often defies easy categorization. This is especially true in situations that eschew the extremes of right-wing dictatorship and left-wing revolution, represented in modern times by situations such as Chile under Pinochet and the Sandinista revolution in Nicaragua. These far more frequent happenings arise in response to the less severe but nearly constant crises that beset dependent relations within capitalism and are related to some problematical aspects of the process of capital accumulation. The problem may be manifested in a number of factors, of economic or socio-political nature, on which capital accumulation depends: resource distribution, a restive and violent underclass, and so on.

In such cases, the state is required to create new mediations to buttress capital accumulation. If these mediations were invariably marked by an increase in the coercive power of the state on behalf of the capitalist class, problems of definition would not arise. Far more common, instead, are state actions steeped in ambiguity, appearing to both attack and buttress capitalism at different turns. Such crises assume a variety of forms that are often flavored with socialist strains of differing intensity. Frequently, they swirl around in the potent radical rhetoric of Third Worldism—a rhetoric made even more attractive and cooptable by the shifting equilibrium of relations between the First and Third Worlds since the late 1960s. Thus the state's role is confounded: is it an agent of mediation on behalf of the existing system or an agent of systemic transformation?

Fascism arose from a parallel crisis in a developed and advanced capitalist economy. Its emergence was linked to the weakening of traditional means of control exerted by the capitalist class over the working class, which threatened capital-labor relations. The state inserted itself powerfully to adjust the situation but did so in a manner that many interpreted as incipient socialism. Here was a state, seemingly independent of class determination, that extended its grasp into industry and finance, was rhetorically antagonistic to bourgeois democratic ideology, and was allied with the popular and middle classes (Dutt, 1935). After considerable review, fascism was given hybrid status, deemed neither capitalist nor socialist. In the end, it came to be correctly understood as an exceptional phenomenon under capitalism. These new conceptualizations focused on the collapse of old patterns of capitalist mediations that occurred when a weakened bourgeoisie could not maintain the traditional means of control over the subordinate classes. In the crisis, the state became more directly and centrally involved in ensuring capital accumulation, which called forth new forms of class mediations, some of them socialist in appearance (Sweezy, 1968; Poulantzas, 1974). It was "a movement in practice, in the conditions of a threatening proletarian revolution, . . . [a] movement supported by the bourgeoisie . . . to defeat the revolution and build up a strengthened capitalist state dictatorship" (Dutt, 1935, pp. 95–96). A state that works closely with labor—which theoretically may be the combination setting off socialist transformation—is not necessarily embracing socialism. In Argentina, Peronism was embraced by the working class;

the result, however, was not communism, as was greatly feared, but populism.

National popularism bears certain similarities to fascism in that both are responses to crises of capital accumulation that center around an activist state. The prominence of the Jamaican state interfaced with the accumulation needs of the capitalist economy, which called for maintaining some balance in class relations as well as promoting quantitative development. The class dynamics of the two models are quite different, however, owing in part to the fact that national popularism is a creature of dependent capitalism. In national popularism, for instance, the working class is relatively weak, thus opening the way for a possible left-leaning posture by the regime. The socialist rhetoric that tends to pervade the political atmosphere needs careful scrutiny, however.

We assert that the distinction between national popularism and democratic socialism is a crucial one. Far from being an exercise in semantics, identifying the various forms of capitalism and the sometimes fine distinctions between these and incipient socialism is an important part of theory building. Democratic socialism as a model of change may be fluid and difficult to grasp, given the state of political practice and, of late, the disarray of the parent model itself. Fluidity does not translate, however, into the inability to make important conceptual distinctions between processes that have the potential for leading to systemic transformation (socialism) and those that announce systemic modification (a different form of capitalism). Analyzing events in Jamaica under the rubric of national popularism rather than democratic socialism allows a different understanding of both the transformative possibilities inherent in the moment and the boundaries circumscribing these possibilities. It also provides some tools that may be useful in approaching the study of similar crises in other Third World countries.

Third World leaders in crisis situations are fond of aligning themselves with socialism, egalitarianism, or anti-imperialism, though everything else about them stands in stark contrast. These alignments are often expedient, if not heartfelt, as Third World expectations almost command them. As Henry Bienen (1977, p. 4) notes, Third World leaders "are seen to be without ideology, a sin in the eyes of those for whom the absence of a leader's writings in anthologies on socialism in developing countries or 'ideologies of the Third World' is equated with the absence of any conception of development or even national interest and dignity or sense of nationhood." Accordingly, these expectations have bred Kenyatta in Kenya, Forbes Burnham and cooperative socialism in Guyana, and many others. And in 1952, the spirit of revolution was betrayed in Bolivia by state functionaries using their privileged position to speculate in foreign exchange (Gomez, 1976, pp. 474–475).

We must, therefore, distinguish theoretical substance from appearance. National popularism is a model of interpretation that explains the socio-political correlates of a crisis in capital accumulation.[4] The particular economic relations are defined by a shortage of capital resources needed to service the interests of the various social classes. National popularism is a response to class conflict; it is not a series of directives and policies handed down from above, however much the vision of the Great Man making history appears to surface. Here the state extends control over resources usually owned by foreign interests and then invests the extracted surplus in the national economy in the service of redistributive policies within the national class structure.

The nature of class conflict plays an important part in defining a phenomenon as national popularism. Of primary importance is the fact that social classes are generally able to define and understand their interests, but the level of class organization and antagonism is not so strong as to prevent them from joining in a working alliance against a common enemy. Unity is undergirded by the pursuit of distinct class interests, often barely hidden under the common banner of economic nationalism and anti-imperialism—the rallying cry and the glue that holds the alliance together. It is, therefore, precarious, since the level of class development militates against any but a temporary and brittle class alliance. Thus national popularism is characterized by shifting inter- and intraclass relations, as events force their basically conflictive nature to the fore.

Also crucial is the role of the state, which needs to create new mediations to overcome the crisis. The state requires a degree of relative autonomy from the class structure, allowing it to apportion resources to the various classes with some sense of balance and to deal with foreign capital. This means that the state may act in ways that run counter to immediate capitalist interests. It remains, nonetheless, a capitalist state engaged primarily in managing systemic contradictions rather than superseding them.

We argue that the conditions for national populism existed in Jamaica in the 1970s. Our study shows that segments of the dominant classes, particularly the industrial fraction of the capitalist class, were eager to undermine the ascendancy of the merchant/commercial fraction and foreign capital, mainly the mining industry owned by U.S. and Canadian transnationals.

The interests of the middle class derived mainly from the ambitions of its budding bureaucratic/entrepreneurial fraction. Its members had substantial control of the political realm (the state apparatuses) since independence in 1962. Strategic political power of this kind had its corollary in the pursuit of economic power; so the fraction began to capitalize on its political power through state expansionism and other financial opportunities from this source.

Within the subordinate classes, two segments falling within the lumpen and sub-proletariats were more significant in the crisis than the others.[5] The Rastafarians were the source of a native Jamaican ideology that merged with other radical ideas and served as an ideological armature for the bureaucratic/entrepreneurial fraction of the middle class and the other subordinate classes. This ideology aided the state in putting constant and effective pressure on foreign capital by bringing to the fore and giving legitimacy to the pro-Jamaican aspirations of the "little people" and capitalists alike. Other members of the lumpen proletariat added significantly to the crisis through the threat they now posed to the system. Importantly, the working class and peasantry were not organized along class lines and did not present a direct or immediate challenge to the status quo. Of the peasantry, it was remarked that they were "invariably asleep, snoring for most of the year" (Munroe, 1981, p. 98). After democratic socialism, they still had not arisen from their slumber.

The PNP regime had little difficulty in marshaling the support of social classes against foreign capital, particularly the mining companies, which were reorganized. Of enormous importance was the bauxite levy unilaterally extracted by the regime in 1974. Distribution of these vastly increased revenues (from $25 million to $200 million annually) momentarily united the social classes against an enemy: the bauxite industry and foreign capital.

The Jamaican state, for its part, had honed its managerial skills since independence. It became progressively the main procurer of investment capital, inviter of foreign investments, and it emerged as the principal source of capital formation. Between 1955 and 1962, the state introduced $420 million in investment; this figure grew to $1.13 billion for the period from 1967 to 1972 (*Daily Gleaner*, February 5, 1973, p. 11). After democratic socialism was declared, this role steadily increased. Jamaican capitalists, on the other hand, remained comparatively weak and did not exercise hegemonic power.

As class reactions and the activities of the state took on certain anti-imperialist forms, they shared many of the outer manifestations of socialism. So did the policies employed by the state. But the crisis was basically a reaction within capitalism—extraordinary but not enough to initiate a structural transformation of capitalist relations. At work, then, was a process of modification rather than transformation.

A closer look at social classes in agitation and alliance is needed, as our competing models appear to have certain common features. For instance, when accompanied by radical and socialist slogans, class expressions may be consistent equally with democratic socialism as with changes more in line with national popularism, that is, induced by capital accumulation within a mode of production not in the least threatened by transformation. The point is readily grasped when we remember that democratic socialism also entails changes in the patterns of capital accumulation. Figure 1 captures the schizoid nature of some of the major class expressions. In addition to stressing the different interpretations each conceptual model would offer for given class expressions, figure 1 highlights the additional class expressions required for the model to succeed. Interfractional rivalry among the capitalist class, for instance, would have to be accompanied by the class's acceptance of state participation and leadership in the economy in order to prepare the ground for democratic socialism. National popularism, on the other hand, would require the capitalists' support for concessions to other classes as well as other accommodations. A review of the major class expressions and their interpretation according to each model lends support to our position that the requirements for the success of democratic socialism were absent while those of national popularism were more often met, albeit in a temporary fashion.

The role of the state also makes interpretation difficult. In its systemic role of class mediation, it has to reflect the contradictions germane to that role. Where the interests of the capitalists are concerned, the role of the state must be, in the end, to protect that class from itself. The state must attempt to ensure that the individual interests of each fraction of the class do not get in the way of the overall welfare of the class as a whole (Miliband, 1977; Carnoy, 1984, pp. 251–255). Getting in the way of the other is the favorite and even expected pattern of behavior of each fraction, as they are so moved by the competition attached to the role. As state actions on behalf of systemic health do not necessarily benefit all fractions of capital, a fraction that is threatened by shifting patterns of capital accumulation may see its impending demise as the end of the system and so define state actions as "anti-capitalist." In addition, the state needs to ensure the adequate distribution of resources to the subordinate classes, particularly at times when such classes pose a potential threat to the system. Thus the spirit of antagonism may appear to reign even though state activities are really aimed at managing capitalism and thus protecting the capitalist class overall.

Capitalist class antagonisms or class unity with the state—such as we saw after the bauxite-led policies in 1974—thus must be measured

FIGURE 1

Major Class Expressions Common to Democratic Socialism and National Popularism and Their Alternative Interpretations

Relevant Social Relations	Class Expressions	Democratic Socialism		National Popularism	
		Interpretation	Requirements for Success of Model	Interpretation	Requirements for Success of Model
Capitalists	Inter-fractional rivalry Open concessions to other classes Visible support for democratic socialism from industrial fraction only	Weakened capitalist base Breakdown of power bloc	Accept and collaborate with state participation and leadership in economy	Jockeying for position in restructuring of power bloc Collaboration with state in temporary broadening of state power to manage the crisis	Continued support for concessions to other classes (even against own immediate interests) Participation in national coalition—power sharing Accept state intervention in economy
Working Class	Increased number of trade unions Increased strike activity	Frustration with present system; readiness for fundamental change Development of class consciousness tempered by commitment to rule of law and common good	Growing commitment to socialism (politicization) Greater participation in economy and society Gains in share of economic resources	Trade union consciousness; narrowly self-seeking militancy Strong streaks of paternalism and clientelism maintain reliance on state to initiate change and redistribution	Temper militancy as result of paternalistic state action Maintain sufficient pressure to ensure continued share of resources

Middle Class (strategic fraction)	Embracing of socialist ideology and agenda Activism within state and its apparatuses (trade unions, government agencies, etc.) to further socialist agenda	Introduction of state-led socialism	Substantial support of party for socialist agenda Control of state and its key apparatuses Forsaking of self-seeking for common good	Search for new mediations to manage crisis (including new ideology) State power being used by class to gain political dominance (forge coalition under its leadership) State power being used to support own economic interests (i.e., join ranks of capitalists)	Availability of resources to distribute within the class as well as to other classes (to maintain stability of coalition) Control of state and its key apparatuses Maintaining balance of power between other classes
Lumpen and sub-proletariats	Increasing violence toward own class and other classes Adherence by significant segment to Rastafarianism	Available to support fundamental change "Oppressed" needed new avenues for social inclusion and justice Incorrigible elements need strong social control	Gains in share of economic resources New avenues for participation (social and economic) Social control	Opportunism Significant element in crisis giving rise to pressure for structural modification	Temper militancy and violence as result of state action Maintain sufficient pressure to ensure continued share of resources

FIGURE 1

Major Class Expressions Common to Democratic Socialism and National Popularism and Their Alternative Interpretations (continued)

Relevant Social Relations	Class Expressions	Democratic Socialism		National Popularism	
		Interpretation	Requirements for Success of Model	Interpretation	Requirements for Success of Model
The state	Aggressiveness toward capitalists and foreign capital Attempts to assert dominance in key areas of economy and society Espousal of socialist ideology and policies so termed, including redistribution of resources	State-led socialism Near autonomy of state from class structure, enabling initiation of socialism State orchestration class relations, since class struggle is neither at issue nor advanced	Acceptance by key state apparatuses of socialist agenda Increasing social equity Democratization and participation of all classes Subordination of capitalist class interests to the common good Foiling of antagonistic reaction by international capital	Relative weakness of "polar" classes allows much play for state action, but within capitalist framework State engaged in forging new mediations to overcome crisis; task involves brittle national class alliance and radical ideology	Maintenance of balance in class relations: weaken antagonists, strengthen allies—with continued state leadership Availability of resources for continued redistributive functions Create and maintain supports within state for new state functions

Note: Each model revolves around different precepts and types of crisis. Democratic socialism focuses on building a democratic and equitable society and emerges from a crisis of transformation. National popularism focuses on resource creation and distribution and emerges from a crisis of modification.

against other factors. The bauxite-led policies of 1974 were accompanied by a marked show of class unity. On the other hand, our analysis reveals the presence of a marked rivalry within the capitalist class (Chapters 5 and 9) and the beginning of a new state-linked fraction—the bureaucratic/entrepreneurial fraction—drawn from the middle class (Chapters 6 and 7). Far from being a sign of impending destruction, however, the rivalry was part and parcel of shifting patterns of capital accumulation within an economic framework that remained capitalist. In order to perceive the differences, one must dig deeper, to the level of the class structure, where changes reveal themselves in economic, political, and ideological terms. In very basic terms, the proper interpretation must be sought in the state of the class struggle that provides the basis of determination from such factors as the level of class consciousness and the role of the state and its apparatuses (Therborn, 1983).

Similar comments apply with regard to state intervention in the economy. Direct state participation may appear in both models, with capitalist reactions ranging from an invitation to the state to intervene on its behalf, to tacit consent to its intervention, to vituperative attacks upon its intervention. In both cases, the weakness of the capitalists is an invitation for a strongly interventionist state, whose agenda will likely include the attempt to relegate the capitalists to the role of junior partners, while assuming a leadership role for itself. The key to distinguishing between the models is again found in the level of the class struggle within and around the state. If we take the PNP at its word, the creation of democratic socialism would depend, in part, on the state having the wherewithal to orchestrate and sustain, with the requisite support from other classes, such a hybrid mixture of capitalism and socialism, a mixed economy in which the capitalist class is no longer the dominant force. What we witnessed instead was a brief alliance sustained by narrow self-interests, with the capitalists retreating and regaining the upper hand once they perceived a threat to their interests.

Our analysis of national popularist class relations shows that they are unstable and brittle. These features can be easily traced to two main factors, capitalism and scarcity of resources. Capitalism, based on the fundamental contradictions between capital and labor, renders any alliance between them necessarily unstable and brittle. Scarcity often means that resources needed by the state to mediate class conflict, whether to create new fractions of classes or to reward one class or social stratum, are usually extracted from another class. This process also makes for unstable and brittle class alliances. We should also note that social classes with a long history of deprivation will compete viciously for scarce resources. In the Sandinista revolution (an upheaval supported by far more revolutionary activity than was the case in Jamaica) "the political allegiance of classes [was] based on their perception of their well-being, not on ideological grounds" (Colburn, 1986, p. 28).

Some timely comparisons must be made here. Class relations under democratic socialism are also unstable. The instability really resides in what Irving L. Horowitz (1982, p. 263) describes as the attempt to blend capitalist "political democratic models" with socialist "economic distributive models." The logic of utilizing capital for productive ends comes into conflict with the nonproductive (social) ends to which socialist regimes often put sizeable portions of their resources. While in both models brittleness and instability will surface in class relations, they originate from different sources. In national popularism, they result from crises within capitalist relations proper, which can be brought about by problems of production or distribution. In democratic socialism they result from the attempt to graft incompatible systems. Our research shows, for instance, that certain sectors of the middle class supported the subordinate classes in their charge of exploitation by the capitalists. However, the middle class withdrew their support when it was felt that the subordinate classes were too favored by programs of social equity, that is, programs placing a heavy burden on economic production. Quite similarly, the industrial fraction was favorably disposed to a section of the middle class, strategically located in politics and the professions (the "strategic fraction") in its quest to wrest power from the dominant merchant/commercial fraction of the capitalists. In the same breath, however, the industrialists attacked this momentary ally for being engaged in non-productive and parasitic forms of activities. If an economy remains essentially capitalist, it is likely that class alliances, especially those flavored by socialism, will be fragile and short-lived affairs. Perhaps the chances of success would be heightened by a measure of isolation, as may have been the case with the "curtains" of the Soviet Union and China. In the case of Jamaica, political traditions and dependency simply forbade it.

What, then, are the main features of national popularism? National popularism is a model of interpretation for particular patterns of socioeconomic change stemming from a crisis in capital accumulation: It is to be distinguished from other models (dependent neo-colonialism, national developmentalism, and national populism) mainly on the basis of the relative autonomy of the state and the level of economic nationalism that is present. Its analytical approach is to focus on strategies that mitigate class conflict within ongoing capitalist relations. Spearheading the class conflicts associated with capital accumulation are those among the capitalists themselves, those attached to the momentary rise of a strategic fraction of the middle class bent on joining the ranks of the capitalists, and the ideology of the Rastafarians. The latter helped to energize the intermediate and subordinate classes relative to the overall confrontation with foreign capital and, in the case of the strategic fraction of the middle class, in the fraction's contest with the capitalists.

Finally, it is a brief and uncertain alliance between various class configurations that was doomed to fail for structural as well as historical reasons.

▶ Testing the Models

The theoretical validity and appropriateness of either model must derive from questions about the state of the capitalist economy. The questions asked of democratic socialism must be drawn from capitalism, the mode it still inhabits, and the system that largely defines whether the capitalist contradictions upon which socialism depends have matured. Even in the case where socialism has made significant progress, capitalist relations will exert considerable influence for some time (Sweezy and Bettelheim, 1971; Foster-Carter, 1978, p. 225). In a real sense, national popularism is the negation of democratic socialism: indeed if one exists in substantial form, the other cannot. This does not, however, make national popularism a residual category. The dynamics of the model point to class action and policy outcomes directed to the specific changes in the class structure claimed to be the basis for new patterns in capital accumulation.

In order to test these models, our analysis focuses principally on the following areas: (1) the state of the class structure and class relations; (2) the role of ideology and consciousness within the PNP; and (3) state apparatuses and their relationship to change.

The first area of focus will involve evaluating class relations relative to the contemplated change. The underlying thesis is that unless the popular classes expressed high or encouraging levels of consciousness to defend the interests sought on their behalf, the democratic socialist project was almost bound to fail. Instead of the required levels of consciousness at this end, we find a class configuration emerging from the middle class and developing around the state that attempted to use state power toward ends that are better understood within the context of capitalist principles.

The second area takes us to the PNP itself. We look at the ideological posture of the party, both historically and in the period in question, to determine the kind of social change the party structure could support. Of course, unlike national popularism, democratic socialism calls for the party to betray the necessary socialist consciousness and ideology to interface with and reinforce popular class support and initiate the policies necessary for the proposed reconstruction.

The most attractive policies, even those enjoying substantial popular support, must rely on a wide range of state apparatuses for their success. The structure and performance of these apparatuses—our third area—are defined and conditioned by the central and dominant organizing principles of the economy. From what script were these apparatuses playing?

Finally, much has been made in other accounts of the important negative factor associated with external opposition to the state's policies, chiefly from the United States (Stephens and Stephens, 1986; Beckford and Witter, 1982). The regime's failure—so the argument goes—had much to do with actions initiated by the United States to undermine the agenda both politically and economically. This is an important issue that we consider in the context of the overall support and opposition the regime encountered. Historically, reformist regimes as well as socialist experiments have been frustrated because of the dependent nature of the economies involved. Jamaica is a dependent economy, with the bulk of its trade, aid, and international borrowing centered in the United States. We can hardly omit from our analysis how this drama was played out and what steps, if any, the Manley regime took to avoid or neutralize concomitant constraints. We do not, however, accord this factor the same level of importance others have given it, not in the least for the reason that it is not central to our thesis. Lack of external opposition may indeed have given the regime more time to attempt to effect its changes, but it would not have altered the basic class dynamics that lend support to either one or the other of our analytical models.

▶ About Political Parties and the State

One might be struck by the importance given to political parties, mainly the JLP and the PNP, and the state in this work. First, the great stress placed on political parties results from the peculiar political history of Jamaica. After the labor rebellion in 1938, the British parliament carefully structured local politics around the major components of modern political society—trade unionism and adult suffrage. The JLP and the PNP were the armatures around which local political activity was encouraged. The central role of these parties was greatly enhanced by a very weak tradition of political organizing within the subordinate classes. Even to this day, there is little in the way of peasant or populist organization, or even the church-oriented "basic Christian communities" that are common in many Latin American countries. In the absence of these mediating structures, the bulk of the Jamaican class structure must depend on the political party. And this dependence is further compounded because organized labor is symbiotically linked to the parties: the Bustamante Industrial Trade Union (BITU), the Trades Union Congress (TUC), the National Workers' Union (NWU), and the University and Allied Workers' Union (UAWU) are all structurally and inseparably tied to the JLP, the PNP, and the Workers' Party of Jamaica (WPJ) respectively. Each union operates as the recruiting agent for each party. To make the practice even more permanent, the JLP and the PNP have conspired to neutralize any efforts to introduce a third party (Chapter 8).

The central role of the state also emerges from peculiar historical circumstances. Like other Third World countries exposed to colonialism and imperialism, Jamaica displays state apparatuses conditioned by the "peripheral" as distinct from the "advanced" form of capitalism (Amin, 1974). Typified by very weak polar classes (that is, the bourgeoisie and proletariat) and sparse capital for development, these economies cannot depend on any developmental impetus that is natural to the antagonisms between capitalist classes, powerful economically and politically, and powerful proletariats, such as those that we encounter in advanced economies.

The process of capital accumulation and the regulation of class relations then falls heavily to the state whose role is distinctly different from that of its counterpart in an advanced economy. In peripheral capitalism the state enjoys significant relative autonomy, that is, it is exposed less to the constraining power of the polar classes (because of their weakness) though the state is no less committed to the continuing existence of capitalism (Poulantzas, 1975). So dominating and intrusive are these states that scholars, albeit erroneously, have begun to refer to a "state mode of production," one that exhibits little dependence on social class relations (Leeson, 1981).

Relative autonomy is shored up by the task of "underpinning politically the articulation of different modes of production, to guarantee, for example, both the reproduction of the relations of appropriation within different modes themselves and the reproduction of the mechanisms for the transfer of surplus from the subordinate modes to the dominant capitalism" (Foweraker, 1981, p. 102). In other words, the state in the periphery arises from the processes by which pre-capitalist forms become integrated into world capitalism, that is, peripheral capitalism and advanced capitalism. The state is, therefore, not automatically for the capitalist class; in a sense, pre- and non-capitalist forms command equal time (Keith, 1988). It is for these reasons that Vergopoulos (1983, p. 4) sees the state in the periphery as the most important social relation there—even more important than the capitalist mode of production. Currently, the states in such countries as Taiwan, Singapore, Korea, and Indonesia favorably attract such a characterization (Amsden, 1979). In South Korea and Taiwan, for example, the state is the prime mover in implementing major agrarian reforms, developing a strong home market for agricultural products, promoting high rates of domestic savings, accelerating industrialization, and initiating policies that support attractive rates of profit.

▶ A Class Analysis in Broad Sketches

It is commonly accepted that 1938, the year of the labor rebellion in Jamaica, represents the beginning of the modern period in the country's history (Chapters 3 and 4). The rebellion laid the groundwork for a modern capitalist state. Within the context of formal political society, the class structure began to emerge more fully in keeping with the capitalist nature of a modern state.

Our time span is divided into three subperiods. The first begins immediately after the 1938 rebellion and ends with the introduction of industrialization and mining policies. It includes the years 1939 to 1952. The second period takes us through constitutional reform and political independence to the realization of economic dependence and barriers to development, that is, 1953 to 1969. The third period, 1970–1980 takes in the eve of the PNP victory to the class and political consequences of the party's policies. Table 2 details the contribution of the different sectors of the economy after World War II and thus provides a partial view of the economic changes that have characterized the modern era.

Figures 2, 3, and 4, herein, have a double focus. They delineate important social classes, their fractions and strata in the process of change, and give their ideological and/or political postures. The focus is mostly on those classes or fractions of classes we contend are central to the crisis. The figures also indicate important events resulting in the transformation (composition, ideology, etc.) of these classes. This second focus captures the forces that define the changing social relations in each succeeding subperiod.

During the first period, the stage was set for the Jamaican economy to undergo a fundamental shift that would spell a diversification of the economic base and a gradual movement away from reliance on plantation agriculture. A combination of factors were at play, in particular the political and economic changes instituted in the aftermath of the 1938 labor

TABLE 2
Composition of the Gross Domestic Product (GDP) for Selected Years, 1950–1985

Year	Agric. (%)	Mining (%)	Manuf. (%)	Constr. (%)	Distrib. (%)	Fin. (%)	Govt. (%)	Other (%)	Total GDP*
1950	24	—	14	8	16	4	6	28	530.8
1960	12	10	14	12	18	4	6	25	1283.8
1965	11	10	15	11	14	4	7	28	1617.0
1970	7	11	16	18	19	4	8	18	2158.6
1975	7	10	18	10	20	5	10	20	2152.6
1976	8	6	19	8	17	5	15	22	2013.6
1977	8	7	18	7	17	5	16	22	1965.7
1978	9	8	17	7	16	5	17	21	1976.4
1979	8	8	16	7	15	5	19	22	1940.8
1980	8	9	15	5	15	6	19	22	1827.8
1985	9	5	16	5	15	7	19	24	1834.7

Sources: Statistical Yearbook of Jamaica (Kingston: Department of Statistics, Government of Jamaica, 1950–1980); Derick A. C. Boyd, *Economic Management, Income Distribution, and Poverty in Jamaica* (New York: Praeger, 1988), pp. 6–7.

* GDP expressed in millions of $ at 1974 prices.

rebellion in the form of self-government and gradual decolonization that were initiated with elections held in 1944, and long-term difficulties experienced even before the rebellion by the plantation sector. Although British policy continued to support export agriculture and discourage manufacturing, impetus was in the direction of the latter. Bauxite mining, tourism, and a program of industrialization through inducements to foreign capital were underway by the end of the period.

The planter fraction of capital, dominant until the labor rebellion, began to lose ground to merchant capital, although its special relationship with Britain shored up its position for a time. For example, Tate and Lyle, the British conglomerate, had purchased 20 percent of the sugar estates in 1938 and did not relinquish them until the 1960s. In the wake of devastating plant diseases in the late 1930s, the U.S.-owned United Fruit Company disposed of its banana estates. The reorganization of the banana industry in the 1940s and 1950s favored local mid-sized and large growers. With the passage of supportive legislation such as the Pioneer Industries Act of 1949, merchant capital began to diversify into manufacturing, a trend that would become more marked in the subsequent period.

The rural-to-urban migration brought on by changes in the rural socio-economic structure had a significant impact on the subordinate classes, although there was less movement into urban industrial em-

FIGURE 2

The Capitalists

Time Period	Socio-Historical Descriptors			
	Dominant Fraction	Other Fractions	Political/Ideological Posture	Transformative Events
1939–1952	Planter (export agric.) Aligned to foreign capital British own 20% of sugar estates 800 farmers own 53.5% of fertile land	Merchant/commercial, some commingling with planter Retail & distributing Dominated by Jews, important Lebanese & Chinese presence Manufacturing (incipient: 1930s—factory production = 3% of GDP, 2,000 workers 1940s—incentive legislation 2,000 manufacturing firms	Class unity JDP contests & loses 1944 elections Growing importance of Jewish merchants & potential tensions with planter & industrial fractions	Labor rebellion (1938) Universal suffrage: Capitalist party loses elections leading to support for JLP Push to industrialization
1953–1969	Merchant/commercial, largely "comprador" Consumer/food imports & "last stage" manufacture Continued Jewish dominance	Export agric. planters declining: rely on gov't. subsidies & purchase of estates in 1960s Mining ("enclave," foreign) Late 1960s: 0.2% of the landowners own 49% of the arable land	Class unity Farmers' Party (export agric. fraction) contests and loses 1955 election Class penetrates influential parastatals & increases influence in government	Industrial incentives legislation (1950s) Independence (1962) Black Power Movement & increasing violence from subordinate elements (late 1960s)

		Industrial (ascendent but mingled with merchant Average import dependence 43.4% (whole manufacturing sector) JMA formed in 1953	Post-1958: growing split with Ashenheim family behind JLP & Matalons behind PNP	
1970–1980	Merchant/commercial, but threatened by state, on the defensive Increasing capital concentration—conglomerates w/expanding infrastructure	Mining (declining post-1974) Bauxite/alumina earnings = US$187.3 million in 1974, US$148.5 million in 1976 Industrial (attempt using state to grow) Mingled with merchant	Merchant/commercial rivalry followed by unity 1972: key industrialist (Matalon) in PNP government 1976: Private Sector Organization of Jamaica (PSOJ) formed Capital flight (esp. post-1976)	Attack on bauxite companies "Socialist" attacks on capitalist structures IMF intervention

FIGURE 3
Intermediate Classes

Time Period	Socio-Historical Descriptors			
	Middle Class	Petty Capitalists	Political/Ideological Posture	Transformative Events
1939–1952	Upper stratum: Jewish attorneys & white colonial officials, few black or brown professionals Middle stratum: brown and black ministers of religion, teachers, mid-range officials Lower stratum: clerks 1943: 4,500 in public sector employment 3% professional, managerial & administrative 11% clerical	Primarily shopkeepers, midsize & small businessmen and midsize farmers 1943: 20,000, or 40% of classifiable labor force	Capitalist values embraced Black/brown professionals & educated support PNP Most petty capitalists support JLP Some reform of system desired	Labor rebellion (1938) Formation of PNP & other modern government institutions Decolonization process begins University of the West Indies founded in 1949
1953–1969	Reduced role of colonial officials More browns and blacks in professions & civil service	Same composition as in 1939–1952 1968: 8% of farms between 10 & 99.9 acres	Orientation to consumption & education Decreasing importance of color	Political independence in 1962 Increasing role of state in demand management & political mediation

	Expansion of class as a whole 1968: 57,000 in public sector employment, 32,000 professional, technical & managerial, 566,400 sales & clerical		Movement away from neo-colonial ideology especially among the young and the intellectuals	Third World decolonizations and Non-Aligned Movement OPEC price increases
1970–1980	Appearance of strategic fraction (bureaucratic/entrepreneurial) around state role Increased black presence (JAMINCORP becomes first black merchant bank) 1980: 110,000 in public sector employment 1973: 42,700 professional technical & managerial, 77,300 sales & clerical	1973: Middle & large farmers comprise 3–4% of the population	Middle class support for PNP wanes between 1972 and 1976 Emphasis on furthering own economic interests (still tied to capitalism)	Declaration of democratic socialism Struggle with IMF Flight of capitalists and others creates economic opportunities

FIGURE 4

Subordinate Classes

Time Period	Socio-Historical Descriptors					
	Peasantry/Agro-Proletariat	Working Class	"Own-Account" Workers	Subworking Class & Lumpen Proletariat	Political/ Ideological Posture	Transformative Events
1939–1952	1943: 45% of labor force in agriculture Most involved in both wage labor (sugar & banana estates) & farming own small holdings	1943: Less than 5% of labor force designated working class, 3% industrial—15,000 1950: 60% of "gainfully employed" in wage labor (usually also involved in other activities)	Self-employed; disguises under- and unemployment; shades into lumpen proletariat	1943: 26% of labor force unemployed; 50,000 would-be entrants unemployed Average worker occupied 60% of the time 19% of the population urban	1938 labor rebellion Semi-proletariat want change—energies channeled into unions & representative government Support for JLP shifts to PNP 1950: 20% of gainfully employed are unionized	Labor rebellion Representative institutions founded Policy of industrialization followed rather than land reform
1953–1969	1960: 39% of labor force in agriculture, i.e., 60% of working males & 30% females	1968: Industrial working class 15%, i.e., approx. 100,000	1968: 36% of employed labor force & 8% of unemployed are self-employed	1959: 17% of labor force unemployed; 32% of population urban	1957: 42% of wage earners are unionized Strikes increase in late 1960s, some	Government contacts with Ethiopia & visit by Haile Selassie (increased accep-

	1968: 78.5% of farms under 5 acres Extensive migration to cities and abroad, ⅔ to England before 1962				radical labor activism High patronage (esp. lumpens) by political parties Increased influence of Rastafarians (⅔ adult males in Kingston slums are Rastafarians or sympathizers)	tance of Rastafarians Black Power Movement & contact with radical intellectuals Rudie Boys
1970–1980	1972: 33% of labor force in agriculture, i.e., 45% of working males and 18% females	1975: Approx. 25% of labor force in working class, 16% industrial—120,000 1973: Approx. 4–7% in agriculture Services are single most important employment source	1972: 35% of employed labor force & 9% of unemployed are self-employed; 12–15% of population is marginally employed	1972: 23% of labor force unemployed; 42% of population is urban	1974: 150,000 paid up union members—60% BITU (JLP), 39% NWU (PMP), 15% "other" 1976: 70,000—100,000 Rastafarians (less other-worldly, more radical); elements of lumpens are militant and violent	Workers' Liberation League formed (later Workers' Party of Jamaica) Increased political participation at various levels

ployment than into un- and underemployment. The peasantry and agro-proletariat were replaced by the lumpen and subproletariats, as "own account" activities (marginal self-employment) became the rule in the new urban environments. The growth of sales and service activities and a civil service entering a period of gradual decolonization provided significant opportunities for a growing middle class.

The years 1953 to 1969 brought to fruition both the promises and limits inherent in the program of economic development. The program brought benefits for the merchant fraction of capital, dominant by the end of the period, the middle class, which experienced significant growth both in size and in purchasing power, and the unionized working class, particularly workers in the bauxite industry. The expanding economy was, however, heavily dependent on imports of consumer goods, semi-finished products, and even foodstuffs, and could not provide sufficient employment. By the late 1960s, 49 percent of the arable land was monopolized by 0.2 percent of the landowners—the bauxite companies being the chief factor in the increased concentration. The threat of growing unemployment was reduced largely through the safety valve of emigration to England and, after independence in 1962, to the United States and Canada.

With decolonization in 1962, the state greatly expanded its functions, with employment in the public sector climbing to 57,000 by 1968 from 4,500 in 1943 (Stone, 1986, p. 42). Behind the visible presence of these "servants of capital" in politics and public office, members of the capitalist class penetrated many of the parastatals (quasi-government agencies) such as the Jamaica Industrial Development Corporation, the Bank of Jamaica, the Banana Board, and so on. The middle class nonetheless began to develop a strategic fraction that displayed considerably more independence vis-à-vis the capitalists. Among the educated professionals, the numbers of those who used state connections to engage in capitalist activities grew.

The limits of the development program were experienced mostly by those at the lowest end of the spectrum—although the economic dislocations it created did not go unnoticed by an increasingly vocal group of radical intellectuals. As the gap in income distribution widened, the possibility of steady employment remained a mirage for most. The annual income of the poorest 30 percent of the population actually declined (in constant 1958 dollars) from $32 per capita in 1958 to $25 in 1968. At the political level, mediating these contradictions became an increasingly intractable problem. The traditional system of patronage (wherein each party provides employment and other resources to its followers when in power) could not contain the activities of the lumpen proletariat, more numerous in this period than either unionized workers or party members. With radical ideologies and criminal activities on the upswing toward the end of the period, concern about these elements grew among the capitalists and middle class as well as among members of the political apparatus.

Since the last period (1970–1980) holds the most immediate interest, we have provided an overview of the class structure in 1973, the year prior to the declaration of democratic socialism:

Jamaican Class Structure, 1973

Dominant Classes:	*Percent of the Population*
Capitalists	0.5
Intermediate Classes:	
Middle Class (Upper Stratum)	4–5
Middle Class (Middle Stratum)	8–9
Petty Bourgeoisie (Upper Stratum)	1–2
Petty Bourgeoisie (Middle Stratum)	2–4
Subordinate Classes:	
Proletariat (Urban Industrial)	22–24
Proletariat (Rural agricultural)	4–7
Lumpen proletariat	22
Subproletariat	15
Peasantry	17–20

This list (adapted from Stone, 1980; Nelson, 1974; Munroe, 1981; and Stephens and Stephens, 1986) conveys graphically the skewed nature of the society after some thirty-five years of modern "development." At the beginning of this period, capital concentration in the economy was increasing. The emergence of conglomerates such as the Industrial Commercial Development Ltd. (ICD) Group, Pan Jamaican Trust, and National Continental Corporation (NCC) was accompanied by an expanding financial infrastructure—a stock exchange, unit trusts, and new commercial banks. The workings of a power structure composed of the famous "21 Families" were revealed. The Matalons, important members of this power structure, were at the forefront of expanding industrial capital and allied to the PNP.

The PNP government installed in 1972 continued to expand the state's functions, attempting to assume a leadership role in the economy. Tensions between merchant and industrial capital sharpened initially, with the state favoring the industrialists in its attempts to emerge from under the tutelage of the merchants. The industrial fraction initially gained some ground over the merchants and became more clearly differentiated, but the merchants were able to withstand the combined attack and retained their dominance. Economic decline and the combined deleterious effects of state policies and activities of the lumpen proletariat in the latter half of the period took their toll on the capitalist and petty capitalist classes, leaving huge gaps quickly filled by members of the black and brown professional/managerial middle class and some of the subproletariat.[6] Through these and other state activities during the period, the contours were revealed of an emergent bureaucratic/entrepreneurial fraction of the middle class that attempted to use state power to join the ranks of the capitalists.

Government programs to increase employment for subordinate classes were stymied by negative growth in the latter part of the period. But income distribution improved significantly as social and economic programs served to reduce disparities. Economic decline caused both opportunities for entry into petty bourgeois activities (such as small-scale import and distribution) and considerable distress (by the end of the period, 40 percent of the work force was self-employed). Increased political activism provided support for organized groups such as the Workers' Party of Jamaica as well as strikes, "captures" (illegal occupation) of land and other property, and so on. By the end of the period, however, no real political alternatives had emerged.

The PNP and Radicalism

PART

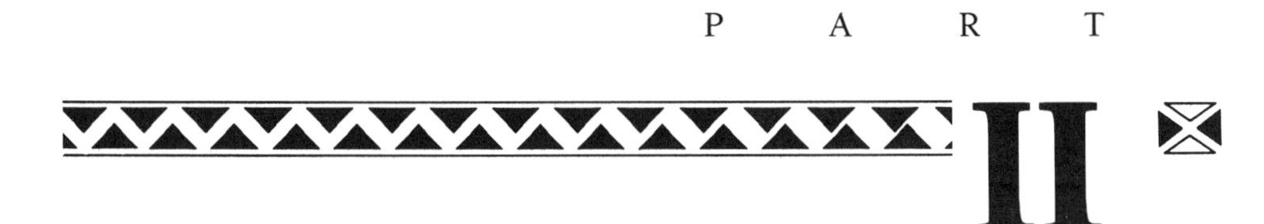

Impulses for Change
The Push from Below

CHAPTER 3

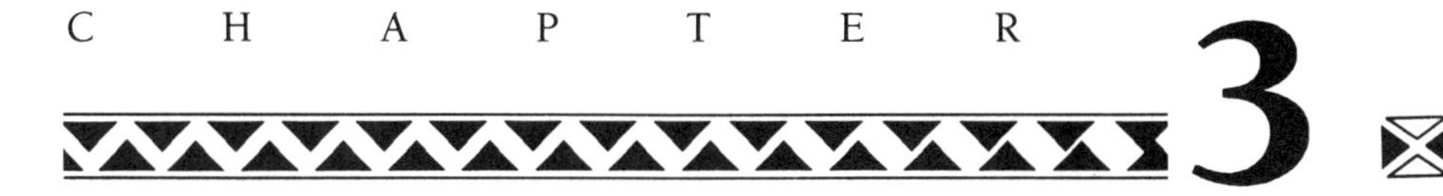

The year 1938 stands as an important threshold in modern Jamaican history. For several months during that year, a labor rebellion engulfed the colony, as a wave of strikes, riots, and violence disrupted normal activities. The immediate spark was perhaps less important than the conditions that preceded it: unstable work, exploitative wages, and inadequate protection for workers. So serious were the implications of the rebellion that the Colonial Office dispatched a royal commission (the Moyne Commission) to investigate and make recommendations to remedy the crisis. (See Ken Post, 1978 and 1981 for definitive works on the rebellion.)

Spurred on by the rebellion as well as the Commission's recommendations, the colonial government took steps to quickly transform the colony into a modern political society.[1] By 1944, Jamaican civic life featured political parties, trade unions, and adult suffrage—the principal ingredients of a well-tried recipe.

Of course, the final product was very much a reflection of local conditions, not the least of which was the perennial need to contain politically and socially radical expressions of the subordinate classes.[2] How did the revamped political institutions fare in this regard? This chapter traces the rise of these new entities, identifying the crucial historical and structural differences between the JLP and the PNP, and focusing in particular on the peculiar role of the PNP as a haven of radicalism. The main features of Jamaican radicalism emerge from a study of four of its more recent expressions: the rebellion itself, and the earlier movements encompassed by Bedwardism, Garveyism, and Rastafarianism.

Political Parties: Origins and Organization

The People's National Party (PNP) was founded in 1938 by Norman Manley, a brilliant barrister and Rhodes Scholar, along with a nucleus of disgruntled middle class intellectuals who were to play an important role in the peculiar politics of that party.[3] In the beginning, the PNP was closely associated with the Bustamante Industrial Trade Union (BITU) formed by Alexander Bustamante, Manley's mercurial cousin, who made himself its president for life. But Manley and Bustamante, a moneylender and adventurer, were temperamentally and philosophically different. Their brief collaboration came to an abrupt end in 1943, when Bustamante formed the Jamaica Labour Party (JLP) after his release from seventeen months' internment for his political activities.

The Jamaica Labour Party (JLP)

Prior to the creation of the JLP as a rival to the PNP, Bustamante's political activities were centered around the BITU, which, by 1943 controlled the bulk of organized labor. This fact may at least partly explain why the JLP was not burdened with a political philosophy: its flanks were covered by the immense and compelling shadow of Alexander Bustamante. Its policies flowed from pragmatism and reaction: a commitment to narrow bread and butter issues, seen as the legacy of the labor rebellion; the protection of business against the socialist threat of the PNP; and the pursuit of law and order. An excerpt from the founding document of the party tells the story:

> The party is pledged to keep within a certain moderate conservative policy in order not to reduce beyond reason, or destroy the wealth of Capitalists to any extreme that will eventually hurt their economical inferiors, but to advocate for the introduction of such measures and Laws that will shorten the terrible wide economic and social gulf that exists today, that almost inhuman disparity between the haves and the have-nots. (*Daily Gleaner,* August 12, 1943, p. 5)

The message was that the JLP would be at one and the same time the defender of workers and of the capitalist status quo.

What can explain the phenomenon of a labor movement with its roots still in a recent rebellion, linking itself after a few short years to a conservative party that defends the status quo? The answer lies in a combination of factors, including the consciousness of the working class, the personality of Bustamante, and the response of the upholders of the status quo.

Let us introduce Bustamante, who stood at the epicenter of these converging forces. The world of Jamaican slaves and their offspring is the world of Anancy, a hero out of African folklore. Anancy is an invincible champion who does not win by strength but by guile. He outwits Brer Tiger, the dull-witted bully, by using tricks that are not overtly demeaning of Tiger, but create a festival of laughter and mockery behind his back (Beckwith, 1929, pp. 219–220; Cassidy, 1971, pp. 275–276). The tale is a secular re-enactment of the myth of David and Goliath, with cunning, stealth, and calculation supplanting the intervention of God. With a keen sense of the psychology of politics, Bustamante joined the elements of David and Anancy, becoming both a secular messiah committed to a sacred cause and a folk tale hero.

In late August 1938, the labor leader took on the might of the *Daily Gleaner* over the question of his political motives. Was it wealth and power that he was seeking? Countered the wily Bustamante: "Yes, I want power, sufficient power to be able to defend those weaker than I am, those less fortunate." A year later (July of 1939), an official from the Colonial Office was subjected to a quip in the same tone: his duty to his people "[was] a sacred responsibility on [his] shoulders and [he would] not sidestep it." He would play the part of "a fox watching everyone," including the Colonial Office![4]

Bustamante's style, however, also included a feisty, confrontational streak. He could, indeed, sound quite subversive, earning from the colonial authorities the label of "rabble rouser" and, along with it, the requisite stint in prison. In 1937, he led the chant "Denham Must Go" (Sir Edward Denham was the governor of Jamaica between 1934 and 1938). In 1938, he attacked the system of taxation in a written protest to the British parliament. Bustamante was given to writing open letters to all and sundry: he challenged the owners of the Worthy Park Sugar Estate with veiled threats made in the name of BITU members; and he threatened to slap Robert Kirkwood's face "and pay for it." (Kirkwood was the general manager of the West Indies Sugar Company and a member of the powerful Tate and Lyle families that commanded substantial sugar investments in Jamaica.) These displays of bravado appealed to the powerless.

The arrogance he displayed toward employers was matched by authoritarianism when the workers were his audience. "If I come back from labor negotiations and tell you to go to work for nothing I expect you to go," he told striking workers in 1938 (*Daily Gleaner,* May 30, 1938, p. 18). Experience proved him right: for much of the time, Bustamante could demand and receive unquestioning obedience from his followers.

In an ex-slave society, fear of forces erupting from below can run rampant, and the colonial authorities certainly feared this messianic labor leader. Yet, perhaps paradoxically, Bustamante also displayed an extreme conservatism that considered change of any kind suspect. For instance, in his testimony before the Moyne Commission, the labor leader all but vowed to keep intact the supremacy of the planters while the laborers would benefit mainly through negotiated increases in wages. Paul Blanchard (1947, p. 96) notes that the "ruling class" liked his political views, in spite of his brusqueness. It was Norman Manley they feared because he constituted a threat to their property rights.

Under Bustamante's leadership, the JLP embraced a Burkean-style conservatism that continues to this day. It was Edmund Burke, one should recall, who warned us against pitying the poor, since it is their natural lot to labor in order to sustain civilization. It will be no surprise that education for the workers was not greatly favored. "The people need bread and butter; education cannot be eaten" was the motto.

It was not long, then, before Bustamante was able to put at ease the hearts and minds of capitalists and colonial governors alike. While not immediately appreciated, the fact was that the JLP agenda fitted snugly into the imperial design. Accommodation was in the cards. The "labor party" became the "capitalist" party" as soon as the elections of 1944 drove home the obvious: organizations representing exclusively capitalist interests could not survive in Jamaica after the institution of universal suffrage (as witnessed in this case by the devastating loss of the planters' and merchants' Jamaica Democratic Party).[5] Following the demise of their party, a number of substantial capitalists joined the JLP, among them Harold V. Lindo, a wealthy planter, and Owen K. Henriques, the wealthy and conservative owner of various interests in agriculture and industry. As increasing numbers of capitalists joined the ranks, the status quo was entrenched by virtue of a quid pro quo: financial support for the party (and Bustamante in particular) in exchange for the party's support of the capitalists' interests (Gannon, 1976, pp. 167–78).

In the end, the 1944 elections seemed to lend support to the JLP's pragmatic interpretation of the labor rebellion and its translation of these yearnings into the politics of bread and butter. These issues, after all, carried a tremendous and immediate appeal for the subordinate classes in circumstances of extreme need. Based on this evidence, it would seem that their consciousness remained steeped in traditional relationships to authority in which appeals to paternalism were buttressed by cunning—the tradition of Anancy. There was another added factor, also rooted in the consciousness of subordinate classes that shored up JLP fortunes more by omission than commission: the PNP was the haven of the middle class, as often detested as grudgingly admired by the subordinate classes.

The People's National Party (PNP)

From their origin to their institutionalization as governing parties, the JLP and the PNP traversed different paths. The JLP, emerging from a pragmatic, narrowly defined labor movement, became a conservative party that favored the capitalist status quo. The PNP's origins were in a nationalist expatriate organization—the Jamaica Progressive League—anchored to the self-serving, reformist aspirations of middle-class intellectuals. These elements were joined by prominent Jamaican professionals of various political persuasions as well as Marxist activists. The organization as a whole was committed to building a broad-based coalition, with the Marxists in particular assuming responsibility for organizing among the subordinate classes and, by the late 1940s, creating a mass base for the party.

Of the two parties, it was the PNP that retained a penchant for innovation and openness to new ideas. While the JLP responded to the letter of the demands of the subordinate classes, the PNP set itself the task of interpreting these demands and giving shape to an agenda of social change. In some respects, therefore, the PNP was the opposite of the JLP. Whereas the JLP could not be said to have a philosophy—unless its naked pragmatism passed for one—the PNP was steeped in ideas.

We have termed "Manleyism" the philosophy that informed the PNP agenda for change, to be discussed in Chapter 4. Manleyism would supplant paternalistic authoritarianism with democratic self-rule and economic justice, in what effectively constituted a redefinition of the "common good." On the political front, the main focus was on self-government and democratic institutions. Social and economic restructuring would emphasize self-reliance (for instance, community development projects) and a shift away from plantations toward a more egalitarian distribution of land to the peasantry.

Even from these scant details, one can see the outlines of a socialist agenda—one that was, however, "bound up as a practical and necessary fact with the destiny of the British Empire" (Nettleford, 1971, p. 98). Partly on account of this intellectual background and the middle class origins of the party, Manleyism was accused of being abstract and removed from the needs of the people. Yet, our analysis reveals that it was not a purely intellectual exercise. In the sections that follow, we will see in fact how what the PNP represented flowed from the ferment of the Jamaican people's traditions, ideology, and expectations.

Our references to Manleyism should not create the impression that Manley dominated the PNP in quite the same way that Bustamante was equated with the JLP. The PNP agenda emerged from a combination of factors including the personality and ideas of Norman Manley, the encounter with the forces of the status quo, and the debate among the various cross currents within the party—a mirror of the broader society. In fact, the party was riven at its core by fundamental ideological differences. Three main cleavages could be discerned: a conservative, pro-capitalist faction, a centrist faction, and a Marxist group, especially significant for its organizational skills (Munroe, 1977).

In part, the internal dissensions were inherent in the PNP's conscious aspirations, as the first truly representative political party, to include all people and be open to all political ideas. As Trevor Munroe aptly describes it, the PNP "sought to span social classes. It stressed that it would work for the people of Jamaica as a whole, for the furtherance of the interest of 'the country as a whole.'" The emphasis was on national reform, but members were not prevented "from believing in any special remedy." Munroe concludes: "to tolerate all the people's views was one way the party hoped to become the party of the people" (Munroe, 1966, pp. 23–24).

The politics of coalition to which the PNP dedicated itself were not without difficulties. Tolerance of "all the people's views" could mean an openness to the suffering and aspirations of the subordinate classes but this openness did not mean the absence of any filters. In the end, the collective vision was forced to submit to the wisdom of rational politics and weighted by an excessive middle class bias. This was perhaps to be expected, since for the subordinate classes themselves, radicalism, paternalism and gradualism were not distinct options but, rather, were intertwined in a world view that contained equal strains of accommodation and resistance. A closer look at the 1938 rebellion and earlier movements from below will give us a better sense of this consciousness and the ways it became incorporated in the PNP agenda.

▶ The Consciousness of the Subordinate Classes

The 1938 Rebellion and Class Consciousness

The wisdom surrounding the program of reconstruction saw the "disturbances" of 1938 as originating exclusively among the workers themselves. There was no evidence to show the role of "any provocative influences from outside." The events were sparked by the "ablest of agitators, hunger."[6] Yet while there were no agents provocateurs around, the rebellion was influenced by external factors, mainly variants of radicalism that were flourishing in parts of the Caribbean and elsewhere, spurred on by the example of the successful Russian revolution of October 1917 (Post, 1978; Karol, 1970; James, 1969). The Bolshevik Revolution effectively dramatized a new and exciting theory and practice of society. It showed that the people can take power and alter, for their collective good, seemingly immutable economic and political arrangements. Though not a socialist, Marcus Garvey himself had clearly recognized the exemplary aspects of this monumental historical event and implored that the black man's "sacred cause of liberty" should take courage from it (Munroe and Robotham, 1977, p. 102).

For the working class in Jamaica, the annual trek of seasonal sugar workers to Cuba—a hotbed of communist activity—became a significant link to socialist theory and practice. From Cuba came a messenger to Jamaica in the person of Hugh C. Buchanan, who was to play a prominent role in the working-class movement and the BITU. An active Marxist in Cuba, Buchanan returned to Jamaica in 1919 to immerse himself in the study and propagation of Marxism-Leninism. He founded the first working class newspaper—The *Jamaica Labour Weekly*—which, from the standpoint of working class interests, stood as the effective counterpoint to the *Daily Gleaner*, an organ of conservatism and the voice of capitalism.[7] Others took up the call: the local tabloid *Plain Talk* carried messages on the evils of capitalism and the necessity for the working class to organize through trade unions (Post, 1978, pp. 132–33; 153).

The example of the Bolshevik Revolution was also important in an interpretative sense. How should the popular will of the labor rebellion be interpreted and what forms should its political organization assume? The JLP, as we have seen, made promises of bread and butter. That approach is not without merit. It was beyond any contention that Jamaica, at the time of the labor rebellion, was plagued by pressing social and economic problems. The Moyne Commission Report as well as the census of 1943 catalogued the dismal state of things: lamentably overcrowded housing conditions, inadequate water supply, poor nutrition, weak education system, high levels of un- and underemployment, and so on. The immediacy of deteriorating social conditions and unemployment could energize a minimalist kind of pragmatism. The BITU, under Bustamante, went after the industries employing the major portion of the labor force. Within bananas (employing 43 percent of the labor force), sugar (18 percent), and stevedoring (6 percent), workers were being starved by wages that in some cases had not increased since 1915. In 1938, male workers in the sugar and banana industries earned the equivalent of US$0.36 to $0.61 daily and female workers earned US$0.21 to $0.36 daily; unskilled male workers in the public sector earned an average of US$0.73 daily, while women did no better than their counterparts in the fields, earning $0.34 daily (Orde-Browne, 1939). A survey of sugar workers revealed that the nominal household expenditure for the survey week exceeded the worker's wages by 12 percent (Post, 1981, I, p. 21). When placed against the banana producers' profits, which were running "at hundreds of percent" (Post, 1978, p. 124), these were starvation wages. And when put in the context of inflation—the cost of living index increased by 55 percent from 1941 to 1942—one can well understand the increasing number of strikes where the chief concern was with wages and working conditions (*Annual Report,* 1948, p. 14).

While objective economic conditions appeared ripe for change, at the ideological level unions and political parties were receiving mixed signals that did not clearly point toward a particular program of action. Take, for instance, the case of Robert Rumble and the Poor Man's Improvement Land Settlement and Labour Association (PMILSLA), an activist grass-roots organization founded in Clarendon parish in the 1930s. The movement resolutely attacked the exploitative practices of the landlords, punctuating each of its denunciations with demands for economic freedom. The difficulty was, however, that PMILSLA's ideology was located within the orbit of colonial (British) paternalism and notions of fair play. Let us listen to Rumble:

> We are loyal true-hearted British subjects and we feel that there has been no Governor who has shown such deep sympathy for the masses. We are the sons of slaves who have been paying rent to landlords for fully many years. We want better wages. We have been exploited for years and we are looking for you to help us. We want a Minimum Wage Law. We want freedom in this hundredth year of our Emancipation. We are still economic slaves, burdened in paying rents to landlords who are sucking out our vitalities. . . . In the name of God and British justice we are asking you to lighten our darkness of we who are the owners of properties that have been leased by Government for ninety nine years.[8]

The subordinate classes expressed pro-status quo sentiments, it would seem, just as freely as they proffered radical and revolutionary epithets.

How should one interpret the message of agitating workers and their supporters singing the British national anthem at the end of meetings? The JLP took them at their word, implementing and supporting policies that would continue, as the anthem went, to "send the King victorious" and allow him "long reign over us." Strikers during the rebellion were heard to sing "Onward Christian Soldiers" after many a fray. The JLP and the BITU interpreted these at expressions of a general and deeply felt disapproval of the prevailing conditions, but the affirmation, nonetheless, of the general principles of British colonial governance.

The position of the British was, of course, unambiguous. The colonial government maintained steadfast allegiance to the precepts of the Passfield Memorandum of 1897, which declared Britain's unwillingness to allow colonial politics to "fall under the domination of disaffected persons, by which their activities may be diverted to improper and mischievous ends" (Harrod, 1972, p. 208). In the 1930s and the 1940s, this was clearly interpreted to include Marxism-Leninism, as Governor Richards gave notice when, in 1940, he denounced communism from the podium and announced a ban on any newspapers that might carry anti-establishment sentiments.[9] The Moyne Commission Report leaves little doubt about the imperial agenda: the colony should continue to produce agricultural commodities, with economic relations essentially unchanged. As Gordon Lewis (1968, p. 92) observes, the Commission failed to address class relations and economic development, appearing more eager to deal with the immorality and sexual practices of the people!

For the PNP, the interpretation of the JLP and the Colonial Office were both inaccurate. Essence had been sacrificed to appearance. Especially for the Marxists, more careful attention ought to be paid to such oft-repeated cries from the workers for "jobs that last long."[10] The JLP read into this slogan the recommendations of the Colonial Office, particularly the Orde-Browne Report, which stressed intensified agriculture to the exclusion of manufacture and industry. Concessions to industry went mainly to pottery and brickmaking (Report of Mission of United Kingdom Industrialists, 1953). The Marxists saw the rebellion as a demand for radical economic reform. The call was for industrialization accompanied by a restructuring of agriculture. Studies conducted by the PNP showed that at the time of the rebellion, 67 percent of all rural males needed to supplement their income by wage labor.

If the demand for "jobs that last long" was the subject of interpretative equivocation, so was the pungent line from one of the popular ditties of the day: "White man ha' de money, but black man ha' de labour." It was an expression that gladdened many a Marxist heart, as it spoke to a dialectical relationship between the exploiters and the exploited, topped off by the withdrawal of labor. Rather quickly though, it was discovered that the fervor of these sentiments should at best be viewed in terms of gradations from lukewarm to warm! There were workers, for example, who preferred to negotiate without union participation. Such was the case with some of the striking sugar workers at Frome, who spurned the representation of the Jamaica Workers' and Tradesmen's Union (JWTU), then under the able leadership of the staunch champion of labor, A.G.S. Coombs. And there were the banana loaders in Portland (at the opposite end of the colony) who openly defied the labor leader, advising him that they were satisfied with "what we are getting" (Post, 1978, p. 266).

To an extent, then, the PNP, especially its Marxist faction, misread the various expressions of the rebellion. The rebellion erupted from a situation of low economic development but the level of revolutionary consciousness required to launch and sustain an emancipatory offensive was not present. Nonetheless, the subordinate classes' apparent lack of enthusiasm and understanding for the beginnings of their own emancipation was not a testimony to their disinterest or ignorance. There was some support for the Marxists, who had been heavily involved in political education and organizing, even bringing such fringe radical groups as the Negro Workers' Educational League within the PNP fold in 1939. But the revolutionary fervor was overestimated.

These examples of working-class consciousness might suggest a singular lack of interest in emancipatory politics. While there is much to support the position, it is noteworthy that most of the examples originate in the rural areas. The Marxists and the PNP as a whole had neglected the rural areas, in part due to financial constraints (Munroe, 1977; Post, 1981, I). Undoubtedly, they would have won more converts had resources and political imagination permitted. It is doubtful, however, that the gains would have been significant. While the labor rebellion certainly fostered political consciousness and awareness, the support for any particular political persuasion, party, or the like, was tepid. Although Bustamante led the JLP to a resounding victory in the first elections under adult suffrage in 1944, the support for his leadership was a mere 23.7 percent of the votes cast. Many voters elected to remain outside the newly constituted political process, but the pattern existed side by side with discrete clusters of unconventional political ideas, which will be discussed more fully below.

Somewhat predictably, the rebellion brought more clearly to the surface certain manifestations with which various interests could identify: Bustamante and his capitalist inclinations, and the Colonial Office and its interest in maintaining the status quo. It also brought into focus and legitimate contention a tradition of local radicalism, steeped in race and religion.

Race and Religious Consciousness

Whether through mainstream Christian churches or ever-present Afro-Jamaican cults, religion has played a dominant role in the political life of Jamaica. This role, which spans history and the class structure, bears some responsibility for promoting democratic principles at the levels of doctrine and practice (Henriques, 1968; Beckwith, 1929; Erskine, 1981; Turner, 1982). Up to the 1950s, the pattern of distribution among competing denominations remained stable. "Respectable" denominations (Anglicans, Methodists, Presbyterians and Roman Catholics) accounted for 50 percent of the Jamaican churchgoers. The Protestant denominations accounted for the greatest numbers (only 5.5 percent of the population was Catholic) and, according to research by Wendell Bell (1964), were more attuned to egalitarian attitudes. The other half of Jamaican churchgoers attended a variety of "less respectable" churches, running the gamut from native Baptists to Pocomania. Actually, it is likely that the figure for native churches is understated, as the stigma attached to them would skew the response (Henriques, 1968, pp. 80, 84). We have it from the complaints of Anglicans that there was a rapid growth of Baptist sects and other syncretic forms derived from that faith at the time of the 1938 rebellion.

The democratic tendencies mentioned above have been demonstrated in practice. Early Methodist and Baptist missionaries delegated a significant portion of pastoral responsibilities to church members drawn from the people. This widespread practice quickly led to a tradition of leadership that resulted in a proliferation of independent, black-led churches (Turner, 1982).

These missionary institutions also inspired the spread of democratic tendencies by their participation in local politics. For instance, they embraced causes that promoted mass enfranchisement and the undermining of the planter hegemony. The Baptists encouraged workers to demand higher wages and to support political candidates favoring peasant interests (Hurwitz and Hurwitz, 1971, p. 130). The roots of these activities were deep in history, dating from the period of slavery. As Mary Turner (1982, p. 153) correctly observes, "Missionaries of all denominations were both essential to the rebel cause and ancilliary to it; leadership lay with the slaves they inspired and trained."[11] This basic fact lies at the heart of the general proposition, applicable even to contemporary Jamaica, that while retreatism and apolitical behavior exist, they were not an impediment to the spread of the democratic ideal. Indeed, as events in Jamaican history attest, religion, embellished with democracy, was the language of political theory and practice.

The influence of Afro-Jamaican socio-religious movements outside the mainstream was at least of equal importance. The history of colonial and post-colonial Jamaica reveals a constant tension between the Eurocentric upbringing of the dominant and middle classes and the Afrocentric yearnings constantly surging from below. The old fear and disdain of the teeming black "masses," with their "creolism," syncretic religious forms, and periodic rebellions, were not diminished with the advent of the new fear peculiar to the twentieth century—communism; if anything, it was augmented by them. For supporters of the Eurocentric status quo—which, in the 1930s and 1970s included middle class reformers—"subversion" had two faces: race consciousness and communism. Often, the first was clad in the raiment of the "native" preacher and his followers who maintained the connection to African culture largely through local religious expressions. As a Catholic missionary to Jamaica noted in the late 1800s, attempts to repress Afro-Jamaican culture failed, inasmuch as "they could not control religion, [which was its] core." He further observed that local forms of religion "continued to develop as a vigorous alternative to mission Christianity and as a source of Afro-Jamaican identity and resistance at a time when European political and cultural hegemony was being asserted" (Stewart, 1984, p. 101).

We will now examine important commonalities between Manleyism and the most influential radical movements of those times—Bedwardism, Garveyism and Rastafarianism. What emerges are the outlines of a tradition of accommodation that typifies the relationship between the PNP and radical movements, even Marxism-Leninism. It is this tradition that explains in part the problems encountered by democratic socialism. It is this dialectic of rebellion and accommodation that so impedes the appearance of the socialist-oriented consciousness that is embedded in Jamaican political traditions and has its origins here.

Bedwardism

Bedwardism was a religious movement that flourished between the 1890s and the 1920s. Established in 1889 at August Town, in the parish of St. Andrew as the Jamaica Native Baptist Free Church, it quickly grew to a membership of thousands, drawn from affiliated "camps" scattered throughout the colony. The preacher-leader after whom the movement was named, Alexander Bedward, has passed into local lore mostly because of two events. First, as a self-appointed messiah, he failed to ascend to heaven in the manner of the prophet Elijah. Second, his many confrontations with the law led to his being placed in the Kingston Mental Asylum, where he died on November 8, 1930.

Beneath the systematic propaganda peddled against Bedwardism—a concerted conspiracy between the courts, the established church, and the main propaganda instrument of the status quo, the *Daily Gleaner*—one discovers a bona fide social movement. Bedward excoriated the government for its racist policies and laws "which oppress the black people. They take their money out of their pockets; they rob them of their bread and they do nothing for it" (Napier, 1957, p. 14). And he placed religious symbolism at the service of radical social change. There was a new wall of Jericho, a "white wall" that must be beaten down by a "black wall" (Bradley, 1960, p. 394).

Bedwardism derived its message and source of support from the poor peasantry and the more precarious members of the working class such as casual laborers and household workers, as well as some craftsmen. It heard these perennial demands for land and justice and refracted them through religious prisms; but the emphasis was on earthly redress more than on retreatism. Land should be taken from the big planters and distributed to the people, Bedward thundered. In what was a comprehensive approach to social organization, he stressed community life, land reform, and political freedom. His communities were organized to provide for the social welfare of their members, and included formal structures designed to cater to the needs of the old and infirm, educate the children, teach self-reliance, and promote some measure of economic self-sufficiency. In short, we are describing the rudiments of community development, which, in an earlier example of the phenomenon—Paul Bogle's community—included internal courts of law to dispense justice and protect fair play (Robotham, 1981; Schuler, 1980).

The mingling of political and religious agendas emerged out of social and political need and oppression as well as deeply entrenched Christianity. The leader effected a symbiosis that transformed the image of Christ through a process of "reincarnation." In so doing he became the twin personification of the religious and the political without committing the fateful error of upstaging the Messiah. Bedward could insist on the observance of heavenly commandments and fulminate against earthly social and political shortcomings. He could baptize his flock in the "soul cleansing blood of the lamb" as well as deliver strong political messages to which the flock reacted with religious fervor. He was cut from the same cloth as Sam Sharpe, the leader of the so-called Baptist War—a slave uprising that took place in the western parishes of Jamaica in 1831.

It is this fusion of the religious and the political that makes Bedward an important figure in local political history. It is, then, erroneous to suggest that Bedwardism "merely promised spiritual redemption" (Henriques, 1968, p. 68). The reality is much closer to Weberian notions of social movements. Max Weber (1964) speaks of prophets and like charismatic figures providing "directions," "paths of development," and "breakthroughs" not available through the more rational forms of religion or politics.

The peculiar symbiosis of politics and religion exemplified by Bedwardism and others (Robotham, 1981) is nicely explained by Leonard Barrett (1977, p. 27): for most Jamaicans, "religion is a total environment" where "folk traditions are nourished and preserved." It is the sphere within which political leaders must pay their deep respect to the "simplicities" of the barefooted.

In its time, however, Bedwardism, constituted a historically novel set of factors, in that it was probably the first global expression of Afro-Christianity as an active molder of the political will. The Morant Bay Rebellion of 1865, of which Bedward spoke approvingly, was also undergirded by Christian counsels, but reflected more the precepts, symbols, and ethos of western Christianity, as transmitted by the Baptists. Furthermore, the rhetorical appeal of Paul Bogle, the leader of that rebellion, though tinged with religious allusions, was steeped in British ideology.

Bedwardism brought into the open, in an organized fashion, many of the unexpressed yearnings of the subordinate classes for radical change. It was also, however, a hybrid creature, still mired in the old social relations. The quest for redress between the races and economic classes did not serve to promote egalitarianism within the movement. Bedwardism retained the paternalistic authoritarianism of the day. Physical chastisement was often the price a flock member paid for disobedience; and like that of Bogle's, the movement had an internal judicial system to promote justice and religious control (Robotham, 1981; Schuler, 1980).

The complex phenomenon of Bedwardism, then, helps us clarify the various strains present in the consciousness of at least a segment of the poor peasantry and lower ranks of the working class. To the extent that the movement was radical (and the reaction of the authorities leaves no doubt that it was so perceived), its radicalism was defined largely through opposition to the racial and economic status quo. The movement furthered democratizing ends only in an indirect way.

In one sense, then, such movements contributed to the growth of a kind of grass-roots egalitarianism, since participation enhanced some members' sense of their own power as actors, enabling them to pursue leadership roles in other movements. There is the case, for instance, of Robert Hinds, who in the 1930s became the chief lieutenant of Rastafarian leader Leonard Howell. Hinds had marched with Bedward in 1921 and was arrested, along with Bedward and 685 followers (Lewis, 1987a, p. 39). To the extent that the movement was authoritarian, however, it gave ample support to the notion of social change as the province of the "Great Man" who was to be followed without question. Parenthetically, a measure of Bustamante's power and influence with the subordinate classes can be traced to these origins. We "will follow [him] 'til we die," sang his supporters, giving him their assurances that, were he to select a dog to run in their constituency, they would vote for it! Such was the degree of devotion he commanded.

The legacy of movements such as Bedwardism, then, is mixed. These movements provide a sense of continuity to the agitation from below that appears as a constant theme in Jamaican history, but are not rooted in the ideal of human equality. The danger thus, is that the quest for redress of racial oppression will produce its own versions of oppression and despotism, as the history of neighboring Haiti bears witness.

In the socio-political climate of the 1940s, neither Manley nor Bustamante would take up the religious and racially loaded aspects of Bedward's message. Afro-Christianity and other local cultural expressions of African origin were still anathema to the dominant classes and much of the middle class, and thus far from acquiring the legitimacy that would allow for inclusion in political platforms. As we shall see in our discussion of Garveyism and Rastafarianism, the establishment reacted with ridicule and with repression to expressions of race consciousness, even if they were posed in rational terms. It was not until the late 1960s and the early 1970s, after the Black Power Movement, that the fear and distaste for these expressions diminished sufficiently to allow their inclusion in national agendas of change.

Garveyism

Marcus Garvey founded the United Negro Improvement Association (UNIA) in Jamaica in 1914. Shortly thereafter, he took it to the United States where the movement eventually acquired nine million members. The UNIA was far-flung, having chapters throughout the Caribbean and the Americas and attracting sympathizers in Africa. While Bedwardism had appealed mainly to the downtrodden, Garvey's Jamaican supporters were drawn from all but the highest reaches of the society, including even some of the budding black bourgeoisie.

Not as strong in Jamaica as in the United States—due to the more repressive nature of colonial society—the local UNIA movement was nonetheless the strongest of those outside the United States, boasting six divisions and many subdivisions by 1919 (Lewis, 1987b; Cronon, 1955). There were also differences in approach and consequences between the North American and Jamaican branches, which arose from the different conditions of black people in the two societies (Vincent, 1976, pp. 171–173). Whereas the former shunned the labor movement on account of racism, the Jamaican UNIA became deeply involved in the trade union struggle and built strong support among the working class.

Garvey's references to a black, redemptive God nicely interfaced with prevailing Afro-Christian orientations. The Rastafarians in particular, whose movement emerged in the early 1930s as an offshoot of Garveyism, responded to Garvey's call to "look to Africa for the Crowning of a Black King" and took up the mantle of black nationalism.

Finally, Garvey's call for the social rehabilitation of the black race encompassed the struggle for political rights and black economic nationalism. These emphases struck a chord especially among members of the black middle class and petty bourgeoisie: the teachers, ministers of religion, farmers, and small entrepreneurs. Among Garvey's more influential middle class followers were H.A.L. Simpson, a solicitor and former member of the Legislative Council; F. G. Veitch, a landowner, physician, and a member of the Legislative Council; and Eustace McNeil, a wealthy landowner and planter.[12] To this list should be added many prominent ministers of religion, among them E. E. McLaughlin, S. M. Jones, S. U. Smith, and Churchstone Lord.[13] Thus the different aspects of Garveyism found audiences among different social groups.

Given a more rational and reformist approach than Bedward's, Garvey could appeal to broad sectors of the society. His approach to politics advanced the process of rationalization: the Garveyite People's Political Party (PPP), founded in 1928, took the unusual step of developing a platform. Not only was the platform, announced in 1929, a vast improvement over the existing "politics by custom" (Lewis, 1968; Munroe and Robotham, 1977); it was also enlightened and possessed significant appeal. The more outstanding features of its comprehensive agenda included land and educational reform, protection of native industries, legal reform, peasant housing, the establishment of a Jamaican university, a minimum wage, workmen's compensation, and land expropriation for public use without compensation (Vincent, 1976, p. 173). The announcement of an election platform was a forerunner of formal party politics in Jamaica. The PNP adopted this PPP innovation in the 1944 elections, which it contested with a slate of candidates committed to a party platform.

Race and color were extremely emotive issues in Jamaica in the 1920s, when ex-servicemen returning from World War I would report freely on their degrading racial experiences in the Mother Country (Elkins, 1978). However, while the race issue was central to Garvey's agenda of change, he was careful at strategic moments to show that he was a man of reason and temperance: "There is no sense in hate; it comes back to you; therefore make your history so laudable, magnificent and untarnished, that another generation will not seek to repay your seed for the sins inflicted upon their fathers" (Essien-Udom, 1966, frontispiece).

On the whole, therefore, the People's Political Party was less aggressive on the issue of white racism than Bedwardism had been. Stress was placed on acquiring power through the self-reliance of the black race, "power in education, science, industry, politics and higher government" that other nations would respect. The message was to get power of any kind (Vincent, 1976, p. 175).

Garvey embraced a kith and kin approach to socio-economic change and, as is well known, was an advocate of black capitalism. This stance was in direct opposition to the radical ideas then being propagated by the communists, with their focus on class rather than race. Should a confrontation be expected? The time did not seem ripe. A celebrated debate that took place in August 1929 between Garvey and a North American communist, Otto Huiswoud, was easily won by Garvey on his home ground (in Post, 1978, p. 2–3). Radicalism in Jamaica was firmly focused on race, although future events would reveal that, far from being over, the debate between race and class was merely beginning.

Garvey's measured reason, organizational accomplishments, and anti-communism did nothing, however, to earn him legitimacy with upholders of the status quo (Vincent, 1976; Stein, 1986). Emphasis on race would not be tolerated from any source; and when it was combined with his charisma and capacity to move people to action it became the object of fear by the Jamaican dominant classes and the brown middle class, which was not immune to Garvey's attacks. For instance, black peasants were left without leaders, Garvey noted, because "the educated class of my own people . . . are the bitterest enemies of their own race" (Correspondence, 1916, cited in Lowenthal, 1972, p. 286).

Like his predecessors, then, Garvey was to "spend time" in His Majesty's prisons. The immediate cause was the tenth plank of his platform, which proposed legislation to curb unscrupulous judges who take certain actions "in defiance of British justice and Constitutional rights" (Lewis, 1987b, p. 210). He was cited for contempt of court, fined £100, and sentenced to three months' imprisonment. The spiritual leadership he unwittingly provided for the Rastafarians also drew suspicion; as we see below, this movement was greatly feared by the dominant classes as well as the middle class, which found its unconventionality too discontinuous with prevailing social, political, and philosophical tenets.

The qualities of messianism surrounding Garvey's figure provided the upholders of the Jamaican status quo with a target for ridicule with which to mount an effective attack. Titles such as "Provisional President of the African Republic" and the "Black Moses" were the equivalent of waving a red flag before a bull. He was given the "Bedward treatment": first, direct mass attention to claims of divinity, expressed or inferred, then institute a systematic program of delegitimation through the use of the church, the law, the press, and other instruments of the state.

Garvey's eclipse was only slightly less ignominious than Bedward's. It included a period of incarceration, highly publicized debt problems associated with the maintenance of Edelweiss, his home, defeat in the Kingston and St. Andrew Corporation (KSAC) elections on January 30, 1930, and self-imposed exile in England.

There is a certain irony surrounding these elections. Garvey had campaigned substantially on the basis of the comprehensive platform indicated above. The winner, George Seymour-Seymour, a wealthy, white planter and businessman, was not afraid to stick to hidebound and outmoded tradition. The customary ad hoc approach to local politics was preferable to Garvey's innovations, he stated, because Garvey's more formal and inclusive program was simply not workable (*Daily Gleaner*, January 22, 1930, p. 1). Of course, this was some fourteen years before much of the natural constituency of Garvey and like reformers would be able to cast their votes in an election, since universal adult suffrage was not won until 1944.

As was the case with Bedwardism, Garveyism contained both radical and conservative tendencies. As John C. Gannon suggests (1976, p. 51), Garvey was an "imperialist." His trappings and entourage were certainly in the finest tradition of royalty and there are many indications that his agenda called for the replacement of one set of masters for another set that might be only a bit more benign and somewhat progressive. His "imperialism" was evident, for instance, during the International Convention of the Negro People of the World, held on August 1, 1920 in New York, when Garvey assumed the title of "His Highness, the Potentate" and conferred peerages and knighthoods upon his "High Executive Council." At the other end of the social spectrum, he was often quite contemptuous of the Rastafarian brethren. (See the *Jamaica Times,* August 25, 1934, p. 21, for example.) Nonetheless, they continued to revere him, stating that Garvey's work is "unfinished . . . and is linked up with the works of Emperor Hyili Silassi I" (*Abeng,* June 27, 1969). The claim is also repeated in contemporary times (Tafari, 1980). Indeed, Judith Stein (1986, p. 273) finds "his ideas . . . rooted in the elite response to economic and racial change." Ken Post (1978), among others, locates Garvey within the middle class ethos.

Nonetheless, from a social-psychological perspective, Garvey provided a powerful model. Here was a black member of the Jamaican subordinate classes, an internationally known figure, who had successfully taken on the might of the white United States and effectively organized an important section of the black population. In a real sense, he was challenging the hidebound notion that black people must surrender to white leadership as a matter of course (Ryan, 1974, p. 30). Even after the movement's decline and Garvey's death, Garveyism kept burning as a beacon of racial pride and black dignity. These emotional referents have minimized the underlying conservatism, middle class orientation, and imperialism of the UNIA leader.

Garvey's leadership, in turn, prompted others to action. In Barbados, one of Garvey's disciples, Herbert Seale, was at the forefront of union organizing. The Barbados Progressive League, of which he was president, acquired more than 20,000 members (Hoyos, 1974, pp. 89–90). In Jamaica, well-known ex-Garveyites such as St. William Grant and Hugh Buchanan (whom we met earlier) became prominent labor organizers for the BITU in the post-rebellion period. The latter became the BITU's first general secretary in the early months of 1938.

Of course, these achievements could (and did) give rise to two major ideological strains. The first was represented by the message that the black race could rise to the heights reached by its white counterpart. Parenthetically, this was Garvey's predominant motive. The second carried revolutionary potential, insofar as some of those who heard and liked the message saw few if any possibilities for accommodation within prevailing socio-economic and political structures. This, in fact, appears to have been the reaction of Hugh Buchanan who became, in the late 1930s and 1940s, a strong proponent of Marxism.[14]

The legacy of Marcus Garvey touched the totality of post-rebellion politics, contributing leaders, organization, and ideas. Perhaps for this reason, the absence of references to race and color in post-rebellion politics is a remarkable omission. We discuss the reasons for Manleyism's shunning of race and color in the next chapter. As for Bustamante and the JLP, the most likely explanation must again be found in the leader's pragmatic and conservative approach: the leader had no use for Africanisms and would not risk arousing the ire of the colonial government and dominant classes whose stance on race was unequivocal. An editorial in the *Daily Gleaner,* for instance, contained the remark, "race questions are not permitted in these columns." (November 14, 1930, p. 12).

With the benefit of hindsight, we can clearly see that one of Garvey's most significant contributions was the push for black economic nationalism. Many middle class entrepreneurs were initially attracted to him, some of them even running on the PPP slate. What Garvey had achieved, particularly in the United States, was not lost on a vibrant group of black entrepreneurs in Jamaica who were searching for leadership and policies on which to anchor their own initiatives. That there was a resurgent consciousness is evidenced by the actions of the newly formed Native Defenders' Committee, which in 1935 actively sought to promote and protect native (black) entrepreneurs from other ethnic incursions.[15] Later, this same interest group would find an outlet in the Pioneer Industries Act, which offered tax inducements and protection from imports to local entrepreneurs (see Chapter 2).

Rastafarianism

With Garveyism in disarray and retreat, the black nationalist movement was splintered. Some prominent Garveyites continued organizing labor and so found their way to the BITU. Many middle class members found outlets in citizens associations, which later became allied to the PNP. For the moment at least, it appeared that race consciousness as a central organizing principle with wide appeal had reached its ideological and political limits and was now in retreat. It retained its centrality only in an offshoot of Garveyism—Rastafarianism—which was perhaps even more marginalized than Bedwardism and found root among the poorest segments of society.

Rastafarianism owes its name to Haile Selassie, Ras ("chief") Tafari, crowned Emperor of Ethiopia in 1930. Dignitaries from many European countries, including an emissary from the British Royal House, attended the coronation—a widely publicized event—and were seen to kneel before the new emperor. Now, Haile Selassie was no ordinary emperor. The name Haile Selassie means "Power of the Trinity" and other titles—King of Kings, Lord of Lords, Conquering Lion of the Tribe of Judah—were rich in biblical allusions. His lineage, it was claimed, was in direct descent from the Queen of Sheba. Ex-Garveyites interpreted literally their leader's quest for a "God of Ethiopia":

> If the white man has the idea of a white God, let him worship his God as he desires. If the yellow man's God is of his race let him worship his God as he sees fit. We as Negroes have found a new ideal. . . . We Negroes believe in the God of Ethiopia, the everlasting God—God the Father, God the Son, and God the Holy Ghost, the One God of all ages. That is the God in whom we believe, but we shall worship Him through the spectacles of Ethiopia.[16]

Through intense study, the book of Revelation yielded the proper interpretation of events. For Rastafarians, Haile Selassie became the God Incarnate, Ruler of Ethiopia, the Promised Land. He was the black messiah whose coming had been foreshadowed: "Weep not: behold the lion of the tribe of Judah, the Root of David, hath prevailed to open the book, and to loose the seven seals thereof" (Revelation 5:2–5).

In line with other Afro-Christian movements, Rastafarians engaged in a rich reinterpretation of biblical symbols, creating a cosmogony in which black people are the lost tribe of Zion of the diaspora, the true children of God, while Jamaica, its white rulers and their brown and black lackeys are Babylon. The new identity was also expressed in personal appearance and lifestyle. Rastas took to wearing their hair in the dreadlocks of Ethiopian Nyabingi warriors. They avoided contact with "Babylon" and its ways, preferring their own society and making a living through farming, crafts or other non-wage labor. With their strange, often fierce biblical language and new-found sense of dignity and power, they were feared and reviled.

Social commentators' characterizations of the movement have ranged from retreatist to revolutionary (Post, 1978; Campbell, 1988; Owens, 1976). In a sense, the movement defies definition, as its various centers exhibit different ideological emphases and political orientations. In its earliest period, which concerns us here, Rastafarianism could best be seen as culturally revolutionary and socially retreatist. Yet the political message held a potential that the British understood as sedition. Rastafarians were told, for instance, that in singing the British national anthem, "they should think of Ras Tafari as he was their king . . . they were only sheltering under the British flag" (Post, 1978, p. 166).

The movement's militant racial ideology and its members' status as social outcasts made Rastafarians the subject of harassment first by the colonial government and then by Jamaican governments after independence (Smith, Augier, and Nettleford, 1960; Morrish, 1982). The now-familiar pattern of repression continued. In 1934, Howell, Hinds, and three other leaders were charged with sedition. One of them, H. Archibald Dunkley, was held in the Kingston Asylum for nearly six months in 1935 (Post, 1978, pp. 165–67).

Historical events, however, conspired to allow Rastafarianism to grow despite repression, since support for Ethiopian resistance to Mussolini's colonial ventures in the 1930s provided it with broader appeal and a measure of legitimacy. Outrage at the fascist war again focused many Jamaicans on Africa and Ethiopianism, triggering an upsurge of black pride. Efforts were made locally to recruit soldiers to go to the emperor's aid. As Ken Post (1978, p. 168) describes it, by 1936 local newspapers were reprinting articles from black American newspapers like the *Chicago Defender* and the *Pittsburgh Courier*.

The excesses of white fascism provided support for Rastafarianism and the more respectable "Ethiopianism" in at least two significant respects. First, the movement could claim, empirically, the fulfillment of prophecies surrounding the Emperor: Haile Selassie's fight and victory over fascism (he returned to Ethiopia as Emperor in 1941) were victories of the "King of kings" over the "beast" (Revelation 19: 16, 20). Second, the destitute condition of the subordinate classes in the 1930s and 1940s, dramatized by the 1938 labor rebellion, made many of the unfortunate search for otherworldly redress. The Rastafarian movement attracted many of the discontented who were being propelled into a wider black nationalist movement. Indeed, by 1939, the Rastafarians represented a broad range of subordinate class elements that shared an interest in social change and the demand for the rights of the black race.

It should be plain from our earlier introduction of the two political parties, as well as from the repetitious reactions of the establishment (like Bedward, Howell also died in a mental asylum in the 1950s), that the subject of race would not command a great deal of positive official regard, even in the ferment of post-rebellion social reconstruction. Among the emergent political structures of the 1930s, race was made an issue only by the PNP Marxists, who raised it periodically through the editorial columns of the PNP newspaper, *Public Opinion*. For the Marxists, however, race was epiphenomenal. In Richard Hart's view, for example, the class struggle in which the Rastafarians were embroiled was "no simple discord of colour" (or race, we might add), but related to the contradiction of capital and labor.[17] The *Jamaica Labour Weekly*, a left-wing tabloid, did not mince words. In its view, it was folly to place undue stress upon race and color; the proper course for "grown men and women [was to join] constructive institutions such as Trade Unions and political parties."[18]

For the rest of the PNP, Rastafarianism was simply outside the pale. W. A. Domingo, writing in the *New Negro Voice* on October 10, 1943, clearly made the point that one should promote opportunities in Jamaica, not Ethiopia. The movement's words and actions spoke of an extremism that must be kept at a safe distance. Its leaders strongly advocated the non-payment of taxes to the government. Claims of divinity were made. At a meeting in Seaforth, St. Thomas, the brethren claimed that Howell was Christ (Post, 1981, I, p. 176). In their turn, some Rastafarian leaders themselves, displaying strongly separatist tendencies, appeared to do everything to discourage an association with the PNP. An excerpt from a dispatch to the Colonial Office provides highlights of a statement Howell made at a meeting in Port Morant on January 7, 1940.

> The white man's time was ended and that soon black men would sit on the throne of England; further that Hitler was in charge of Europe and that all European powers would be overthrown in 1949, and that at the end of the war the white nation would be utterly exterminated. (Post, 1981, I, p. 95).

The centrists and conservatives in the party, for the most part, identified with the posture of the *Daily Gleaner*. In a nutshell, the movement was the epitome of barbarism and the dregs of society were threatening to turn civilization on its head. It was tactless, and worse, dangerous, the position continued, to employ race and color except in the sense of assimilation. The antidotes for this habit of mind were performance and enlightenment captured by a favorite saying: "It is not the color of the skin/But the true heart that beats within." All this, as well as their unconventional dress and appearance, placed Rastafarians beyond the established canons of taste and civic responsibility. For the time being, at least, the movement was destined to remain a counterpoint, growing on the margins nonetheless, to resurface during more auspicious times in the late 1960s and 1970s.

Subordinate Class Movements and Modern Politics

We have already alluded to some continuities between the movements highlighted above and the politico-ideological structures of modern Jamaica. Messianic claims apart, Bedwardism was remarkably in tune with Manley's ideas. Bedward's far-flung camps, for instance, were in the tradition of community development dear to Manley, while his insistence on land for the peasants differed from Manley's only with regard to the means of acquisition—one advocating expropriation, the other compulsory acquisition.

The continuities also embrace Garvey's movement. The *Election Manifesto of the People's Political Party* outlined the plans for a new society to replace paternalistic and oppressive socio-economic and political structures. Again, land reform was central to the program. Of course, Garvey's attempt to fit his agenda into the mainstream, as opposed to the marginality of Bedward, meant that the former paid closer attention to political structures. Like Manley (although with less aplomb), Garvey advocated self-government and the reform of existing institutions. "I read *Up from Slavery*," wrote Garvey, "and I asked: 'Where is the black man's government? Where is his king and kingdom? Where is his president, his country, and his ambassador, his army, his men of big affairs?' I could not find them, and then I declared, 'I will help make them'" (Garvey, 1967, II, p. 126).

Bedward's quest for justice and salvation through religiously based organization bears similarities with what is, for Manley and Garvey, a humanistic quest. Sydney Webb's assertion that Fabian socialists (of which Manley was one) are interested in better housing and living conditions for people "as a means to an end—the development of individual character,"[19] can be an apt description of Garveyism. Garvey, in fact, placed a great deal of emphasis on character as the foundation for genuine emancipation.

On the whole, one is struck by the similarities in the high moral tone of Garvey and the PNP leader. Garvey spared no effort in appealing to the creative wellsprings of the black race. His consistent message was that the white race did not have a monopoly on the laudable achievements of civilization. With a renewed belief in the race and assisted by astute and affirmative leadership, black people would prevail. This theme finds an echo in Norman Manley's recurrent exhortation to Jamaicans that "man stands tallest when he rules himself." Both stressed that freedom carries a responsibility for rational self-development, which should be largely self-directed.

In turn, Bedward's sense of biblical justice could meld, in crucial ways, with the very British notion of fair play that undergirded Manleyism. Again, what emerged for Bedwardism as religious necessity was dictated for Manley by secular ethics. Viewed carefully, Bedward's famous "wall" speech, which effectively commenced his difficulties with the authorities, is both vengeful, in the Old Testament sense, and secularly moral in ways that would appeal to those believing in justice and fair play, two bywords of British ideology.

Garveyism and Manleyism also shared the conviction that education provided the strongest impetus for the rapid development of the masses. Garvey's elaborate and grandiose ideas on the founding of a Jamaican university, an academy of music and art, and the general promotion of education were quite Fabian, differing with Norman Manley's only in terms of scale (Hurwitz and Hurwitz, 1971, p. 191).

There was, of course, a major discontinuity between these movements and the new political leadership, as they did nothing to make race part of the discourse. If for Bustamante the causes were probably rooted in pragmatism, Manley's reasons were located in ideology: race questions were considered divisive and antithetical to the goal of uniting all under the banner of the common good.

Nonetheless, the similarities underscore the fact that Manleyism, except for its rational and secular roots, had much in common with the substantive aspirations of the Jamaican subordinate classes. At the same time, Manley's approach was worlds apart from the leadership style that appealed to these classes, introducing a discontinuity that would somehow have to be rectified. In fact, one could argue that two different cognitive orientations were at work.

Afro-Christianity had produced two politico-cultural strains. One, nurtured by the Baptists and the Moravians, could more easily be attuned to the rationality and competition associated with liberal democracy. The other, derived from the Pentecostal Church and the Church of England, abhorred political involvement, directing its adherents to avoid its corruption and store up their treasures in heaven. Though it is not possible to state it with any certainty, people in the rural areas were probably more likely to be affected by the latter. This folk culture might have resonated in a realm that was not easily accessible to modernization (Alleyne, 1988). Differing definitions of certain aspects of reality would then place the two cognitive orientations at loggerheads; from the newly emergent Western-centric perspective represented by Manleyism, the folk culture existed at a lower level of political development. From the opposite perspective, Manleyism could be seen as too removed and abstract. This would partly explain the initial difficulties experienced by the PNP in 1944: its system of narration did not find common ground with the prevailing culture of the subordinate classes.

Bustamante's approach to politics was far more in tune with this folk culture. Like Bedward, Bustamante subscribed to what we might call "politics as revelation." Bustamante, the secular prophet-politician, would mesmerize workers some fifty years after Bedward held his flock spellbound. The revivalist preacher, of course, infused his politics with religious sensibilities, while the labor leader played upon the basic instincts of survival and political psychology.

Education and Consciousness

If "native" religion did not provide much support for rational theorizing and abstract notions of the polity and the common good—such as were pursued by the PNP—education represented much more fertile ground. We will focus specifically on the importance the subordinate classes placed on education and the ideological strains that gave buoyancy to Norman Manley's rationalism and abstract principles.

Three factors account in large part for the importance of education for the subordinate classes: denial of economic opportunities, a reaction to the disdain of the dominant classes, and mimesis (adopting the affectations and outlook of one's "betters"). On the first count, Jamaica had a static, caste-like class structure that did not encourage upward social mobility through capitalist activities (Smith, 1970, p. 54; Jones, 1978, pp. 285–86). Aspiring brown and black entrepreneurs were largely prevented from pursuing entrepreneurial activities, since loans could not be easily gotten from commercial banks. This economic limitation channeled the creative offspring of the subordinate classes into the professions and the civil service (Holzberg, 1977). It is impossible to say with precision what influence this early structural limitation had on subsequent educational agendas. It is clear, however, that denial of economic opportunities was one of several social factors that contributed to the creation of a special place for education in the West Indian mind.

Traditionally, common people had always respected the Chinese, the Jews, the Syrians, and Lebanese for their economic strength and purpose. But they would not want their children—if they had a choice—to "sell salt fish behind counter" or "cut pants length." They would want their children (especially their sons) to be "a great man not only in music but also in sociology and economics," or "a man of international fame, a man who by virtue of his political genius has acquired much respect" or "a diplomat and to do wonders for the nation." Unlike the son of a white man, their boy would not pursue a life of business, working himself up the ladder through apprenticeships.[20]

This strong association with a world defined, often bombastically, by the verbiage of grandiloquence, is a trait of the subordinate classes on which others have commented (Ayearst, 1960; Henriques, 1968). Traditionally, the contours of power are configured from the magic of sonorous words and the political medium. Words define power, real and intended, and even appear to set the course for concrete change. We must wait until the decade of the 1970s for these educated children of the subordinate classes to start turning their attention to the power inherent in property (see Chapters 9 and 10).

The result was that the subordinate classes placed great stress on the power of education to promote upward social mobility and equality. For the most part, their educational priorities centered less upon inculcating "habits of industry" (Olivier, 1936, p. 366) than on patterns of instruction of a more sophisticated kind. As early as 1895, parents objected to their children's participation in practical work; they wanted them to receive solid book learning and nothing else (Hurwitz and Hurwitz, 1971, p. 185).[21] In fact, many had strongly supported Crown Colony trusteeship because it had markedly improved education, and wanted the trusteeship to continue until the system of education improved significantly (Will, 1970, p. 29). Between 1880 and 1903, the thirst for education made the electorate "curiously tenacious and sensitive" with the result that members of the Legislative Council made every effort to comply with these demands (Will, 1970, p. 66).

The quest for protection through the colonial government is easily understood in the context of the legendary disdain expressed by the ruling class for the subordinate classes. This disdain is the second factor accounting for the importance accorded to education. Contempt gave rise to two related sets of positive reactions from the people: a stiffening of the collective backbone from the direct affront to mass dignity and pride, and the intensification of efforts to promote education. The reactions of mothers who would stand for nothing short of "decent" book learning and those of the Teachers' Federation who strongly supported the adoption of British-style curricula as opposed to "native" curricula testify to the mindset. The dogged pursuit of elected officials and governors to increase expenditures for education lends support to the point.

That this attitude was merely mimesis is at best a half-truth. It is true that socialization may have some negative effects, but it can never be so all-encompassing as to produce mirror images of the socializers. The essence of the process of socialization is not to ape fully and totally, but to do so against some standard of self-directed notions and expectations. The colonial master was, to a large degree, the model. Organically, this stereotype had been reinforced by the patterns of educational instruction and the content and kind of materials employed in the process. Of course, this agenda bred a certain dependence on colonial influence and counsel but suspicion and distrust were equally real reactions. This explains, in large part, a general native watchfulness in matters of curriculum content.

Mimesis may have been a factor in the opposition to a native curriculum, but it was undoubtedly true that such a curriculum would be inferior to that of the colonial masters. The idea of a Jamaican university advanced by Governor Probyn in 1919 fell out of favor because the notion of a predominantly native curriculum brought this issue to a head.

The subordinate classes were also immeasurably influenced by those from their race who appeared to be the equals of the colonial masters; these scions served as exemplars. The relationship (as was the case with Norman Manley) was however steeped in ambiguity: admiration could turn to suspicion, as the successful all too often adopted views that ran counter to the wishes of their less fortunate black "brothers." The case of J.A.G. Smith underscores the point. A fiercely proud black man with an immense racial pride and a commitment to the betterment of the subordinate classes, Smith balked nonetheless at the idea that such an undeveloped electorate should be immediately granted the responsibilities of government. His stance has been labeled reactionary (Post, 1978; Munroe, 1972, Chapman, 1938), and yet his attitude was, in a sense, invited by the very standards the subordinate classes, through their educational pursuit, demanded of themselves.

The "curiously tenacious and sensitive" quality that was so intimately a part of their thirst for education made them particularly susceptible to the influence of the products of an education they desired or were counseled to emulate. It also rendered them gullible to the wiles of such an elite, whose capacity to manipulate was heightened by this form of devotion and dependence. It is easy to conclude, then, that the PNP's position on education would meet a favorable response, especially as education had been sadly neglected. Equally, however, we begin to understand the complexity of the feelings subordinate classes had for their educated "betters" such as Norman Manley: they were underscored by a grudging admiration that earned Manley their respect but not their full trust, and certainly not their love.

The gap between these aspirations and reality was enormous, as the sad state of education in the 1930s was beyond belief. The illiteracy rate was 60 percent, with literacy loosely defined as the ability to sign one's name. This was merely a carry-over from the post-emancipation days when plantocratic politics and sections of the Church of England conspired to block the path of the ex-slave. In fact, so deep was the neglect that the salary of the elementary school teacher was not large enough to be taxed. The state of secondary education for blacks was also appalling. Census data from 1943 show that only 3 percent of the population attended secondary schools—an increase of 1 percent over the 1929 figure

(Hurwitz and Hurwitz, 1971, p. 184). All the secondary schools were monopolized by the whites, which meant that colored teachers had few opportunities for employment. In 1921 the Jamaica Imperial Association (JIA) bitterly opposed the education bill, complaining that its cost had tripled since 1911. The very next year, one of its members advised against excessive spending on education, since this could result in "overeducation" and thereby give subordinate classes a misleading picture of their place in society. In fact, their pattern continued until the eve of the labor rebellion (Carnegie, 1973, p. 31).

The Politics of the PNP
The Emergence of a Rational Practice

C H A P T E R

The task confronting us here is to trace the patterns woven into political party structures by the social forces that had been excluded from power prior to 1938—the subordinate and middle classes. Since our concern is with the antecedents of the democratic socialism of the 1970s, we focus primarily on the PNP and the ways it incorporated and transformed radical ideas in the course of institutionalizing its role in post-universal suffrage Jamaica.

▶ The PNP Project: From Authoritarian Paternalism to the Common Good

The project for which Norman Manley had been preparing the party and the colony was the transformation of the outmoded political philosophy undergirding colonial structures into a form incorporating the philosophy of the common good more representative of the times. The common good, as we define it, is a theory of political obligation compelling a citizenry to obedience. The theory sustains its legitimacy from a process of historical interpretation that competitively argues for the use of certain phenomena for achieving goals in the general interest. These phenomena, which maximally influence legislation and the formulation of policies, include race, social class, community, and so on.

There were clearly two versions of the common good jostling each other at the time. The first—the pre-rebellion version—was informed mainly by notions of racial superiority. It was the white race, or at least its cognitive style, its sense of morality, and its political wisdom that set the agenda for the colony's development. It was a body of ideas instructed by Thomas Carlyle's injunction that the enlightenment of the ex-slave was the white man's burden. The white man was born wiser and must play the role of parent to the misbegotten. It was a version of the common good delineated by paternalism.

On the other hand, Manley's version of the common good embraced egalitarianism, which is evident in his statement at the founding of the PNP on September 18, 1938:

> As I see it today there is one straight choice before Jamaica. Either make up our minds to go back to crown colony government and have nothing to do with our government at all, either be shepherded people, benevolently shepherded in the interest of everybody, with as its highest ideal the contentment of the country; or have your voice and face the hard road of political organisation, facing the hard road of discipline, developing your own capacities, your own powers and leadership and your own people to the stage where they are capable of administering their own affairs (Nettleford, 1971, p. 16).

It was clear that he saw the future of Jamaica as one based on communal needs that no single group—least of all the plantocracy—could satisfy on its own. This explains in part his consistent efforts to bring even those considered by the people to be insincere allies—prominent capitalists, for example—into a coalition of some kind. This version of the common good was, however, fragmentary and we have had to piece it together from a variety of contexts. It could at times be inconsistent and did not always make clean breaks with the competition. It was, for example, influenced by paternalism.

At its inception, this version of the common good was most clearly enunciated as a political philosophy linked to an agenda for refashioning political structures. PNP official statements, such as the 1940 *Statement of Policy,* outlined reforms based on some form of socialist principles such as nationalization of public utilities and sugar factories, a state banking system, and land reform. However, much of the platform remained at the level of pronouncement, as the agenda became more immediately translated into a call for land distribution and the adoption of Manley's successful experiment in community development, the Jamaica Welfare Limited. We will briefly review the latter before engaging in the longer discussion of the Manleyist call for political restructuring.

Until the 1930s, the Jamaican economy was dominated almost

exclusively by British economic interests and economic ideology. Neo-mercantilism was on the ascendant and the Ricardian formula counseled that one's colonies should produce agricultural goods to be exchanged for the mother land's manufactures. Predictably, the British answer to economic distress in Jamaica was to strengthen plantation agriculture. Even small-scale manufacturing was opposed, ostensibly because of local limitations.[1] Initiatives directed to industrialization, the consolidation of inward-oriented agriculture (that is, agriculture for local use rather than export) and an independent peasantry were discouraged.

With notions of the common good at the forefront, Manleyism proposed the reorganization of economic and social structures. The demise of plantations and their replacement by a class of independent peasants and small farmers was, according to this agenda, an essential part of post-rebellion economic reconstruction. Equity required not just better wages for workers and peasants, as the JLP would have it, but the ownership of viable pieces of agricultural land and new opportunities for economic activities. In the 1950s, these concerns would be translated into the program of industrialization, as Jamaican leaders attempted to create a closer match between their concerns and the prevailing interests of North American capital.[2] Both agendas were a far cry from what British paternalism considered suitable for Jamaica.

Manley's immediate concerns, however, were with issues of taxation and land. In the 1940s, the PNP lobbied to introduce more direct forms of taxation, a system of land valuation, and the principle of eminent domain, allowing compulsory acquisition of land for public development. As the records show, the peasants' common complaint about the monopoly of land by the big landowners had been one of the underlying themes of the labor rebellion. The principles of land valuation and eminent domain addressed the peasant's need for assurance that the land was no longer the sole preserve of the plantocracy.

The Jamaica Welfare Ltd., founded by Norman Manley in 1937, also addressed the issue of social reorganization. Begun as a non-political entity to foster self-reliance and undercut traditional patterns of inferiority and dependence, the organization had among its programs cooperatives and self-help educational efforts in the rural parishes. These endeavors met with initial resistance, but in time the organization scored significant successes, ranking them among Norman Manley's truly lasting contributions (*Public Opinion,* September 6, 1969). As we saw in the last chapter, agrarian communities and self-help projects were not foreign to the Jamaican tradition.

The Democratic Ideal in Governance

As we noted above, activities on the economic and social front took a back seat to the PNP's political agenda. Reform in this sphere was made more urgent by the heavy-handed paternalism that imbued governance in Jamaica, both before and after the rebellion, in spite of the increasingly laissez faire attitude of Britain toward her colonies.

These paternalistic tendencies were expressed through distinctive structural forms. In the political realm, there were a governor, a legislative council, and, later, an executive council that were monopolized by British-born personnel or powerful white members of the plantocracy and later of the merchant class. The first black man was elected to the Legislative Council in 1901. By 1935 this body had eight black members, but the records suggest that they shared the paternalistic attitude shown by the whites (Campbell, 1976). Until the grant of adult suffrage in 1944, political involvement and representation from below were marginal at best.

The same patterns prevailed at the level of ideology. Education, religious instruction, and morality were heavily infected by paternalism from outside Jamaica. Public officials in education were all recruited from Britain, with instructions to mold Jamaicans in their own image. In like fashion, the Church of England, with its archbishop and heads of dioceses recruited in Britain, sought to instill patterns of morality consistent with the British ethos. It was a regimen hostile to native forms of religion, and one that, as we saw in Chapter 3, was not averse to utilizing the courts to its advantage.

At the lower reaches of the colonial society, paternalism shaded into its darker side—the peculiar denial of humanity associated with the colony's experience with slavery. The depravity of the plantocratic mindset had tainted successive historical periods, as evidenced by the long record of denunciations from varied sources such as Lord Olivier (1936) and the Moyne Commission, down to the pre-emancipation Baptists—who had observed that "the old methods [were] the rule—improvement the exception" (Hurwitz and Hurwitz, 1971, p. 37).

Politically, the most glaring denial resulting from paternalism was that of the franchise. Between 1901 and 1935, not more than 15.4 percent of the population over twenty was registered to vote in Legislative Council elections. In 1935, that figure was 9.9 per cent. Over the same period, the level of voting increased: in 1901, 27 percent of the registered voters exercised the franchise; by 1935, 49 percent did so (Gannon, 1976, pp. 17, 19). The nature of the deprivation is more telling as one realizes that for the period in question, blacks—among whom were the largest number of the disenfranchised—accounted for 80 percent of the population, while "coloureds" made up 18 percent and whites accounted for only 2 percent. It should be noted, however, that political power for all was limited by the strict trusteeship of Crown Colony government, instituted in 1865 in the aftermath of the bloody Morant Bay Rebellion.

Thus, the political project facing enlightened leaders in the post-rebellion era was to transform authoritarian paternalism into a more equitable philosophy of the common good. As we saw in Chapter 3, Manleyism was better equipped for this project than Bustamanteism, which was closer to the old paternalistic tradition than the new philosophy of the common good. Bustamante, as a conservative source told it, saw self-government as a formula for Jamaicans "to work out their own destiny according to the collective unwisdom of the greatest number of the inexperienced" (Gannon, 1976, p. 31). His appeal was not so much in his ability to respond to circumstances by fashioning new structures and discourses but in finding new language to address the psychology of subordinate class oppression and aspirations. Thus his 1944 campaign slogan, "Self-government means slavery," took the newest political current and linked it to the history of oppression by the local power holders. In a sense, he was the epitome of a brand of conservatism that resonated with their own ambivalence toward change. The general notions of the common good embraced by Manleyism, while not problem-free, were more progressive and comprehensive and sought to effect changes at the very core of the people's concept of their own humanity.

The concept of the common good oscillates conceptually between two extreme positions: the broadly ethical and quite abstract Rousseauist rendition and one that functions principally as the means to the good of individuals (Maritain, 1951; Simon, 1980). The Manleyist version assumed a position somewhere in between these two. While it was first evolving between the mid-1930s and 1944—the year of adult suffrage—this formula was very abstract and severely hampered by voluntarism. A more practical approach was developed, however, between the PNP's dismal failure at the polls in 1944 and 1949, when the party prepared for the next elections.

The PNP's initial failures raise the issue of the degree to which the subordinate classes were ideologically attuned to the PNP's call for democracy—the very kernel of the PNP philosophy. As we saw in the last chapter, the subordinate classes did not insist on egalitarian or democratic structures and leadership in their movements. Their organizations reflected alternative social, racial and religious ideologies that asserted the humanity and economic aspirations of the ex-slave. They reacted sometimes strongly against oppression—by planters and merchants in particular—and spoke of freedom, but, to the best of our knowledge, did not challenge the paternalistic or authoritarian role of their own chosen leaders. In the case of working class movements, there were instances of workers refusing to allow Bustamante or A.G.S. Coombs to negotiate for them in 1938 and 1939, but these seem to be the exception rather than the rule. Thus, while relationships to their own leaders may well have been ambivalent, emancipation from paternalism from within did not seem to be an ideological priority among the subordinate classes. What is more, these classes may actually have enjoyed bestowing the familiar trappings of power on their own, recreating in a sense the potential for the same oppressive structures from which they sought relief.

Their paternalism notwithstanding, many of the organizations of the subordinate classes, including especially the Baptist churches and Rastafarians, did instill in their members a sense of personal power and confidence. In so doing, they enhanced their potential for social action and leadership. As they developed their decentralized and non-hierarchical organizations between the 1940s and 1960s, Rastafarian groups asserted an egalitarian and democratic ethos founded on the ability of each of the brethren to "reason" and contribute to the recreation of culture and a common worldview (Owens, 1976; Campbell, 1988).

Support for liberal democracy also surfaced in the appreciable body of petitions to stipendiary magistrates, colonial governors, and the British monarchs. Their basic formula was a combination of democratic epithets and religious supplications surrounding a call for justice, freedom, and the rule of law.

We recognize that our position runs counter to that of other usually well-informed commentators, who have depicted subordinate classes as powerless and apathetic masses, the victims of disorganization, and easy prey to manipulation. Orlando Patterson, for instance, saw British ideology as a force that met little resistance from local culture and institutions since they were in a state of disintegration (Patterson, 1967, p. 178). Such positions certainly reflect the dominant views of the Jamaican middle class and Jamaican professionals and intellectuals up to the 1960s but are not, in our view, justified by historical facts.[3] The educated would almost naturally take advantage of their privileged position in a largely semiliterate society, but it is grossly inaccurate to suggest that subordinate classes were like putty in their hands.

Of course, the effects of British ideology, especially education, could not be entirely staved off. However, as other accounts reveal, there was in the retreat from Western forms a new syncretism, a vibrancy that kept working in subterranean fashion, to produce, say, the creativity of Rastafarianism, reggae, new art forms, and so on. The fact that the activist tendencies that surfaced at the time of the labor rebellion were not pursued and developed further had just as much to do with "apathy and dependence" (Munroe, 1972, p. 17) as with the implementation of national policies designed to redirect activism from below into other channels.

While it is true that the message from Jamaican social movements is mixed (democratic leanings go hand-in-hand with authoritarian leaders) it should be recalled that this period of Jamaican history constituted the epochal transition to a formal political society. It is not uncommon for such early forms to be imbued with a degree of authoritarianism, which both Bustamante and Manley displayed in different ways and to different degrees. However, while dependent on their leaders for the definitions of the desirable and the possible, the subordinate classes were not devoid of strong ideological preferences.

This helps clarify why Bustamante, the autocrat, could be accepted as a leader by workers in 1938 and newly enfranchised voters in 1944, while Norman Manley, the democrat, suffered a resounding defeat personally as well as for his party. It also suggests, however, that Manleyism could indeed speak to the consciousness of the subordinate classes, showing the PNP the way from the defeats of the 1940s to the victories of the 1950s. Notions such as universal suffrage and self-government were not too abstract and removed from pressing physical needs, as Bustamante would have it. Support for the democratic ideal could be built. It would need, however, to be melded into the "system of narration" of these classes. Somehow the notions of liberation expressed by folk culture and the Manleyist democratic ideal must be integrated, thereby bringing democracy closer to a concrete, achievable reality.

This insight helps us see the importance of symbolic pronouncements, leadership styles, and the like—the wrapping, in other words, without which the kernel of political ideas would not find easy acceptance. We may postulate that for subordinate classes to support the very abstract notions of democracy and democratic political structures they would need some prior experiences, such as the movement for self-government, to bring these abstract ideas into the realm of political practice. Of course, this injunction has nothing to do with their understanding of and belief in democracy.

The movement toward self-government and independence by Manleyism as part of the common good (as well as by the Marxists, who played a crucial organizing role), appealed at least to the latent political sensitivities of the people. The push for adult suffrage gathered pace in 1938, in spite of resistance by most of the power holders. At various times, the Colonial Office, the Legislative Council, and Bustamante voiced opposition to the agenda. The Colonial Office, mainly through the Moyne Commission, appeared sympathetic but was concerned about the readiness of the colony to assume the requisite responsibilities. In the Commission's view, adult suffrage should be extended only on a trial basis.

The recalcitrance of the Legislative Council was partly explained as a concern about preparedness. Some black and colored members (those of the Elected Members Association of the Council—J.A.G. Smith and Harold E. Allan among them), for instance, opposed the measure on the ground that their poorer counterparts were ill-prepared to shoulder these responsibilities. Yet, underlying the more acceptable explanations ran themes shared by the Jamaican dominant classes: conflicts of interest and the unwillingness to improve the political stock of the black subordinate classes, as the following dispatch to the Colonial Office attests:[4]

> I should hesitate to put much power into the hands of local politicians. They are totally untrained for any responsibility. Manley, the leader of the PNP Party, is anti-English and would as his first act, if he ever got the power, release that firebrand "Bustamente" [*sic*], who in a speech once advised cutting the throats of all the white population. Yet in all probability, when the next election is held, a large number of his supporters will be elected.

The dispatch further stated that self-government was dangerous and asked why democracy must always be linked to it!

Bustamante, meanwhile, continued to view politics pragmatically. As befitted his opportunism, he played the various interests against each other: workers and peasants against the middle class, against Manley and the PNP, against the governor and Legislative Council, and so on. His interest in self-government can be gauged by his confident predictions that he would be the first governor of Jamaican stock. And he remained adamantly against self-government even after universal suffrage—which was part of the reforms he had fought—catapulted him into the position of chief minister in 1945.

In spite of resistance, events pressed toward the consolidation of the democratic ideal. Between 1938 and 1944, when constitutional reforms were enacted, it was Norman Manley and the PNP who occupied center stage. Based on the preparatory electioneering that allowed the party to take the message throughout the rural areas, Manleyism was able to provide subordinate classes with a more concrete feel for its philosophy. The PNP organization dovetailed into the rapidly growing citizens associations, which proliferated in the rural areas under the direction of teachers and ministers, the natural political leaders of the rural areas. Some 140 action groups were organized colonywide by 1939 (Post, 1981, I, p. 66). Through the Harmony Division of the local United Negro Improvement Association (UNIA), Manleyism also received the endorsement of Garveyites, who promptly traced the self-government movement to their leader.

The new constitution, which went into effect on November 20, 1944, established formal power sharing through the popularly elected House of Representatives, the Legislative Council consisting of nominated and elected members, and the Executive Council, which was designated "the principal instrument of policy." The governor wielded enormous power but the expanded electorate could now have a degree of influence on government policy, since membership in the Executive Council was partly controlled by the House of Representatives. These changes, we should recall, took place against the backdrop of a dramatically increased electorate: registered voters increased from 6 percent of the population in 1935 to 52 percent in 1944 (Gannon, 1976, p. 97). Nonetheless, the new constitution began to demonstrate some of the attractive achievements of the democratic approach.[5]

With the self-government movement, something new had begun creeping into the political mix. Members of the subordinate classes were gravitating towards the PNP and beginning to make alliances with the middle class and Manleyism, which were at the forefront of the reform movement.[6] In particular, the Rastafarians signed the party's Self-Government Resolution, testifying to the triumph of timely abstract principles over the immediate growling of the stomach. The brethren, it will be recalled, were among the poorest strata in the society. Other sections of the population reacted with apparent enthusiasm. The *Daily Gleaner* reported that a crowd favoring the call for self-government "swarmed forward to a man to affix their names to the document" (Hurwitz and Hurwitz, 1971, p. 203). The British House of Commons was also taken in by the people's identification with the movement, observing that it "was largely suggested by the people of Jamaica themselves" (Hurwitz and Hurwitz, 1971, p. 205). This short list of instances of Manleyism in commendable action is, however, dulled by some of the drawbacks of the common good and its Manleyist version.

Limitations of the PNP's Notions of the Common Good

Inasmuch as the common good is about brotherhood, equality and other rather abstract rights and principles, it can obscure or make light of the real divisions and conflicts within the body politic. In embracing the common good, Manleyism found grist for existing tendencies within the PNP toward abstractness, rationalism, and an unjustifiable reliance upon voluntarism. Two major difficulties the common good posed for the PNP were the limitations inherent in the philosophy itself, namely its "abstract" nature and the resulting tendency to obscure the real divisions in society; and the delegitimizing effects of the obvious class connections of the PNP leadership.

The abstract penchant of the PNP greatly favored the role of ideas in the evolution of the political agenda. The PNP's platform was built on the fundamental premise that the colony's socio-economic reconstruction should start with the politics of unity: all the contending social classes, groups, and interests should immediately put aside their partisan interests and work to remove the divisions that had "degraded the political sense of the country and [were] partly responsible for the degeneracy and gross corruption of our quasi-democratic Institutions and for the decay in leadership."[7]

For subordinate classes, the problematical nature of this agenda was compounded by the mindset and practices it carried. First, the rational bargaining suggested by Manleyism was historically suspect, as the more significant gains of the subordinate classes were literally won on the battlefield (such as in slave riots, the Maroon wars, and the Morant Bay Rebellion, to cite a few examples). Manleyism represented a pattern of bargaining that was unfamiliar and without a convincing record of success.

Second, while Manleyism effectively gelled with certain features of subordinate class consciousness, there were lingering suspicions about the Eurocentric affectations of Manley himself. Indeed, this reaction flowed from a natural ambivalence inherent in the historical situation: Manley, the urbane barrister, epitomized the new, enlightened Jamaican many aspired to be, if not themselves, then for their children. It was not easy, however, to forget the past and adopt new permutations whose guarantees could not be assured. Of course Bustamante, with his truculent language, presented a far more agreeable posture in the role of watchdog.

Before proceeding, we must deal with the omission of race and color issues from the Manleyist definition of the common good, which might well be seen as a third—and quite serious—limitation. It seems clear to us that while race and color issues were matters of some concern, the bulk of the subordinate classes saw them less as swords than as shields. It is true that "complexion, social position, and money have always counted for almost everything"; but, arguably, this was a source of greater concern for the aspiring middle class, which, unlike the subordinate classes, was pressed more by social and political deprivation than economic necessity.[8]

Before the Moyne Commission, educated black members of the middle class complained about the "mulatto" (brown) members who were showing "no appreciable difference from the whites."[9] The blacks—so the position went—were as able as anyone and, after all, color should neither matter in an abstract sense nor provide unfair and unjust advantages in the quest for power. We will see that the Jamaica Progressive League, a founding organization of the PNP, betrayed similar dispositions in its definitions of the role to be played by the subordinate classes in modern Jamaica. Often race and color would be used selectively to promote the power-seeking of the educated middle class who would quickly translate them into abstract rights. W. M. Macmillan (1938, pp. 60–61) notes that some were clamoring for a House of Lords as part of the new constitution of 1944—an index of Afro-Saxonization if ever there was one.

From the standpoint of the subordinate classes, both factors tend to be couched mainly in terms of brotherhood. There are notable examples, of course. There was little doubt that Bedward with his white wall/black wall analogy spoke of black leadership from a narrow, racist position. Nonetheless, the subordinate classes were in search of equality, such as the Bible and British principles of justice and fair play enjoined as a right. Such tabloids as the *Worker and Peasant,* published by the Negro Workers' Educational League, and the *New Negro Voice,* published by the local UNIA division, often published articles in that vein.

A notable contrast seems in order. While the crisis of the late 1960s and the early 1970s has much in common with the dynamics of this period, the meanings of race and color have changed dramatically in the present. Brotherhood and fair play continue to be very much a part of the change agenda but unlike in the 1940s race and color factors now provide the basis for restructuring Jamaican society. In a sense, these factors can be likened to the legal term, *primus inter pares* (first among equals). During the 1940s, a form of associate status was being sought and was requested.

It is clear that, for the PNP, the choice to construct a consensual political agenda meant the avoidance of politically volatile issues such as those involving color and race. These issues were the preoccupations of the less enlightened. True leadership and a just society should repose in a context that does not allow "the skins of candidates" to become an issue in political elections.[10]

But issues of race and color could not be wished away. Although, as we have seen, their expression was steeped in ambiguity, they were important facets of the consciousness of the subordinate classes. It can be argued, for instance, that Bustamante's early victories were as much due to his more concrete platform and successful labor organizing as to his greater willingness—albeit purely in a rhetorical vein—to speak to the reality of class and race oppression in the society. The equation of self-government with slavery (as expressed in the JLP slogan in the 1944 election), for instance, sent a message that he, unlike Manley, understood the oppression meted out to subordinate classes by the local power holders. With its strongly middle class orientation, the PNP would perhaps have had greater political difficulties in raising such issues, because they were deemed divisive. Yet, given its philosophy, one might argue that this task should have assumed the stature of an ethical imperative. Instead, correcting colonial political ineptitude took precedence over race and class issues just as bread and butter issues should bow before the greater importance of remedial political thought. Thus spake Manleyism.

The failure to confront racial discrimination robbed Manleyism of a crucial ethical dimension. This inconsistency was glaringly exposed by the PNP's avoidance of the Rastafarians who, by the late 1930s, had become the dominant expression of black nationalism. Undeniably, certain features of the brethren's creed together with their retreatist political doctrines provided the PNP with a justifiable reprieve. However, by 1942 the attitude of the Rastafarians appeared to have undergone some encouraging modifications. For one, the Ethiopian World Federation's signing of the PNP's Self-Government Resolution spoke of their commitment to certain aspects of the Manleyist version of the common good. These overtures were followed by a donation to the party's Internees Fund (Smith et al., 1960). The Rastafarian movement's identification with Manleyism was also seemingly prompted by the PNP's active role in human rights organizations, such as the Jamaican Council for Civil Liberties, formed in 1941 to protect the constitutional rights of individuals who were made targets of indiscriminate detention.[11] During the following year, Rastafarians took to providing guards of honor for PNP politicians at political meetings.

In the end, however, the party rejected these overtures. While some members were vituperative in their rejection (in their opinion Rastafarianism was the epitome of barbarism) the episode highlighted two related strands of thought that are representative of the PNP's position on race and color. From the conservative side, the potentially subversive nature of racial rhetoric was feared:

> [Rastafarians] were attempting to resurrect the causes that led to the unfortunate incident of 1865 [the Morant Bay Rebellion] by the abominable doctrine of skin for skin and colour for colour being

> inculcated in the minds of the ignorant and hot-headed masses . . . who for the most part can be driven to any extremes at the present time owing to the evils of unemployment and privation.[12]

Closer to Manleyism was the understanding that racial rhetoric (and class rhetoric, we might add) was divisive and inimical to the nationalist project. This position would naturally seek alliances not among strident advocates for the black race but among those who remained steeped in the ideology promoted by education and mainstream religion, as exemplified in Una Marson's famous poem (Scobie, 1972, p. 152).

> *God keep my soul from hating such mean souls,*
> *God keep my soul from hating*
> *Those who preach the Christ*
> *And say with churlish smile*
> *"This place is not for Niggers."*
> *God save their souls from this great sin*
> *Of hurting human hearts that live*
> *And think and feel in unison*
> *With all humanity.*

Though race and color may differ, in brotherhood we stand! While nationalism was coveted, it should be seamless. It should not be riven by the emotive connotations of color.

Class issues did not fare much better than race and color under the banner of the PNP's common good. If subordinate classes were not always subjected to middle class disdain, the class implications of their social and economic difficulties were nonetheless glossed over by epithets steeped in moralism. The stance of the party on socialism and its relationship with its Marxist wing provide ample evidence of unwillingness to take positions on class issues.

From its inception, the party's position on socialism was ambiguous. Especially in the early 1940s, mixed messages were the rule, with Manley himself shunting awkwardly between Fabianism and Marxism-Leninism—a philosophy with which he was never quite comfortable. The Marxists would stick to a fairly orthodox communist line, while Manley himself would court equivocation. By his somewhat convoluted definition, the PNP was "a Socialist Party by declaration but its programme [was] the straightforward expression of an expansionist policy with socialist safeguards."[13] Though replete with inconsistencies, scattered here and there in the party's 1940 *Statement of Policy* were references to class conflict that were clearly influenced by the Marxists. Yet the statement is also evidence of the protean nature of the PNP's socialism:

> In proposing this declaration of policy which opens before those who are honestly progressive and who believe in the principles of equality and justice in the world, I am certain we propose a practical course which will one day prevail. At the same time, whilst asking you to accept so much, I ask you to remember that you are not being committed either to revolution or to godlessness or any matter which might conflict with your conscience to which you owe a duty. But you are being asked to accept a policy which will search both your hearts and your minds, your courage and your understanding, if it is to develop and prosper in the island.[14]

This was hardly a statement to put all doubts to rest. The party's relationship to its Marxist wing, to be discussed below, elucidates further its position on social class as a political and ideological factor. Precious little was made of the existence of class divisions, let alone class conflict. We shall see in later chapters that the modified notions of the common good adopted by the PNP under Michael Manley admitted the existence of class divisions but refused to accept class conflict. As was the case with its predecessor, extreme reliance was placed on voluntarism.

It can be argued, of course, that the level of consciousness and organization of subordinate classes allowed the PNP (and the JLP) to postpone or give short shrift to the issues of race and class in developing its political agenda. Yet, in an indirect way, as we have seen, these issues surfaced in the suspicions of subordinate classes about the PNP's one-dimensional emphasis on long-term structural and political change—an approach we have termed "abstract nationalism."

History showed that the aspirations of the subordinate classes were multifaceted. The labor rebellion and its aftermath suggested that the most immediate needs of the people were material: how to cope with food shortages, rising prices brought on largely by World War II, and non-existent jobs. This was the side that Bustamante understood so well. What was required, however, in terms of both needs and the level of development of the consciousness of subordinate classes, were approaches that kept all their needs—material and ideational—in tolerable balance.

While not developed politically to the point of being able to generate their own agendas, the subordinate classes were nonetheless sufficiently conversant with the phenomena of color, race and class to be wary of approaches that glossed over their immediate needs, favoring instead the future good of all. These niceties of reason were especially suspect when voiced by those who kept uncomfortably close company with light skin and wealth, as Manley himself found out at first hand. Being a member of the Jamaica Imperial Association and holding briefs for a perceived enemy of the worker, the Standard Fruit Company, drew suspicion.[15]

The suspicions of subordinate classes vis-à-vis the middle class were not unfounded. Even the Fabians had to admit that they were "unsympathetic to the aspirations of the barefooted man" (Nettleford, 1971, p. xiv). Almost uniformly, middle class organizations had barely disguised their lust for power, using the self-government movement as a launching pad. For example, representatives of the Jamaica Progressive League wrote in 1938 that the people should align their support to the leadership of the middle class and so help indirectly to improve their standard of living. The "masses" were expected to contribute only "indirectly" to their economic and political development! Earlier, a prominent middle class Jamaican, R. W. McLarty (1919), had stated the case with undisguised directness. The subordinate classes were "ignorant, inarticulate and helpless; must be thought of and for; and can be lifted only by wisely conceived and carefully applied legal and social methods, and by the interested and sympathetic personal association and service of those who have risen out of the mass." Robbed of their dignity and patronized by their betters, the people who "know no better and desire no better" had a few redeeming virtues—among them, they "could be taught and inspired" (McLarty, 1919, p. 11). This characterization stands at the more acerbic end of a range of negative perceptions of the subordinate classes. Nonetheless, between the extremes of the hostile and the sympathetically benign, the vilification maintained a constant refrain, shared even by Garvey himself: "the common people of Jamaica will sell their mother for a morsel of bread and a drink of rum."[16]

How did this affect the fortunes of the PNP? In truth, the party's search for an affiliation with "every organization that seemed to have anything like root in the people" (Munroe, 1966, pp. 24–25) ended with an excessive middle class bias. The membership list included such organizations as the Jamaica Progressive League (JPL), the Jamaica Union of Teachers, all other teacher associations, the Women's Liberal Club, the Kingston Federation of Citizens' Associations (KFCA), the National Reform Association (NRA) and other "progressive" organizations. These entities, which stood for specific issues or programs such as adult suffrage and self-government (like the JPL) and cultural, civic and charitable undertakings (like the KFCA), contributed much energy and many ideas to the PNP. However, they displayed a strongly elitist streak, exemplified by JPL statements that the subordinate classes should leave politics and government to the educated middle class or that the KFCA members were not expected to associate with just any worker (Carnegie, 1973, p. 114). Needless to say, this posture neither invited nor won the support of the subordinate classes.

In the end, the PNP's terms of reference might have been sweeping but the subordinate classes were neglected at the level of party organization. Most of the peasant- and worker-based organizations that later joined the party—the PMILSLA and the Negro Workers' Education League, for example—were brought aboard by the Marxists.

The PNP and Marxism

Our account thus far has painted the PNP largely as a middle class party with a liberal ideology but influential conservative tendencies relative to the subordinate classes. This picture must be rectified by recalling that one of the three factions of the party was decidedly on the left. As we stated above, it was the Marxists who were responsible for much of the labor organizing and political education undertaken by the party. While their official presence was uneasy and relatively short-lived (they were expelled from the party in 1952), they contributed much to diluting the middle-class influence within the party and preserving a place for the needs of subordinate classes.

The Marxists were frequently at odds with the rest of the party over doctrinal issues, to be sure, but also over organizational ones—for instance, the need for the PNP to become more like a political party and less like a political lobby or national movement. There are, nonetheless, explanations for the fellowship between the PNP and the Marxists. We find two explanations to be compelling: first, they were linked by a reciprocal need and, second, there were ideological affinities that allowed their connection to continue over the period of institutionalization of the PNP.

In the Jamaican context, both Marxism and Manleyism originated in the quest for democracy and nationhood. They had, however, different definitions of the task at hand. The Marxists were compelled by the underdeveloped state of the economy and the working class to subscribe to some practices of liberal democracy, but they persisted in the orthodox notion of the dictatorship of the proletariat. As an adherent, writing in *The Workman* in 1939, stated it, the position was that within "the ambit of the capitalist system the workers' burden can be lightened only by a united working class two-edged [industrial and political] weapon: a Strong Labour Union and a militant political party."[17] Manleyism, on the other hand, embraced liberal democracy with a very strong Lockean flavor. It tended to treat class relations as though they involved equal and economically stable individuals who were prepared to recognize and respect each other's rights and privileges. This quality is presupposed by Manleyism's abiding reliance upon voluntarism. For the Marxists, the prevailing conditions in Jamaica were quite the opposite. Socio-economic reconstruction must embrace the reality of class inequality.

In spite of these differences, the Marxists were an asset to the party, since their skills at labor organizing were well known. The Marxists and their cadres were responsible for organizing and maintaining the stability of the BITU during the period of Bustamante's imprisonment in 1938. After 1943, when Bustamante withdrew the BITU from its association with the PNP and so removed the mass base upon which Norman Manley had counted, mass organizing became vital. Again, the Marxists' accomplishments were impressive. Skilled workers and civil servants were quickly organized, forming the first trade union arm of the PNP—the Trades Union Congress (TUC). From its inception in 1943 to its ultimate replacement by the National Workers' Union (NWU) in 1952, the TUC was the PNP's equivalent of the JLP's BITU. During the first nine months the Marxists worked with the TUC, union membership grew from 9,000 to 14,000. Clearly, the Marxists owed their continuing existence and efficacy within the PNP to their substantial role in this practical organizational thrust.

There was a self-serving component in the Marxists' approach to the PNP as well. At the level of rhetoric, the expansiveness of Manleyism implied the ability to win over the entire party by force of argument and the efficacy of theory and practice. From a Leninist perspective, the opportunity existed for infiltration; in fact, the Marxists seemed to have attempted to create a Marxist cell within the PNP sometime between 1947 and 1948, which contributed in part to their ouster in 1952.

The second crucial explanation for the accommodation between Manleyism and the Marxists is to be found within the inner recesses of ideology. Marxism labels the common good an epiphenomenon, as part and parcel of the ideas of the ruling class; but at the basic levels of its definition of human nature, subtle agreements are established. More specifically, liberalism sees human beings as creatures of good will, influenced by self-gain and fear but also inordinately susceptible to reason and humanism. From the perspective of historical materialism Marxists see the common good aspects of humanism as deriving from the naturally productive, cooperative, and self-actualizing characteristics of *homo faber* (human beings as purposive makers).[18] These two positions, though fundamentally different, contrast starkly with the conservative position that sees human beings as naturally individualistic, selfish and unproductive, induced to act principally by self-gain and the fear of unpleasant results.

There was also a "reason dimension" within the Fabian/liberal bent of Manleyism that was genuinely open to the rough and tumble of debate, however awkward it might be. We merely have to cast our memory back, on the one hand, to the fundamental rationale for the founding of the PNP and, on the other, to Norman Manley's *Statement of Policy* that proclaimed the PNP a socialist party in 1940. This discursive gambit anchored itself to two Lockean theorems: first, that the natural harmony of interests existing between individuals enables them to live together peacefully and productively; second, that where action is required in the name of society, rational discussion is the best approach for determining agreeable courses of action. Again, the approach can be contrasted with the autocratic style of Bustamante who brooked no opposition even from his cabinet ministers.

The general openness of Manleyism to some measure of debate meant that doctrinal constraints did not initially reduce the Marxist faction to the role of disgruntled bystanders. While their role in political organizing gave them some clout, it is also a fact that the democratic thrust to Manleyism was consistent, to some degree, with Marxist principles. It should be remembered that Marx, unlike Lenin, lauded civil society for its freedom, private property and other features that would promote the decline of capitalism.[19] Of course, as Lenin stated it in *The Proletarian Revolution and the Renegade Kautsky,* the question would ultimately have to be faced: "It is natural for a liberal to speak of 'democracy' in general; but a Marxist will never forget to ask: 'for what class?'" (Borodulina, 1972, p. 606). At all practical levels, then, these two systems could not be expected to coexist without deep rivalry.

The socialism of the PNP finally made good its threat to desert the Marxists. Naked expediency had run its course. Apart from the obstructionist policies employed by Norman Manley and the PNP to prevent the spread of communism within the party—the debate surrounding the forging of the *Statement of Policy* bears this out—the attack upon the Marxists bore in from the authorities, the Colonial Office, and private interests, such as the *Daily Gleaner*. In October 1942, Richard Hart was interned at Up Park Camp (the headquarters of the military), on the charge that he and his associates were plotting to control the PNP, had been promoting racism and class animosity, and were fomenting extreme revolutionary doctrines (Munroe, 1977, pp. 66–67). This reaction followed upon the heels of a prominent showing of the Marxists at the Third Annual Conference in 1941. Here that faction won a majority of the delegates in support of its communist agenda (Munroe, 1977, p. 15).

Starting with Governor Richards's imposed restraints on communism and the banning of newspapers carrying related sentiments in 1940, Norman Manley began to pay close restraining attention to the faction. From February 1944, the year of the electoral loss, the time was ripening to begin taking decisive action. In 1947, a resolution was unanimously passed by the PNP's Executive Committee to effectively cease revolutionary work within the party (Munroe, 1977). By 1951 it was time to expel the communists. The main target was Ken Hill, who, it should be recalled, in the elections of 1949 defeated Hugh Shearer, Bustamante's protege, for the labor leader's old Western Kingston seat. The very popular Hill also captured the mayoral seat of the KSAC in 1951 and was the president of the TUC.

Strategically, the PNP targeted the TUC as the object for its principal attack. In a bipartisan effort, both political parties passed regulations to debar the participation of trade unions in polls, as long as they were procommunist (Munroe, 1977, p. 56–65). As a result, the TUC was outlawed and replaced by the National Workers' Union (NWU) on April 1, 1952; it was to become the trade union arm of the PNP, one heavily subsidized by the United States (Harrod, 1972).

So the PNP finally let go of the proverbial tiger's tail. Indeed the process of dislodgement remained consistent with the traditional approach to radical political strains within the polity. Like Garveyism and the Rastafarian movement, Marxism had to be prepared to toe the conventional line protected by a now enlarged status quo (presided over by the Colonial Office, the JLP and now the PNP) or suffer recriminations. The structural contradiction had emerged and matured. However, the impact was minimal at the concrete level of politics. The strong delegate support for the Marxists within the PNP did not follow them into exile: in the elections of 1955, Richard Hart's People's Freedom Movement received 0.4 percent of the vote; Ken Hill's National Labour Party secured 1.2 percent. The latter result is staggering when it is remembered that the TUC then controlled a little over 20 percent of unionized labor.

The gradual institutionalization of the PNP in the developing two-party system favored Manleyism. By the early 1950s, the necessary foundations for parliamentary democracy and the onward march to political independence had been laid. There was no immediate need to support a revolutionary agenda nor its adherents. Furthermore, the pro-democratic tendencies in radicalism writ large, in widespread religious ideologies, and in the pursuit of education, could now have an umbilical relationship with this new socio-political formula.

And so the eventual ouster was predictable, although the temporary alliance between Marxists and liberals in the PNP had produced mutual advantages. The Marxists' organizational needs and those of liberal democracy, to which the rest of the PNP subscribed, had counseled cooperation against Bustamanteism, which, for all its hostility to the plantocracy, was infected by political and philosophical backwardness. The compatibilities between Marxism and liberal democracy, however weak, strengthened the capacity of the PNP to promote democracy. For all practical purposes, the Marxist project was inserted into a more conjuncturally appropriate liberal democratic agenda.

The relationship of the PNP and Marxism in the 1940s holds important insights for events of the 1970s. Then, a Marxist organization (the Workers' Liberation League, precursor of the Workers' Party of Jamaica) engaged in political work on its own, with some success. Nonetheless, the conclusion was that the ground was not ripe for Marxism. So it was that, after a bitter campaign of denunciations, the Workers' Party of Jamaica made overtures to the PNP in 1976. Again it was the PNP rather than the JLP that was seen as a haven for radicalism.

However, back in the 1940s the task for the PNP was to overcome the difficulties that dogged its organizational efforts. The disjuncture that existed between thought and action, between political and economic ideas and practical economics, had to be bridged.

▶ The PNP Moves to the Center

By the mid-1940s the strategies of the PNP began to shift and consolidate in favor of capitalism. The need to score an electoral victory in the next elections (1949) prompted the party to concentrate on its grass-roots organization. Such members as Wills O. Isaacs and Manley himself took to the streets and even showed at times that they were prepared "to meet violence with violence" (Eaton, 1975, p. 123)—quite a departure from the 1944 campaign when Norman Manley thought elections could be won with the force of personality and intellectually reasonable positions.

The new labor union arm of the PNP—the Trade's Union Congress (TUC)—matched the BITU stride for stride from 1944 to 1952. It successfully organized such branches of the civil service as public works, prisons, hospital services, and telegraph services (Munroe, 1977, p. 36). Strikes were also led against the *Daily Gleaner* and the Jamaica Utilities in 1948. Indeed the union increased its share of unionized labor from 10 percent to 52 percent between 1945 and 1950. The coup de grace was perhaps the winning over of the influential Independent Port Workers' Union from the BITU in 1949. To this growing base was added the Jamaica Welfare Limited, with its powerful rural bases, and the Jamaica

Agricultural Society—an organization with vital links to the agricultural community.

The PNP did not win the elections of 1949 but it pulled itself into the middle ground. It won thirteen seats to the JLP's seventeen, but the index of popular vote told another story: the JLP received 42.7 percent, the PNP 43.5 percent. Of course, the fact that the JLP was now the government, with a record that was far from pure, could account for at least as much of the shift as did the changes in PNP theory and practice. The JLP had, for instance, introduced anti-labor legislation to restrict picketing and strikes (Post, 1981, II, p. 511). The effect of such actions was such that Bustamante himself decided to vacate his seat in Kingston for a safe one in the sugar belt. His old seat was captured in 1949 by Ken Hill, a central figure in the TUC and a PNP radical.

Undaunted, the PNP continued to move closer to the center. It consciously shed its agnostic image; hymn-singing and Bible reading soon became a regular feature of PNP group meetings (Bradley, 1960, p. 386). Indeed, by the 1950s both political parties amply used established religion in their political work (Waters, 1985, pp. 75–76). Truth had been learned the hard way: most Jamaicans live, move, and have their being within religion. Unconventional ideas and radicalism were receiving less direct attention.

The party was slowly but unmistakably moving toward the moment of reckoning between its socialism and the pressures associated with governmental politics. A central issue revolved around the agenda for reform and modernization. In this regard, it is instructive to consider the party's manifesto, the *Programme for Action Now,* unveiled at the PNP's annual conference in 1945.

This document marked a significant turning point in the entrenchment of the party's liberalism. It bore only vestigial resemblance to the famous *Statement of Policy* issued in 1940. In that document, the PNP had declared itself a socialist party and spoken of nationalization. Scattered here and there, albeit in a flush of inconsistency, were references to class conflict that were clearly influenced by the Marxists. The *Programme for Action Now* was less insistent on the more basic features of the common good. It was not that the focus was abandoned, as it was shifted to a more practical plane: securing the electorate's vote. The main issues addressed were employment promotion, the development of resources, security, public health, and workmen's compensation for factory and industrial workers. The issue of eminent domain was deemphasized, with the party concentrating on policies related to protection against eviction and adequate compensation for improvements on rented lands or leaseholds.

The PNP did return to the issue of fundamental land reform in 1949, again calling for cooperative farms, with the government providing houses, roads, water, and other assistance. Each farm would encompass 2,000 acres and include 400 farmers, who would also be provided with a private tract for their personal use. Farm managers, aided by committees of workers, would provide guidance and expertise. However, by now, such strongly leftist pronouncements seemed to be more anomalies than directives for future government action.

The intention of the PNP was unmistakable. It was certainly not moving closer to the narrow conservatism of the JLP. In fact, Bustamante took pleasure heaping scorn upon land cooperatives; he cautioned the people that this would be the first step to communist-style expropriation. Nor was the party sacrificing its notions of the common good to expediency. It was merely moving up the curve of consolidation; the business of practical politics is the business of winning elections.

The PNP's plan was definitely capitalist but differed crucially from the JLP's version. The JLP's continuing resistance to the politics of self-government illustrates its conservative bent both politically and ideologically. During House debates in March 1949, Bustamante informed the opposition that the JLP would continue to resist self-government because the colony lacked the financial soundness to undertake the responsibility, and because Jamaicans were not fit and proper persons to govern themselves. He actually implied that self-government would usher in a reign of terror: "It is the consensus of opinion among certain elements that follow the P.N.P. that if the party got into power whatever it did the great lawyer [Manley] would defend them and so even the honour of our women not alone in houses but even on the waysides would not be respected" (Eaton, 1975, p. 140).

Such recalcitrance lingered until the publication of the JLP's revised constitution in 1951[20] in which the party committed itself to allowing Jamaicans to participate more directly in the general affairs of government, to reorganize the party structure to reflect these policies, and to work for self-government and dominion status. While by 1956 the JLP and the PNP shared a common agreement on the self-government issue and even betrayed similarities at the organizational level of the party, the competing subsets of capitalist ideology kept them significantly distinct. The liberal democracy of Manleyism and the autocratic conservatism of Bustamante both contributed to the entrenchment of capitalism locally but, for all its rhetorical dedication to democracy, the JLP remained an essentially conservative party ruthlessly guided by an autocrat who maintained an iron grip on his party even in retirement.

For Manleyism, consolidation at the economic level was demonstrably different, with issues of equity and social justice remaining at the forefront. However, as we explore in more detail in Chapter 8, the tension between socialist and liberal policies had to be addressed. Plans for cooperatives were not the only target of detractors. By 1949 the PNP had started to remove the ambiguities attached to nationalization, which had been addressed with characteristic obliqueness in the *Statement of Policy* of 1940. In December 1949, Manley reassured Jamaicans that the PNP had "absolutely no intention of going any further than the four public utilities: . . . Bus Transportation, Light Power, Broadcasting, and Telephones. All talk about our nationalizing land or ordinary business and industry [was] absurd and untrue."[21]

The goal of gaining control of the means of production, a prospective policy supported by Manley himself and widely shared and propounded by the Marxists, would be pursued in ways that were not incompatible with the principles of capitalism. After all, public ownership of the utilities can and has been cogently defended as a sound public policy. It would have been quite a different matter had the list included nationalization of large plantations and sugar estates, which the Marxists advocated. In a real sense, then, the policy of land reform, that is, collective agriculture of the size and character mentioned above, suffered a setback. Only poor and mostly non-arable lands would be made available to the peasantry—a policy not consistent with socialism and Norman Manley's own testimony before the Moyne Commission. That testimony had shown him pressing for a slow but definite movement away from plantations and toward an economy based on the peasant mode of production, as demonstrated by the following exchange, "Q: You suggest, gradually, I admit, the division of Jamaica into smallholdings? A: Looking far ahead when you and I are both dead, yes."[22]

The PNP's policies moved sharply toward the support of private capital. Several of its position papers, pamphlets, and its developmental blueprint *Plan for Progress,* published in 1949, succeeded in pushing a reluctant Bustamante towards industrialization. The first significant result of this goading was the Pioneer Industries (Encouragement) Act passed in 1949, designed to promote employment in light industries targeted to create some 3,000 new jobs. Also heavily courted was "direct investment of private foreign capital," with the PNP promising to enact legislation to create a favorable climate and "to give [foreign investors] security and protection in all proper ways."[23] As direct extensions of this cumulative initiative, the PNP successfully forced into existence statutory bodies such as the Industrial Development Corporation, which attracted, helped to finance, and generally assisted investors in the industrial sector. This endeavor was followed by the party's own initiatives when it took office in 1955. The most noteworthy efforts were two pieces of legislation: the Industrial Incentives Act and the Export Incentives (Encouragement) Act, both enacted in 1956. The former was directed at promoting import substitution while the latter provided incentives to industries manufacturing exclusively for the export market.

There had been ample warnings from at least 1949 that the party would court capital. The PNP's socialist program, Manley had announced, would be based on "an orderly political development and genuine business sense" (People's National Party, 1949, p. 15). And, as confirmed in later communications, Manley was already then favorably disposed toward development programs such as Puerto Rico's Operation Bootstrap (Hart, 1972, p. 289). By 1957, then, the the PNP's transformation was complete. The plethora of centrist policies beginning with the *Programme for Action Now* in 1945 succeeded in transforming the economy into a modern capitalist one by 1962. Industrialization by invitation would lead to relatively high rates of growth but brought with it its own share of contradictions that matured in the late 1960s.

By the early 1950s the economic policies of the JLP and PNP were becoming indistinguishable, as the modernization theme of the PNP pushed aside the traditionalism and ad hoc recipes of the JLP. For this initiative, Manleyism was rewarded in the elections of 1955 and 1959 with eighteen of thirty-two, and twenty-nine of forty-five seats respectively. In drawing itself to the center, the party captured the support of the people for two principal reasons: its economic policies and the concrete advances made through the application of democratic principles. Although its economic policies had very serious deficiencies, as we mentioned in Chapter 1, they held out some hope to the people who could not remain indifferent to consumerism and the progressive changes in the occupational structure. These structural developments went some distance to meet the demands for "jobs that last long" made by the strikers during the labor rebellion.

The tide of change introduced by the institutionalization of modern political society beginning in 1944 (trade unionism, formal representative political parties, and adult suffrage) had given rise to constitutional advances in 1953 and 1962. The first of these, the new constitution of May 5, 1953, was uneven in its effects, in that considerable power still rested with the governor, but it corrected some of the major deficiences that had plagued the local exercise of power. In particular, the new constitution increased the numbers and powers of elected versus appointed members; it severely limited the power of the governor to remove the (still appointed) chief minister; gave ministers the authority to direct their departments and initiate proposals and policies for reference to the Executive Council; and created separate, decentralized ministries staffed mainly by local personnel and headed by local permanent secretaries (Barnett, 1977; Munroe, 1972).

The governor was still the chief minister but responsibility was shared and was delegated even more by the passage of the Jamaica (Constitution) Order-in-Council of 1962. Here the structure of responsible government was formally complete with the introduction of largely autonomous political institutions: a head of state and a head of government (a governor-general and a prime minister); an executive branch comprising the prime minister and ministers appointed and dismissable by him; a bicameral parliament; and an independent judiciary. Overall, the independence constitution effectively completed the installation of a British form of government (that is, a Westminster-type democracy) in the local political structure.

It is difficult to fault many of the compelling strictures about the decolonization process raised by critics such as Munroe (1972, chap. 5). These constitutional drills are definitely flawed with respect to the formal correlations between substance, form, and execution. For instance, political emancipation without economic autonomy of some kind makes a mockery of the use of newly acquired power. However, we must make an important observation, often missed or trivialized by critics, relative to the perception of the subordinate classes. For the scholar and the trained political scientist, the analysis takes the form of an examination of the inner logic of political phenomena in relation to proffered or implied theory and practice. For the subordinate classes, the immediate effect of this march to greater political self-determination has to be gauged at the level of political psychology. For the electorate, the efficacy of politics depends inordinately upon the politician's ability to appeal successfully to the electorate's irrational and subconscious prejudices. If, as is correctly stated, practice informs theory, then the theorist will have to be guided in large measure by the voice and actions of the people.

We suggest that, in fact, between 1944 and 1962, the subordinate classes focused considerable critical attention on the exercise of power. First, it is useful to recall the high rates of participation in the political process. Between 1944 and 1962, 49 to 54 percent of the population was registered to vote—a high number, considering the large proportion of young people in the population. During the same time, the percentage of registered voters who cast their votes increased steadily, from 59 percent in 1944 to 73 percent in 1962 (Gannon, 1976, p. 97). The effects of this newly won power were expressed in clearly observable ways. Planters and merchants, for one, withdrew from active politics, their bitter complaints a sign of waning power. As Bell (1964, p. 119) states the matter, "Some of [the old elites] no longer attempted to participate in the major affairs of the society but complained bitterly of recent political developments, heaped invectives on the heads of both the electorate and the new politicians, and in general deplored the present democratic system in Jamaica." In addition, given the party/union symbiosis, for subordinate classes political and labor organizing victories were not distinctly separate. In fact, they would stage demonstrations at the legislature to demand jobs. And the years following 1944 were years of intense labor organizing, considerable strike activity, and—for unionized workers—significant gains. JLP-BITU members were even to witness the gradual change of mind by their leader on the independence issue. On July 10, 1947 Bustamante belatedly declared after many years of resistance that "England should rule [Jamaica] no more" (*Daily Gleaner,* July 10, 1947). After that, he wavered, until his final acquiescence in 1953.

By now the democratic ethos had taken substantial root among the subordinate classes and the democratic process was evolving a more expansive form of political morality. A noteworthy example was reflected in the new JLP constitution of 1951, which removed the extreme authoritarianism of its leader. Proceedings in the House also betrayed this quality, suggesting in the end the formative role of the PNP in these developments (Barnett, 1977; Phelps, 1960).

▶ Concluding Note

Our analysis strongly indicates that Manleyism was the most agreeable carrier of fundamental change, though not by conscious design. Its underlying principles of the common good and Lockean philosophy could appeal at once to the existing strains of both radicalism and liberal democracy. Indeed, Norman Manley himself gave more than a gentle hint of this accommodation in the debate on the *Statement of Policy* in 1940. He stated that the PNP "had contained the elements common to Liberal and Socialist thought. There was, for instance, the principle of public ownership of utilities—a Liberal idea sixty years ago—and the Public ownership of monopolies—a socialist principle but one which might be accepted by a purely Reform movement."[24] Unlike the JLP, with its deep affection for traditionalism, Manleyism could speak to a sense of collective consciousness, especially through the use of the more enlightened aspects of the democratic tradition. It is in this respect that it can be strongly suggested that Manleyism inherited some of the galvanizing rhetoric of the global radical tradition while managing to remain faithful to capitalism. But there were vital capitalist checks and balances built into this apparently pliable structure. The Lockean invitation is extended to liberal and Marxist to reason together but in the event of an impasse, Machiavelli will prevail.

It should be clear now why the PNP occupies the contradictory positions of pillar of capitalism and haven of radical tenets and practices that are inimical to the capitalist mode of production. Most of the determinate causes are related to the peculiar juncture between the party's politics and the dynamics of the class structure. Of enormous importance was the blend of political philosophy and economics whose strongest suit was perhaps the ability to offer realistic promises, generate political progressivism through the advocacy of the common good, and thereby contain the forces of discontent kept alive by a long and seemingly unending tradition of want.

The Progressive Accumulation of Capital

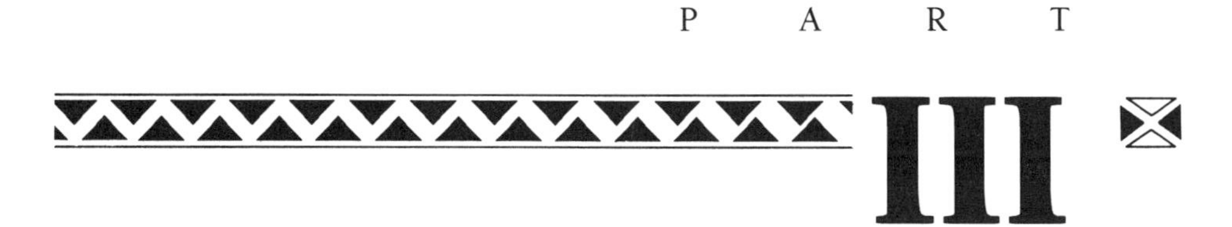

The Capitalists

CHAPTER 5 

Some analyses of events in Jamaica have tended to depict the capitalist class as an undifferentiated whole—an approach that creates serious obstacles to a proper understanding of the events under study. Doubtless, the class often presents a united front, especially when its core—capitalist property relations—appears threatened either through political action or ideological challenge. Such was the case in the aftermath of the 1938 labor rebellion, when capitalists formed the Jamaica Democratic Party in a show of unity that successfully hid inner tensions even from Bustamante and Manley. The appearance of unity may be even more notable when ethnic and kinship ties provide an added binding factor, as is the case with the 21 Families that, through their interlocking networks, controlled much of the Jamaican economy into the 1970s and beyond. And yet, capital has its own requirements that do create tensions even in so-called oligarchies. Overall, the responses of the class to crises reveal more ambiguous and subtle patterns, with intraclass tensions and conflicts common and seemingly inevitable. The process of capital accumulation, we would argue, automatically stirs up competition between the fractions of capital, which ebbs and flows with the overall requirements of capitalist stability and expansion. These fractions must compete as a function of their inherent nature (Marx, 1975) and "on pain of elimination from the ruling class" (Sweezy, 1968, p. 339).

Through the history of Jamaica, three main fractions of capital have competed for dominance: the planter, merchant/commercial, and industrial. The respective strength of each vis-à-vis the others has alternatively grown and waned as changing patterns of growth or the attempted expansions of capital brought intraclass tensions to the fore. In the main, however, dominance was conditional on mediation—chiefly by the colonial state and intermediate social groupings—as all fractions have been beset by systemic weakness. These two characteristics, fractionalization and systemic weakness, help put into perspective the emergence of the new politics of the 1970s.

▶ Planter versus Merchant

> Friday, April 9, 1920 . . . Fred L. Myers of Kingston offered me £72 per ton for my sugar, and the controlled price in England for April was fixed at £67. I considered this a good offer. I however was a bit suspicious about it and went up to Kingston yesterday to enquire of Mr. Rose, who is Thompson Hankey's agent, if he had had any news from them of the rising prices. He said no, so I went to Myers and told him that I would accept £75. Rather to my surprise, he accepted my offer and I sold 100 tons. Today I have been offered £80 per ton, so I dropped £500 on the sale of 100 tons yesterday. Anyway, the price is a splendid one and I have been very fortunate in my sales of sugar this year as I did not sell beforehand as some other planters did. Several of them sold their crops, or the greater part of them, at prices ranging from £30 to £45. So far I have sold none below £70 save the 8% for island consumption which the Food Controller has commandeered at £35.10.0. (Craton and Walvin, 1970, pp. 272–275)

Fred Clarke, owner of Worthy Park Sugar Estates (St. Catherine parish) and author of this diary entry, was justifiably pleased with himself. This time around, at least, he had been more fortunate than many of his fellow planters. But the precipitous decline in the price of sugar in two years—from £100 to £15—caused many estates to fold. Throughout the 1920s and 1930s low prices would continue to create havoc for the beleaguered planter, while the merchants' profits remained high. Nor were the merchants' inroads only in the economic sphere. Planters had to contend with the likes of H. G. deLisser, commonly known as the "special ambassador for Jamaican commerce in Britain" (Roberts, 1951, pp. 113–14). DeLisser, scion of a well-known planter family, was also the editor of the *Daily Gleaner* (the veritable "voice of Jamaica") as well as the secretary of the Jamaica Imperial Association—a conservative and powerful organization which saw itself as the mechanism to "supplement government because of government's inadequacies" (Carnegie, 1973).

And yet these details, which seem to demonstrate merchant dominance over the planter, convey only one side of a far more complex reality. For instance, the planter-dominated legislature managed to defeat the Native Industries Bill in the 1930s. Here the planter, presuming to act on behalf of the capitalist class as a whole, prevented merchant/commercial interests from securing protection for proposed industrial operations. Planters would not stand idly by—as the position was stated—while others tinkered dangerously with the country's imports (Carnegie, 1973). This ability of the planter to set limits to the avenues for the accumulation of merchant capital, to pass off the interests of the fraction for those of the class as a whole, is revealing.

Sugar plantations suffered in the aftermath of the abolition of slavery in 1838 and during the entire free trade period of the 1840s to 1930s: the number of plantations declined from 600 in 1838 to 140 in 1896 (Beachey, 1977). The export of bananas, championed by small farmers and plantations owned largely by local merchants, came into its own by the 1890s and so filled part of the gap, while sugar exports only revived in the 1930s with the turn to neo-mercantilism in Britain. The merchants were, for much of the period between abolition and the 1938 labor rebellion, the most dynamic force in the economy. Given this turn of events, the continued ability of the economically weakened planters to retain their dominance over the class structure needs to be explained.

Two historical circumstances dating to the crisis of plantations in the aftermath of abolition helped shape future class relations. First, the colonial government intervened, at times through selected mediators, to support the threatened plantation economy. Second, local merchants, among whom Sephardic Jews were the dominant force, managed to consolidate their position, becoming full-fledged (but not dominant) members of the capitalist class. The ascent of this particular ethnic group is linked closely to the race and color hierarchy of the society as well as to economic and cultural factors. These patterns speak of two constants: the role of the state in shaping and supporting capital accumulation and the class and race or ethnic rivalries swirling around merchant capital.

The State's Role in Capital Accumulation

While the planters were losing ground, the local merchants were on the ascendant, having benefited from entrepot (warehousing and shipping) trade with South America, trade with the North American market (both illegal and lucrative during mercantilism), and, later, trade with the local peasant economy that developed and grew between abolition and the early 1900s. In 1850 the merchants' income was the highest of all occupational groups in Jamaica (Hall, 1959, chap. 2). The fraction managed to consolidate and improve its economic position despite the decline of planters by shifting its activities to the more dynamic sectors of the economy, allying itself, along with the coloreds, to forces pushing for economic innovation. A comparison of national income figures for 1832 and 1850 reveals a significant shift in merchant income—which increased by 50 percent during the time—toward greater reliance on the peasant sector (Eisner, 1961, pp. 28, 46). There was increasing diversification of trade, especially with North and Latin America, in the latter part of the century (Eisner, 1961). For instance, the records reveal that many traditional merchant families (Jews, Lebanese, and so on) owned banana estates around the turn of the century (*Handbook of Jamaica*, 1890 to 1910). In the 1930s, banana exports were valued at £2 million and represented 50 percent of the value of all exports (*Annual Reports* 1946, 1948). There was also considerable penetration of allied economic activities. From the first decade of the twentieth century, the cross fertilization of the Jewish business community began to increase, expanding to encompass insurance and commission agency businesses and representation of several airlines (Holzberg, 1987, p. 158). A review of the legal profession shows Jewish lawyers highly represented throughout the years leading up to the labor rebellion (*Handbook of Jamaica,* 1920 to 1936).

There were limits, however, as the rising merchant fraction found several major obstacles in its path in the form of British economic interests, the continued hold of planters on key organizations, and ideology. The institution, in 1865, of the authoritarian Crown Colony Government, which reduced the Legislative Assembly to an advisory body and severely limited suffrage, put a stop to the inroads Jewish merchants had been making into direct political power. This forced a rechanneling of their efforts into more indirect avenues of influence and greater involvement of the merchant fraction in economic activities directly and indirectly tied to the plantation. The government during the Crown Colony period succeeded in propping up this weak and withering dominant fraction that retarded Jamaica's development. The merchants, in allying themselves to it rather than to industrial capital, gradually came to represent a retrograde force.

With regard to economic interests, it was clear that, planter weakness notwithstanding, the Crown would not consider alternatives to plantations as the key elements for the organization of economic life in the colony. Industrialization had to be curtailed at all costs: legislation proposed in 1902 to guarantee interest on capital investments to companies "manufacturing or preparing the products of the island" was defeated (Eisner, 1961, p. 204). By the 1920s there were only fifty-two manufacturing concerns in Jamaica.

The sugar industry declined fitfully between 1914 and 1939, with sugar prices falling some 40 percent between 1927 and 1937 (Verity, 1938, p. 18). Banana cultivation, initially frowned upon as "backwoods nigger business," did eventually provide an acceptable alternative to sugar for many planters. However, more important for the continued dominance of the planter fraction was the support the sugar industry received from an external source: the International Sugar Agreement of 1937. The agreement set in place a quota system for producers in the British empire along with a guaranteed market and an arrangement whereby shortfalls by one producer would be made up by others. This agreement coincided with the world-wide shortage of sugar induced by World War II, and for many years up to 1951 the British Ministry of Food provided the industry with a guaranteed outlet and stable prices. Powerful British interests also had a big stake in the fortunes of the industry. This partially explains constant British support for the Jamaican export agriculture fraction, in spite of their decline relative to the other capitalists.

The industry's own efforts to lobby in England were, however, substantially responsible for these supports. Jamaican sugar organized its own defense, with JIA sponsorship. The Jamaica chapter of the West India Committee, formed in 1904, was essentially a lobby for West Indian sugar in the British Parliament. In 1923 this Committee organized the West Indies Parliamentary Committee consisting of members from both houses of Parliament; its declared aim was to watch over the interest of British West Indian affairs in Parliament (*Handbook of Jamaica,* 1923). In 1942, the British West India Sugar Association was formed. This body helped to establish the Commonwealth Sugar Agreement, which first came into effect in 1953. These agreements set limits upon the volume of exports but relieved the industry from the uncertainties of free trade.

Alongside these powerful lobbies came the Royal Empire Society of 1938; its aims included the promotion and "the preservation of a permanent union between the Mother Country and all other parts of the Empire" (*Handbook of Jamaica,* 1939). There can be little doubt that this last organization carried direct implications for the planter fraction as a whole. Economically, it interfaced with the introduction of British capital into the sugar industry as effected by the 1937 purchase by Tate and Lyle of seven local sugar estates. Politically, it provided another instrument for combatting the anti-plantocracy fervor that was spreading throughout the Caribbean in the late 1930s.

The resilience of the planter fraction was also due in part to ideology. The main political and economic organs of the state were weighted in favor of the plantocracy. For instance, up to the late 1930s the membership of the JIA reflected the various fractional interests but was dominated by the planters. Such power and influence filtered out into the wider society. Let us take, for example, the Parochial Boards, the centers of political power in the parishes. Each board, which operated from the main town in each parish, consisted of the member of the Legislative Council for the electoral district, the custos (chief justice of the peace) of the parish (nearly always a planter), and nine to fifteen persons elected by taxpayers qualified to vote in Legislative Council elections (*Handbook of Jamaica,* 1910 to 1933). This formula ensured the planter's power and influence in a single stroke: the custos as a member of the Parochial Board also sat in the Legislative Council formulating laws and policies at the highest reaches of government. Further, the challenge from the merchants and would-be industrialists was offset by the fewness of their numbers and the concentration of their members in the larger towns, especially Kingston, the colony's capital. These examples by no means exhaust the supportive mechanisms at the planter's disposal, which provided the fraction a solid ideological and organizational base that neutralized their economic weakness.

Intraclass Rivalries

Ideology is usually a reliable index of inner tensions within class relations. We see this truism reflected in the distinctly anti-merchant and anti-Semitic biases of the Jamaican capitalist class. Historically, merchant trade was tainted by immorality: "trade was perilous to the soul and avarice a deadly sin" (Tawney, 1926, p. 37). The merchants' lowly origins in the itinerant peddler moving from town to town, village to village, won them the dubious title of "pies poudreux" (dusty feet). The Jamaican planter class continued such traditions, buttressed by a long-standing conflict with British merchant interests that, throughout the plantation era, exercised control over trade and commerce and prevailed in matters of legislation and government policy.

This antipathy was pervasive (Hurwitz and Hurwitz, 1965). In the early 1800s, the few merchants who had occasion to visit King's House (the governor's mansion) were received in the kitchen (Wright, 1966). This, in spite of the fact that some eighty of them had accounted, in the 1700s, for the bulk of Jamaica's taxation revenue (Holzberg, 1987, p. 19). Jewish merchants were begrudged and reviled by whites, coloreds, and blacks at all levels of the society (Hurwitz and Hurwitz, 1965, p. 41). Until 1826, they suffered the same legal disabilities as the coloreds. For instance, they could neither vote nor give testimony against whites. The prejudice revolved at least partly around their success and the nature of their business practices. Jewish merchants were known to charge rates of interest fully 15 percent above prevailing rates (Pitman, 1917, p. 136). Considered exorbitant by local businessmen, these rates were nonetheless allowed since the British treasury benefited from the steady flow of hard currency they generated (Marcus, 1970, p. 112). While these practices were open to others, the taint of the trade, combined with the formation of alliances among the Jewish merchants, meant that they soon constituted a nearly monopolistic force (Stewart, 1808, p. 151).

Ideology does sometimes give way to need, and in the crisis years of the mid-1800s the political position of Jewish merchants started to change. Beginning in 1835, Jews were elected to the Assembly; eight of the Assembly's forty-seven members were Jewish by 1849, the number increasing to thirteen by 1865.[1] In time, they penetrated key political apparatuses, occupying important positions in the Vestries, as justices of the peace, and as members of the Privy Council. George Solomon, a Jew, was the de facto finance minister between 1860 and 1863 (Andrade, 1941, p. 31). By the time of Crown Colony (1865), they had evolved into a powerful political force that was not always allied to the status quo. According to Governor Eyre,

> There are forty-six members of the House exclusive of the Speaker—of these, twelve are Jews and eleven are colored gentlemen. There are always joined with them those members who are always in violent opposition to authority and those who consider themselves to have some personal grievance. Together the combination is a formidable one and represents fully half of the entire House.[2]

Crown Colony closed the door to further direct threats against the status quo by the Jewish faction and rechanneled somewhat their activities but did not decrease inter-fractional antagonisms. As already hinted in Clarke's diary entries and further supported by the literature of the period, intense friction between merchant and planter continued.

Diary entries of the period describe the growing financial interrelationship between the beleaguered planter and the Jewish merchant. As commission agents for overseas buyers, as suppliers of goods and equipment, as part owners when sugar enterprises fell on hard times, Jewish merchants took increasing advantage of their strategic position. Among the prominent commission agents are the names Alexander and Myers, now household words in Jamaica (their offspring are currently members of the "21 Families"). These agents were known to use privileged information, much like insider trading, to gain advantage and extend the range of their interests. The firm Lascelles deMercado, for instance, went from commission agents and wharf owners to serving as creditors to the industry. Jewish merchants, as the data show, were not heavily involved in sugar and landownership (Andrade, 1941, pp. 136–142): this is explained in part by the strong merchant traditions of the group and the earlier exclusionary practices of a hostile white population. They did, however, enter into export agriculture through salvaging operations arising from liens and mortgages on planters' crops and property. The prominent merchant/agents would in time own interests in several sugar estates. Up to the 1940s, Owen K. Henriques owned New Yarmouth Estates in Clarendon; Lascelles deMercado, until recently one of the leading commodity distribution houses, was part owner of Vere Estates in Clarendon; the Ashenheims, Henriques, and DaCostas controlled Appleton Estates in St. Elizabeth and Bernard Lodge in St. Catherine. We should note that these were investments made principally from a merchant/commercial base. The field was laden with profits. In 1916 a prominent merchant realized profits that amounted to one-tenth of the government revenues for that year (Reid, 1976, p. 42). In 1933, 50 percent of the government's revenues came from customs duties. In 1938 it was said that Jamaica's overseas trade of £11,517,961 had been built up "practically with Jewish Co-operation and initiative, [and that] almost every important article of trade [was] represented by some prosperous Jewish firm" (Andrade, 1941, p. 37).

Converging and divergent interests kept the planter and merchant fractions in a precarious balance between collaboration and conflict. Progressively, the Jewish faction became the upholder of the ideological status quo. The *Daily Gleaner,* whose board of management included the Ashenheims, Levys, DaCostas, deMercados, and Delgados—all drawn from Jewish commercial and professional families (Carnegie, 1973, p. 168)—served to entrench and consolidate the ascendant Jewish interest within the capitalist bloc. It also generalized the broader interests of the class throughout Jamaican society. H. G. deLisser, editor between 1904 and 1944, played a masterful role in the realm of opinion management, keeping political radicalism at bay. As we have seen, the *Daily Gleaner,* with support from the Jewish business community, plotted the undoing of Bedward and threatened to legislate against the proprietary rights of such movements, hounded Garvey from Jamaica by effectively ensuring his political failure in the 1930s, and doubtless influenced Norman Manley's own decision not to incorporate race and color into the PNP's agenda. The faction had well earned its place in the power bloc.

It is wrong, however, to speak of an oligarchy with interests more coalescent than conflicting (Reid, 1976; Phillips, 1976). Clearly, the respective capitals were in competition—a competition made keener by clashing ideological perspectives and anti-Semitism. Partnerships between planter and merchant were uneasy compacts. The merchant was aggressive and the planter took steps to counter the trend. In 1929, for instance, the planters formed the Sugar Manufacturers' Association (SMA), which was charged to rehabilitate the industry but had the hidden agenda of fighting the merchants. The association was empowered to fix the prices of sugar consumed locally, to guarantee the export of 80 percent of annual production, to secure government subsidies and other supports, and to regulate commission payments to merchants. Regulating the merchant's commission was a clear enough rap on the knuckles, but government subsidies partially removed the planter's dependence on the likes of Fred L. Myers and eased the credit restrictions imposed by banks such as the Canadian Bank of Commerce.

▶ Shifts within the Alliance

Poulantzas (1980, p. 127) has noted that the process of capital accumulation requires a fraction to be dominant within the alliance of the capitalists (the "power bloc"). While the planter fraction had managed to hold this position through difficult times, its grip was becoming increasingly tenuous. The fragmentation of interests into competing and often hostile entities diluted the main organizing principles of the fraction. We will view these changes in the context of the Jamaica Agricultural Society (JAS), formed in 1885, and the Jamaica Imperial Association (JIA), which were the principal organizations around which the leadership of the planter fraction revolved.

During the early years, the planters operated within the JAS relatively free from destructive conflicts. By the early 1920s, key positions on its board of management were monopolized by prominent members of the JIA. Exercising tight control over agriculture, the management of the JAS organized a number of successful conferences on bananas, oranges, and cotton. But this act of organizing the producers of other crops carried certain immediate dangers, since once created, these new organizations might slip out of one's control.

Problems developed around continued support for sugar. The Sugar Industry Aid Board of 1921 initiated support by providing a low-interest loan of £40,000 that assisted in restoring the credit-worthiness of the industry. The industry continued to receive support through sugar control boards, the first of which was created in 1929 to protect the sugar planters. But strong resistance within the Legislative Council at the end of the decade made traditional government loans and subsidies no longer a matter of course. Bananas and the newly emergent industries assisted into being by sugar—coconut, citrus, and cocoa—had their own representation in the Legislative Council. These interests, as we shall see, were not entirely synonymous with those of the sugar producers, although there was some common involvement.

The first major anti-sugar force was really the Agricultural Produce Advisory Board of 1926, formed as the supportive arm of the Jamaica Producers Association, with the task of consolidating producers (except sugar) into cooperatives. Its goals indicate a high degree of independence from sugar: to develop agricultural resources with the full cooperation of other entities; to develop overseas promotional campaigns to aid local exports; to negotiate on behalf of its members; to advance loans to its members; and to generally protect its members against all and sundry.

Within four years, the Association created four major producers' associations: banana (1927), citrus (1929), coconut (1930), and dairy (1931). Two marketing and distribution subsidiaries were also created: the Jamaica Producers Marketing Company and the Jamaica Direct Fruit Line. The gathering strength of these associations can be measured by their impact on production and marketing. Citrus producers, boasting a membership of 850 by 1929, erected sophisticated marketing facilities. By June 1934, 272 members of the Coconut Growers' Association had signed marketing contracts for annual sales of 30 million nuts. Later, unrefined and refined coconut oil and soap were introduced. The banana cooperative was perhaps the most impressive of all. For one thing, it challenged the monopoly of the United Fruit Company and succeeded in capturing a fair portion of the British as well as the European market throughout the 1930s. At the end of November in 1929, the association had 76,000 members with some 40,000 acres under cultivation; between 1920 and 1930, the number of plantations increased from 329 to 505 (Eisner, 1961, p. 287).

These changing patterns are significant in that they reveal that the center of power was shifting away from old-fashioned arrangements within the capitalist alliance. It is important to grasp the subtleties exhibited by the alliance. First, it is true that the merchants and merchant capital had penetrated sugar and bananas and had filtered into the lesser agricultural crops coming into their own (Eisner, 1961, p. 207). Nonetheless, we should in no way speak, as Ken Post does (1978, p. 84), of a fusion of interests. We have provided sufficient evidence to support our contention that the very logic of capital is raised upon the foundations of competition. While in less-developed economies antagonisms may appear more as tendencies than as full-blown fractional conflict, there was enough conflict present to reveal these basic processes at work behind any appearance of fusion.

As a second point, while the sugar sub-fraction was losing ground, the export agricultural fraction still held sway within the class as a whole. The various industries within the agricultural sector had been released from the tyranny of sugar and by 1935 were jostling for position in the Legislative Council. Sugar interests, which had earlier monopolized government attention and resources, now had to compete fiercely with the newcomers. For example, banana growers argued successfully for an increased subsidy (over the prevailing sum of £20,000) to support an operation that stretched to Rotterdam and included the transport of passengers and freight (Eisner, 1961; *Handbook of Jamaica*, 1931).

How real, it might be asked, was the fragmentation of power if the sugar planter was most likely a banana producer, the owner of coconut groves, and a producer of oranges? Available data suggest that ownership and production of bananas, cocoa, citrus, and sugar was quite heterogeneous and occurred in whatever combination climate and soil would allow. It appears also that the level of capital concentration was not of the kind to create monopolies. Rather the emphasis was on self-development spurred on by attractive prices. Indeed, over this period, the profit margins on sugar were about 10 percent while those on bananas rocketed to 43 percent, with some sugar planters quickly abandoning ship (Eisner, 1961, pp. 207–208).

▶ The Merchants Flex Their Muscles

The gradual erosion and displacement of plantocratic domination provided the merchant/commercial fraction with the space to expand. Social ostracism ceased. All the exclusive clubs, once the sole domain of the planters, now admitted merchants. By 1923 the elite Jamaica Club not only admitted merchants to its membership but had six of them (mostly of Jewish background) on its management committee of seventeen. The membership of the Royal Jamaica Yacht Club, with the governor as Commodore, included ten merchants (nine of whom were Jewish) on its management committee of thirteen officers (*Handbook of Jamaica*, 1923, pp. 550–551; *Handbook of Jamaica*, 1931, p. 479). By the 1930s, Jewish politician/businessmen like George Seymour-Seymour were indistinguishable from their erstwhile detractors as they donned the affectations of wealth and high status (Gannon, 1976). This insouciance was evident in Seymour's contest with Garvey for a seat in the Kingston Corporation in the 1930s.

The merchants were also prying loose from the grip of the planters. What is now the Jamaica Chamber of Commerce, the principal lobby of the merchant/commercial fraction, was once under the control of the planter fraction. Improbably, yet understandably, the Chamber began as the Royal Society of Agriculture and Commerce in 1840. Indications of the merchants' recalcitrance were gradual yet perceptible. In 1886 the Society, owing no doubt to the prominence of the merchants, formed the complementary Merchants' Exchange, which was an effective weapon against the planter.[3] By 1922 the process of severance was complete, as "Royal" and "Agriculture" were both dropped from the organization's name making it the Jamaica Chamber of Commerce and Merchants' Exchange.

On the economic front, the greater share of merchant capital went into the developing tourist trade (Eisner, 1961, p. 287). The penetration of the non-banking and insurance spheres was also noteworthy. Building societies, the principal instruments of non-banking finance, were under the exclusive control of the merchant/commercial fraction. In nearly all cases, the directors of the major societies, the Victoria Mutual Building Society, the Jamaica Permanent Building Society, and the Jamaica Mutual Life Assurance Society were exclusively and predominantly of merchant/commercial background. Directorships were often interlocking. It was not unusual to find the names of prominent businessmen on several boards of management at the same time. These monopolistic tendencies, in turn, led to the rapid absorption of lesser incorporated private institutions.

In the insurance field, prominent members of the fraction were in command. Such noted commission agents, sugar and rum producers, and wharf operators as Fred L. Myers and Lascelles deMercado were heavily represented here. So, too, were Sir Alfred DaCosta and O. K. Henriques, who had interests in sugar. Indeed, O. K. Henriques was a principal director in the Insurance Company of Jamaica in 1931. Other important figures were George Seymour-Seymour, a commission agent with interests in sugar and bananas, and the Lyons family, connected with wharf operations among other commercial enterprises.

The merchant/commercial fraction gradually welded itself into a formidable entity that would continue to hold sway in Jamaica well into the present. Genealogical and other forms of research on interfamilial and interlocking relationships (Reid, 1976; Phillips, 1976) have matched all the merchant/commercial interests mentioned above with contemporary structures of economic power in Jamaica. Isolated by that process is a nucleus of twenty-one families whose impact on the present economy has been formidable.

In order to grasp more fully the dynamics of planter decline and merchant ascendance we must consider the special demands of each fraction relative to its own growth and reproduction. Colonial policies, after all, explain only part of the tension between planter and merchant. The production of sugar involves a proletariat: to the extent wage labor is central, this form of production sets in motion a body of industrial capital whose growth and reproduction are tied to the increasing exploitation of wage labor and patterns of investment sustaining or increasing profitability. By extension, capital invested in sugar is technically involved in a pattern of investment associated with high levels of productivity, large-scale production, and growing industrial expansion.

In a theoretical sense, sugar production embodied most of these dynamic characteristics, coupled with the additional advantage of preferential treatment by the Colonial Office. But the oft-repeated story is that any number of obstacles, not the least of which was that very Colonial Office connection, impeded the self-propelling elements built into this form of investment.[4] The net result was a gross asymmetry. As an industry requiring a large pool of surplus labor, sugar had progressively displaced the peasantry, making its members more and more dependent on wage labor. More disastrously, most sugar workers are employed for only six months of the year and have to seek alternative outlets for their displaced labor. On the other side of the equation, sugar's industrial capitalist potential was stillborn. Table 3 tells part of that story.

The data speak of an undeveloped economy, with manufacturing making a marginal contribution to the national income. Trade, especially if taken together with distribution, was impressive, testifying to the stability of the merchant/commercial fraction. In fact, profits from trade were more than twice those of the planters, constituting 22 percent of all profits. What may surprise is, on the one hand, the low contribution of agricultural wages (21 percent of total wages and salaries) and, on the other, the high planters' profits (10 percent of all profits). In fact, agricultural laborers' wages were barely higher than domestic wages. It was peasants and small farmers that made the most significant contribution to the national income.

The picture presented by these data is unmistakable. The volume of capital and the pattern of its organization could not sustain capitalist social relations. A capitalist infrastructure had been installed but the planter fraction and its practices were out of step with what was

TABLE 3
Composition of the National Income, 1930

Sources of Income	% of Total (rounded)
Agriculture	
Wages	7
Peasants' and small farmers' profits	35
Planters' profits	7
Non-Agricultural	
Domestic wages	7
Construction	5
Manufacturing	1
Distribution	4
Other	10
Other Profits	
Trade	15
Professions	2
Self-employment	9
TOTAL	102 (£19,950,000)

Source: Adapted from Gisela Eisner, *Jamaica, 1830–1930: A Study in Economic Growth* (Manchester, Eng.: Manchester University Press, 1961), p. 99. Used with permission.

implied and required by capital accumulation. We should now understand more fully why the labor rebellion erupted at about the time when these contradictions matured. The explanation for the demand for "jobs that last long," the rebels' steady refrain, lies also in these conditions. Equally, we can appreciate Norman Manley's early support for peasant producers. While their fortunes were deteriorating at the time, they had demonstrated their viability by generating more than half of all profits.

Post-rebellion conditions beckoned to a more developed market with a larger role to be played by distribution. In a structural sense, this fundamental break in traditional social and economic arrangements favored the interests of the merchant/commercial fraction that would definitely benefit from the severance of domestic helpers, yardboys, and washerwomen from their pre- and non-capitalist ties with the land. But the script, if their demands for "jobs that last long" were heeded, also called immediately for industrialization. So the tensions between planters and merchants would barely be resolved before another skirmish took its place: the merchants versus the industrialists.

▶ The Planters' Continuing Decline

Throughout the chapter, we have been discussing the forces that were undermining the dominant position of the planter. By the 1940s the general tide had taken an irreversible turn. The decline was hastened by several factors. First, pressure on wages, caused by inflationary trends during World War II, was far more problematic for the planter than the merchant, who was not a large employer of labor. In turn, the war signaled the decline of Britain and its colonial economic priorities, that is, its emphasis on export agriculture, and the beginning of a closer relationship with the United States and its economic priorities, which did not favor traditional agriculture. To these difficulties were added the effects of bad weather and plant diseases—especially the disastrous leaf spot disease that practically wiped out bananas.

The combined effect of these fundamental shifts, as Post (1981, II, p. 351) correctly indicates, highlighted the need to reform the power bloc. The cohesion that had promoted the domination of the planter fraction was loosening. The political instruments that had held the alliance together—the Sugar Manufacturers' Association, the JIA, the Jamaica Chamber of Commerce, the Jamaica Agricultural Society, and the Citrus Growers Association, to cite the more powerful—had lost touch with that unifying ingredient. As for the merchant/commercial fraction, it is perhaps a significant index of the shifting alliances that by 1948 the JIA, once the most prominent organization of the capitalists as a whole, represented planter interests almost exclusively. Such powerful representatives of the merchant/commercial world as O. K. Henriques, Sir Alfred DaCosta, and Sir Noel Livingston seemed to be paying greater attention to their business interests (*Handbook of Jamaica,* 1948, p. 667).

Overall, these reverses contributed to a substantial, all-around weakening of the planter fraction as a whole. Sugar had rebounded from the earlier misfortunes, but the sub-fraction was now the hub of a wheel with defective spokes. In fact, the setbacks experienced by the other sub-fractions led to a rash of self-serving withdrawals. Government subsidies and other financial assistance would now be vital for the process of rehabilitation and the various support associations seemed to have gone after scarce resources in a self-serving way. This observation is borne out by the increased number of control boards and other devices that nestled parasitically around the central organs of government.

Undergirding the decline were important structural factors. The incompatibilities between merchant capital and export agricultural capital, touched on earlier, deepened with maturing contradictions. First, merchant capital avoided the export agricultural sector; second, it had to contend with industrial capital, a nemesis that it was partly responsible for creating. Let us take them individually.

If, as indeed was the case, the merchant possessed the bulk of the investment capital, then the new configurations within the bourgeoisie were predictable. To begin with, there was no plausible reason to plow capital into a moribund and problem-ridden industry whose average net profits for 1939 through the mid-1940s was less than 4.0 percent (*Report of the Jamaica Sugar Industry Commission, 1944–45*). It was evident that the process of consolidating sugar estates, which was undertaken in the 1920s, had not fulfilled its promise. That merchants preferred to invest their surplus funds in British and United States stocks and bonds was quite understandable (Post, 1981, II, p. 382).

The demands of merchant capital were out of harmony with sugar and, for that matter, all of export agriculture for structural and conjunctural reasons. Structurally, merchant capital abhors risk taking, shuns a slow turnover on investment, and prefers to restrict operating costs to a minimum by employing as few hands as possible. Sugar production, in particular, demands most of what was abhorred by the merchants and it remains a mystery that they should have penetrated the industry in the early years of the twentieth century except that it was one of a limited range of options and provided the most accessible route to financial growth and political prominence.

There were yet other negative elements. The market fashioned by sugar (less so, bananas, as production continued year round) was static. Seasonal production played havoc with market expansionism, because merchants could only look to the harvest season for their profits. Merchant capital's logic of growth is aimed at luxury consumption, which was in no way consistent with the growing number of gardeners, domestic servants, and the like, that is, part-time, low-paid workers. Merchant capital was enmeshed in the socio-economic arrangements demanding fundamental change within clearly defined, structural limits. As Karl Marx notes (1975, III, pp. 326–327), "within the capitalist mode of production—that is as soon as capital had established its sway over production and imported to it a wholly changed and specific form—merchant capital appears merely as capital with a specific function." When that stage arrives, merchant capital, in spite of its wishes, falls under the sway of industrial capital whose logic is based upon a completely different set of calculations.

It is important to put this situation into proper perspective, since it serves partly to illustrate the conjuncture, that is, the effect of these developments on the alliance of the capitalists. Quite directly, the period of the late 1940s and the 1950s witnessed a tug of war between the three main fractions of the class: the declining export agricultural fraction, the merchant/commercial fraction, and the incipient industrial fraction, whose gradual rise stemmed as much from the internal contradictions of merchant capital as from the requirements of internal class relations.

Let us look more closely at the conflict between the merchant and the budding industrialist. Of course, the rule of capital accumulation is that the merchant's capital will in time seek qualitatively new investment formulas to ensure its growth. More specifically, the merchant would come to expect greater opportunities with the widening of the market suggested by the dislocations and population movements following the events of the late 1920s.[5] But the prospect of an expanded internal market and the anticipated good fortune of the merchant had to confront socioeconomic factors that begged for more favorable state policies. Domestic servants, yardboys, gardeners—many of whom at the time of the labor rebellion had "neither land nor agricultural tradition" (Macmillan, 1938, p. 169)—were incipient class formations cut loose from their moorings within pre- and non-capitalist relations that were progressively undermined by export agriculture. In addition to the pronounced rate of rural-to-urban migration mentioned earlier, the number of agricultural workers (including unpaid family members) had been steadily declining. Between 1954 and 1961, the overall decline was just over 37 percent (Jefferson, 1972, p. 83). Thus, the dislocated social groupings—separated from agricultural labor, the peasantry, and artisan occupations—were forced more and more to cast their lot with proletarian pursuits or some form of petty capitalist and marginal (own account) occupations. These developments would significantly alter the pattern of state mediation as new social forms and relations were inserted into the process of capital accumulation.

The reaction of export agriculture to its declining power and influence assumed political form: the Central Committee of Primary Producers (CCPP) was formed in 1947 as an independent lobby. The CCPP was dominated by large farmers who sought to directly influence policy on the behalf of agriculture. The lobby appeared to have had only marginal success but its general sentiments led to the formation of the Farmers' Federation (FF) on February 3, 1951, which aimed to agitate on behalf of the small farmers. At the helm was Robert Kirkwood, the chairman of the Sugar Manufacturers' Association, who had expressed concern for the plight of the sugar industry and the rise of merchant/commercial interests.[6] The effort swelled into the formation of a political party, the Farmers' Party (FP), on March 14, 1953. Among its ranks were large landowners such as A. S. Campbell, the general manager of Innswood Sugar Estate and a member of the Legislative Council between 1944 and 1954; J. P. Gyles, a successful large farmer of St. Catherine who was later to become a minister of agriculture under the JLP; Roy A. McNeil, a lawyer and farmer from St. Catherine who later held the portfolio of minister of home affairs under the JLP; and Oscar deLisser, a wealthy farmer of distinct capitalist origins. The FP's entry into competitive politics was disastrous. In the general elections of 1955, none of its thirteen candidates captured a seat in the House of Representatives. In fact, the party could muster no more than 3.9 percent of the vote, despite the high caliber of its candidates and its sound financial base. It is true that the party's effectiveness was greatly impeded by highly publicized quarrels within its higher echelons and by defections at the most inopportune moments, but these developments had little impact on its ultimate failure.

The industrialization agenda, on the other hand, was securely based and withstood the highly publicized pro-agriculture recommendations of the World Bank in 1952. At the heart of what was objectively required (and subjectively demanded during the course of the labor rebellion) lay industrialization. For the dislocated, it could give at least a partial solution to their demand for steady employment. For the capitalists, there would be the necessary reordering of the alliance to meet the new patterns of accumulation introduced by these new formulas. More concretely, intraclass conflict between the merchants and the budding industrialists would heighten as the state could not favor both equally. This is so because both forms of capital are mutually exclusive in a structural sense and, while merchant capital excels at amassing investment funds, it had to be converted into industrial capital to undertake the transformation required by the Jamaican economy. Industrialization would open up markets (a middle class explosion) from which the merchants in their distributive and/or commercial roles benefited; but once the industrializing impetus was set in place, the conflict and competition between these fractions had to intensify.

Of course, this thesis, as expounded by Marx (1975, III, p. 336), is immediately applicable to a context more dynamic than was the case in Jamaica in the 1940s. Marx's model is the advanced metropolitan economy, where hegemonic relations within the capitalist alliance and the state made the transformation of merchant capital into "the servant of industrial production" compelling. In our case, none of the fractions clearly dominated in the dual senses of hegemony within the class and with regard to the state. Nonetheless, the antagonism between the fractions is structurally imprinted in the relations they collectively bear to capital accumulation, though the invariable results are crises in production and distribution throughout the system.

▶ Enter the Rebellious Child: The Industrialist

The industrial fraction was a late arrival on the scene. During the 1920s and the 1930s, the economy's rudimentary industrial base was understandably linked to agricultural products—tobacco, copra, sisal and rope, beer and ale, and the processing of hides. There were only eight factories in production in 1890; this number swelled to sixty-two by 1910, then declined to fifty-two in 1928 (largely through consolidation) (Eisner, 1961, p. 174). Other industrial indicators, electricity usage, for example, suggested gradual industrialization. In 1939, 4.323 million kwh were generated for thirty-nine registered companies that employed 8,840 workers. Only eight years later, in 1947, 9.083 million kwh were generated for sixty-nine registered companies employing 20,123 workers (Central Bureau of Statistics, 1949, p. 6).

Government also provided some measure of protection. In 1931, the passage of the Coconut Industry Aid Law prevented merchants from importing edible oil. Four years later, in 1935, the Safeguarding of Local Industries and Trades Law prohibited the manufacturing of matches without government approval. This latter legislation mitigated somewhat the effects of the defeat of the Native Industries Bill by the planters in the same year. The action had been prompted by the desire to stem the potential rise of monopolies as well as by anti-Semitism, as the predominantly white planter fraction verbally deplored the fact that industrial development was falling into the hands of "aliens." A Jewish outfit, Henriques Brothers, was spearheading local industrial efforts in producing coconut oil, citrus, and matches. It is suggested that a prominent member of the Henriques family, O. K. Henriques, chairman and managing director of the match factory and himself a member of the Privy Council, was "quite instrumental in persuading the colonial government to pass [the Native Industries Bill]" in 1935 (Holzberg, 1987, pp. 140–141). It testifies to the growing importance of the industrial interests at this time. Other Jews who were branching out into industry were the DaCostas and the DeMercados, who had established the Jamaica Biscuit Company and other light industries as early as 1909.

By the end of the 1930s, there were many voices arguing for the importance of secondary industries for economic development. The case was made, for instance, for such industrial activities as canning, containers, confectionery, footwear, alcohol production, and textiles. Starting with World War II, the *Public Opinion* (September 9, 1939, p. 1) struck a blow for the industrialist/manufacturer. The high cost of imports, the tabloid reasoned, should provide the impetus to can local beef and process bacon.

Gradually, pressures clustered around the adverse terms of trade between agricultural exports and the imports from metropolitan suppliers. Jamaican exports, it was noted, could buy less abroad in 1947 than they could in 1938. The adverse movement in the terms of trade "had been against the exporter of produce more or less primary and in favour of the exporter of items of secondary production" (Central Bureau of Statistics, 1949, p. 7). Indeed, the call for import-substitution was quite distinct: "All of these are related to local developmental possibilities such as fishing and livestock raising, flour milling, motor vehicle assembly, textile and alcohol manufacture" (Central Bureau of Statistics, 1949, p. 9).

Unquestionably, this general pro-industrial mindset promoted the supportive legislation introduced over the period, perhaps the most far-reaching of which was the Pioneer Industries (Encouragement) Act of 1949. This piece of legislation was wide-ranging in scope, embracing a total of nineteen industries projected to employ 1,000 workers and increasing to 3,000 workers when maximum production was established (*Annual Report—Jamaica,* 1953, 1956, p. 182). The impetus to industrialize received added strength from three sources. World War II created shortages of consumer goods that local entrepreneurs supplied; the dollar crisis of 1947–48 provided manufacturers with the opportunity to produce goods previously supplied by the United States and Canada; and Jamaicans who had served in the armed forces overseas returned with new technical skills suited to the industrial thrust. Between 1948 and 1957 impressive increases were registered in the levels of capital formation and there was a significant inflow of capital between 1950 and 1966 mostly from North America (Girvan, 1971, p. 13).

These impressive values are reflected in the performance of the relevant sectors of the economy, as Table 4 shows. The marked industrial thrust to the economy at this time can be appreciated by comparing these data with figures from 1930 (see Table 3) when agriculture dominated the economy, accounting for 48.7 percent of the national income. As Table 4 reveals, the decade of the 1950s saw profound changes, the trend continuing into the 1960s. Agriculture had declined by half between 1950 and 1960, and by half again by 1970. During the 1950s, the sector did not fare well against construction, distribution, and manufacturing, all of which increased significantly, contributing together 34

TABLE 4
Contribution of Economic Sectors to the Gross Domestic Product, 1950–1970 (% of total)

Sectors	1950*	1960†	1970‡
Agriculture, forestry, and fishing	24.0	12.1	6.7
Mining and quarrying§	—	9.7	12.6
Manufacturing	11.3	14.0	15.7
Construction and installation	7.6	11.9	13.3
Distribution	15.1	18.1	19.0
Public administration	6.1	6.1	7.8
Other (incl. services)	35.9	28.1	24.9

Source: Department of Statistics, Kingston, Government of Jamaica.

* Total GDP of $5,308,000 in constant 1974 prices.

† Total GDP of $12,838,000 in constant 1974 prices.

‡ Total GDP of $21,586,000 in constant 1974 prices.

§ Bauxite mining operations began in 1953.

percent of the GDP in 1950 and 44 percent in 1960. Between 1950 and 1957 alone, net agricultural output increased by a mere 17.5 percent while manufacturing, still then in its early stages, increased by 65.4 percent (*National Accounts Income and Expenditure, 1950–1957*, 1959, pp. 2, 8).

Industrial capital had been forcing itself into the interstices of the economy. The conflict and competition between the merchant and industrial fractions (and between them and the export agricultural fraction) began to surface within the apparatuses of the state. Up to 1938, almost all members of the Legislative Council who were not ex officio members came from agriculture. By the mid-1940s there had been a noticeable increase in the number of the "newly ascendant manufacturing and merchant interests" (Phillips, 1976, p. 18). This fissure got much wider and the issues surrounding it grew even more complex. An interesting feature was the extent to which internal economic factors aided clearcut fractional distinctions. Peter Phillips (1976, p. 21) saw the 1940s as a period favoring the rise of "an urban manufacturing group" whose interests would "benefit from the closure of Jamaican markets to foreign goods" and could entertain support for "the national stirrings" of the times.[7]

But the rivalry assumed more overt forms. In our earlier discussion of the planter/merchant rivalry, we indicated that the Jamaica Chamber of Commerce was first under the control of the planter fraction in the guise of the Royal Society of Agriculture and Commerce. As we further indicated, conflict and competition between the two fractions led to the merchants' own organization, the Jamaica Chamber of Commerce. Under like conditions of conflict and competition, the Chamber of Commerce gave birth to the Jamaica Manufacturers' Association in 1953, with Aaron Matalon as its first president.[8] At this time, the conflict between the two fractions was an open secret. R. W. Youngman, the president of the Chamber, vehemently denied charges that his organization was opposed to industrialization. However, he felt that locally made goods, for local consumption or export, should be marketed through the same channels as imported goods of a similar kind. These goods, he added, should also be fully competitive with imports in quality and price (Chapman, 1954, pp. 25–26).

In short, the mercantile community wished to keep the new industrialists under its thumb. Self-interest clearly played a crucial part, since the undisputed control of import and distribution channels had enabled the fraction to keep its profits high. The community also opposed the idea of protection for these industries: there was perhaps no surer way of destroying the new industrialists than to heed the Chamber's recommendations. For it should also be noted that the mercantile community opposed industrial subsidies, whether as protective tariffs or restrictions on imports. Indeed, the key issue, namely, fractions of capital in fierce competition, was readily grasped by the enlightened opinion of the day: manufacturing industries should take precedence over "the purely distributive agencies" whose role was important "but not the highest" (Chapman, 1954, pp. 71–72).

The incipient industrial fraction attracted significant bi-partisan support over the succeeding years.[9] In 1969, the ongoing tug of war with the Chamber of Commerce necessitated the intervention of the minister of trade and industry, Robert Lightbourne, to defend the fraction and reassert the government's protectionist and import restriction policies, and its condemnation of those (mainly the merchants) who had been complaining about the quality of local manufactures. Investigations revealed that the complaints might not reflect the true situation, thereby suggesting some mischief on the merchants' part. In the end, the JMA, in a show of conciliation, acknowledged the support of merchants and consumers, in spite of the occasional problems in their relationship, and invited the presidents of the Chamber and the Consumer League to sit as honorary directors on the JMA Board of Directors.

The fraction's fortunes improved. In 1970 the Jamaica Development Bank promoted industrial production and manufacturing over distribution, as a matter of specific policy. The Bank had approved business loans of $2.9 million for production and manufacturing, a figure that represented 60 percent of all its loans (*Economic Survey of Jamaica,* 1970, p. 104). In 1969, the Bank also instituted a new system of classifying installment credit that clearly favored the industrialist. New business was divided into two groups according to whether loans were for consumption or productive purposes. Loans for consumption purposes, especially the consumption of imports, were subject to restrictions, while credit for productive purposes was exempted from all restrictions (*Economic Survey of Jamaica,* 1969, p. 106).

A note of caution is in order, however. Up to the late 1960s and very early 1970s, we may speak of a tendency at best toward the gradual development of an industrial fraction. Export agriculture, as we mentioned earlier, lost its dominance in the late 1960s, but the merchant/commercial fraction maintained a strong position in the economy. In fact, it seems that a measure of accommodation was reached that enabled merchants to maintain their traditionally high profits through distribution of locally made products. Thomas Balogh (1970, p. 306), a consultant to the government, warned: "By charging extraordinary profit margins at a number of consecutive stages of the merchandising process . . . [the merchants have] not only [secured] unduly high incomes for themselves, [but seem] to have encouraged industrial inefficiency and indeed made it profitable." Later, it was found that profit margins for locally manufactured items were higher than those for imports, suggesting that "the imported price [was] being used as the price leader in the local market" (Stone and Brown, 1976, p. 362). In turn, such practices impacted negatively on the industrialization drive: in the hands of the merchants, much import-substituting manufacturing became assembly operations, often tied to earlier, well-established commercial traditions.

Three main causes can be easily identified. First, the dynamics discussed above were caught in the process of transforming one species of capital into another and depended a great deal on the resources of the merchant/commercial fraction. It should be remembered that the economic base of the economy was essentially weak, making the transition from one species of capital to another a gradual one. Second, the policies implemented by the state, particularly against the distributive sector, could be only partially enforced. For one, the openness of the economy made monitoring difficult. Such was the case with the Prices Commission's efforts to curb consumption in 1970 where openness of the economy was compounded by the ratio of imports to the total supply of goods and services.[10] For another, the mindset of the appropriate organs of state (the Trade Administrator's Office, for example) betrayed a "comprador" bias, thereby frustrating the process even further.[11] Third, while the economy was being transformed, agriculture provided the most effective employment outlet. As noted in Chapter 1, even at the height of industrialization, the modern economy could absorb only a few new entrants into the labor force.

At the time of the crisis of the late 1960s the relative position of the fractions had shifted. The export agricultural fraction was maintaining a tenuous position by relying on state subsidies and, in the case of sugar, state purchases of unprofitable operations. Distribution and manufacturing occupied the commanding heights of the economy. Interfractional rivalry was increasing but the merchant fraction gradually but perceptibly entrenched itself.

In the 1970s capital concentration revolved around securities trading. Old, family-operated businesses were now becoming public corporations. A stock exchange established in 1969 provided the mechanism for deepening the control the class exerted over the economy. Assets that were once tied up were released for investment in other ventures. Original owners acted as directors of new public corporations. Management fees, allowances of all kinds, and even practices in flagrant conflict of interest were used to consolidate their power (Holzberg, 1987, p. 173; Thomas, 1988, p. 381). In 1970 fully 70 percent of Jamaica's Gross Domestic Savings were in the form of undistributed corporate profits and depreciation allowances. When democratic socialism was declared in 1974, public companies controlled J$500 million in assets (Holzberg, 1987, p. 172).

Prominently represented here is the Jewish subfraction, the core of the 21 Families mentioned earlier. Exercising a level of control far in excess of their numbers—Jews numbered less than 0.025 percent of the Jamaican population but accounted for roughly 23 percent of the entrepreneurial elite—the fraction dominated. Seven of the 21 Families were Jewish, of this number, four—Ashenheim, Henriques, DaCosta, and Matalon—accounted for twenty-two of the forty-seven corporate boards for the 1971 and 1974 stock exchange companies (Holzberg, 1987, p. 187).

As accomplished managers, class members could not be kept out of the activities of the state. Traditionally, class members avoided popular elective office, having learned the lesson in 1944 that elections to Parish Council and Parliament were the domain of the middle class. Instead they became heads of parastatals, commissions of enquiry, and diplomatic missions. The pattern continued during the government of Michael Manley, which appointed members of prominent Jewish families to important administrative posts. Notable examples are O. K. Melhado, a member of the

"21 Families," who became the managing director of the State Trading Company (STC) in 1977, and Eli Matalon, who served as minister of state in the Ministry of National Security and Justice in 1974. Another was the formidable and very talented Mayer Matalon who distinguished himself as a tough negotiator against the bauxite industry in 1974.[12] Attorneys from the fraction also figured prominently in the reorganization of the bauxite industry.[13] In fact, the industrial fraction of the class was targeted by the Manley regime to spearhead future growth and economic development. This preferential treatment created tensions within the class but nothing of a rupture signaling decline.

▶ Concluding Note

The pivotal argument of this chapter is that the widely held notion of the capitalist class as a tightly-knit oligarchy is false. Competition, which is intrinsic to capital accumulation, deepened rivalry within the capitalist class as Jamaica's economic development progressed.

In the 1920s, the planter fraction, if not hegemonic, was dominant, receiving an appreciable measure of support from the Colonial Office. The gradual but perceptible decline of sugar and bananas, and thereby the export agricultural subfraction, signalled the rise of the merchant/commercial fraction. However, this rise to prominence lacked support of the kind received by its antagonist, the planters. It possessed substance but had limited power at the level of the state apparatuses.

Events did not allow the merchant/commercial fraction to entrench itself. No sooner had the planters and the merchants crossed swords than the industrialists signaled their readiness to push for ascendence. In fact, there was much in the structure and the needs of class relations to support this species of capital. Indeed, the transformation of the class structure, mirrored in large part by the events of the labor rebellion, placed a growing burden on the state both to mediate these grievances between and within social classes and to ensure that such undertakings complied with the requirements of capital accumulation. By definition, merchant capital could not be placed in a favored position; it was industrial capital that could provide the "jobs that last long," create rapid economic activity, and promote the kind of economic expansion to transform the domestic servants, the yardboys, and the "scufflers" into productive workers.

But the problem of securing adequate capital resources remained stubborn. These resources had to evolve from merchant/commercial coffers in relation to capitalists among whom a clear-cut fractional ascendancy was impossible. This lack of ascendancy is structurally endemic in the class and accounts for its ongoing instability and brittleness—features that go hand in hand with the process of capital accumulation.

It should be noted as well that while the various fractions of capital were jostling each other, that process resulted in the consolidation of a capitalist state and concomitant relations. The intracapitalist rivalry surfaced with greater impact during the crisis.

The Strategic Middle Class

CHAPTER 6

The history of Jamaica has always foreshadowed the rise of a strategic fraction of the middle class, a fraction that was destined to play a pivotal political role. Circumstances rooted in the history of slavery and developing capitalist relations placed members of the class in the crucial political role of mediating between the (white) dominant class and the other (black) social classes. Increasingly, middle class power became concentrated within the apparatuses of the state. Over time, these circumstances would ensure the fraction's attempt to translate political clout into economic power. The possibility of arrogating economic power, of penetrating the capitalist class, results from the chronically weak economic base of this class. All the significant periods, starting with the pre-Emancipation era, through Crown Colony, down to democratic socialism, testify to the occurrence and consolidation of forces favoring the rise of this strategic fraction of the middle class.

▶ Evolution of the Middle Class

Jamaican history can be divided into three periods: from the eve of Emancipation in 1838 to the beginning of Crown Colony government (1820–1865); Crown Colony government itself (1865–1938), ending with a transitional period (1938–1944); and the modern era, that is 1944 and beyond. In each period, crucial social relations of intermediate elements had mainly to do with mediating the antagonisms of the classes above and below them.

Pre-emancipation to Crown Colony (1820–1865)

Slavery created a climate of distrust and animosity between the dominant class, the plantocracy, and the slaves and their emancipated offspring, which later compounded the intrinsic antagonism between capital and labor. During slavery and progressively thereafter, political rule came to depend more and more upon the strategic intervention of mediators drawn from an intermediate element, the "coloreds," who constituted nearly 10 percent of the population in the early to mid-1800s. Their bi-racial ancestry, intermediate status, and social aspirations continued to make them appear, at least for a time, as the natural ally of the planter and a barrier against the ambitions of the black population.

The planters were not alone, however, in seeing a mediatory role for the coloreds. Particularly after Emancipation, the British government sought to increase their political power in order to temper the gross excesses of a plantocracy not taking too kindly to the loss of power over its slave population. Of course, if the humanitarian motive was sincere, it was heavily sprinkled with the need to maintain the economic health of the colony—which the planters' folly was rapidly destroying—and "keep up, if not increase, the consumption of British manufactures."[1] Historically, the planters had won the right of self-rule in local matters, which now greatly hampered direct intervention by the governor. With tutoring from the Marquis of Sligo, the governor, in 1836 the coloreds formed a "liberal clique," the Town Party, using this vehicle to gain increasing influence in the Legislative Assembly. The Crown gave further support by appointing educated coloreds to "Superior Offices": by the mid-1850s, one from their number was appointed to the prestigious and influential office of Attorney General (Campbell, 1976, p. 236).

The coloreds fulfilled the vision—and more. While these elements were not of uniform economic status, they had been united by the social and political ostracism that affected them equally (Hall, 1972b). And, while some were planters and had been slave owners, overall, they questioned the wisdom of equating the health of the colony with the health of the plantation sector, supporting instead economic alternatives, chiefly the locally oriented economy of the "small settlers," crafts, and the like. We can begin to trace a pattern—though hampered by political liabilities and steeped in ideological ambiguity—of a class-in-formation disassociating from the weakened economic structure and attempting to supplant the dominant class by championing economic innovation (Keith and Keith, 1985, pp. 80–83).

The coloreds executed their political charge with distinction and several of their more prominent representatives strongly supported the cause of the beleaguered freedmen. Overall, colored politicians were the only ones who brought legislation to improve the conditions of peasants and laborers. While not altogether successful, the coloreds opposed the various motions of the Assembly to introduce income tax and levies that would further affect them; instead, efforts were made to shift this burden onto the estates (Heuman, 1981, p. 147). Equally, the new legislators displayed a keen sense of their own interests, rapidly acquiring the cunning needed to entrench themselves. For instance, they enacted measures in the Assembly to reduce the salaries of all public offices, thereby making government service unappealing to the expatriate. Colored-owned newspapers like *The Watchman* and the *Morning Journal* were part of the machinery that ultimately broke the strangle hold of the planters.[2] In fact, by the mid-1850s the planters no longer controlled the Assembly: a coalition of coloreds and mainly Jewish merchants were the new caretakers.

The planters fought back. An early ploy, attempted in 1839, involved a most improbable coalition: planter and ex-slave.

> *For now their tricks are all found out,*
> *And they must go to the right about,*
> *For white and black the day is now,*
> *And Brown and the brown have fallen low.*
>
> *Bravo White! and well done Black!*
> *To the devil send the greedy pack;*
> *And with a pull both strong and long.*
> *We'll sing perdition to the Watchman's throng.*[3]

Alliance with the Imperial government was clearly a more rational approach, and one that became more probable as the power of the coloreds increased beyond all expectations. The strength of ideology was such that the British had expected the coloreds to follow the lead of the whites as a matter of course. Most prominent coloreds did, in fact, strongly favor British values and policies. Some were even likened to prominent English intellectuals. However, even those values, in the local situation, required action that ran counter to British expectations. Take Richard Hill, the stipendiary magistrate of Spanish Town. According to Anton V. Long (1956, p. 27),

> [Hill] was to Jamaica what Sir James Stephen at the Colonial Office was to the British Empire. Both of them were men of high principle who employed their talent in subordinate positions out of a

> conviction that their obligation to humanity outweighed any rewards which either might easily have obtained in another capacity. For both men there were times when they felt that duty required them to become involved in political affairs. For Hill this was such a time. He could not remain silent while preparations were being made to restore the harshness of slavery.

So the realities of politics had thrown both camps together in a startling example of forced fellowship. The prevailing voting patterns were such that "all candidates of European extraction [were being] sedulously rejected" in favor of their colored counterpart.[4] As W. P. Morrell (1966, p. 261) put it, the coloreds and the peasantry "were showing themselves eager for consequence and power."

This eagerness was quickly throttled by a series of modifications to the franchise that reduced the small freeholder vote by two thirds and, by 1860, translated itself into a dramatic drop in the numbers of coloreds in the Assembly (fourteen out of forty-seven). The planters and the British Parliament now acquiesced in the belief that the rise of the coloreds to sudden prominence was "retrograding" civilization and "subverting . . . the natural order of society."[5] As Governor Grey (1846–1853) comments somewhat superciliously, the coloreds lacked "that common understanding which results from a long continuance of homogeneous renewals of a people."[6]

The Morant Bay Rebellion of 1865 provided the immediate rationale for replacing representative government with autocracy ("Crown Colony" government). In the end, the rebellion was about the inability of the political system to mediate the conflict between planters and freedmen. If the latter were about creating an independent peasantry, the former, in alliance with the Imperial government, would never countenance such a project. Nor were the coloreds securely on the side of the ex-slave. The property qualifications required of the electorate severely restricted any inclinations to the watchdog role and counseled appropriate alliances with power holders. Humanitarian inclinations themselves, when present, were tempered by ideology. In Jamaica, as in other colonies, coloreds strove to win planter approval by shedding the so-called "nigger yard" culture in favor of the preferred British noblesse oblige. In fact, some of them accepted the existence of planter privileges that they "did not expect to enjoy" for themselves (Delson, 1981, p. 85–86). In the end, the coloreds' pursuit of equality was personal and not troubled by considerations of social justice and fair play for the black population. Indeed, many had been committed slave owners (Hall, 1972b, p. 198; Cox, 1984, p. 150).

Crown Colony Period (1865–1938)

The future tone of relations between dominant and subordinate classes was set by the time of the Morant Bay Rebellion. We would not speculate that, had the coloreds been allowed to entrench their instruments of mediation, the conflict between planter and peasant would have been significantly minimized. We will volunteer, however, that most of the structural indicators point to inevitable mediation by appropriate strata or fractions of the middle class, especially as relations between the planters and the peasants and laborers were now more deeply strained by the spilling of blood on both sides.[7]

In theory, there is support for the contention that Crown Colony government dealt the severest blow to the coloreds. Graham Knox (1965, p. 142) argues that the coloreds, by joining forces with the planters around and in the aftermath of the Morant Bay Rebellion, hampered their chances for political self-determination. Their ambiguous political and ideological positions, combined with the control of the Colonial Office would probably not have allowed a complete alliance with subordinate elements. Whether the coloreds could survive politically as an independent force was doubtful. It is, however, beyond doubt that their structural role in the development of mass politics was assured: between the extremes of the plantocratic cooptation and pursuit of naked self-interest there exists an impressive number of postures and activities at once consistent with the coloreds' structural role and agreeable to the subordinate elements. In particular, their overall predisposition to and respect for education led to the rise of a strong tradition that paved the way for their counterparts, the black middle class, in later times. Limited in their access to the planters' sources of wealth and power, the coloreds controlled the sphere of education: they were the parsons, teachers, and civil servants, professions that could serve to turn the rational frame of reference of capitalism upon itself. It was Joseph Shumpeter (1942) who remarked on capitalism's propensity to give rise to a critical frame of mind that, after destroying many traditional institutions, turns upon capitalism itself.

The coloreds played a particular role in economic history as well as in political history. It appears that only a minority of their members established their financial base by buying into plantation agriculture; in fact, they were generally criticized for not plunging enthusiastically into acquiring bankrupt and abandoned sugar estates. The more dynamic among them selected an alternate economic base. The spread of entrepreneurial efforts away from plantation agriculture was impressive: along with the more adventurous blacks, the coloreds pioneered small scale industries, mining operations, the production of silk, and the like (Hall, 1959, p. 33; Post, 1978, pp. 36–37; Brathwaite, 1971, pp. 80–95).

With market forces operating against these new investment patterns and the Colonial Office stubbornly propping up sugar and the planter, the coloreds were, however, thwarted in their efforts to establish a sound economic base. These constraints persisted in the period under review and well beyond. Indeed, the coloreds' economic assault was halted by the rise of the Jews whose trading and commercial activities and subsequent emergence into the capitalist class were discussed in Chapter 5. Their gradual growth and consolidation stifled not only the economic advance of the coloreds but that of other politically strategic fractions of the middle class (Eisner, 1961, p. 314). The coloreds gradually lost their footing in the mercantile field partly—and not insignificantly—owing to their inability to secure adequate financing for such ventures. This disqualification was common throughout the West Indies. In Guyana, for example, coloreds found themselves ousted from business by newcomers, the Portuguese and Chinese, from whom commercial credit was not withheld (Jeffrey and Baber, 1986, p. 42).

Education, in the end, contributed more to increasing the political power of colored and black elements than economic activities. Education had structural specificity. The Church played an important role in promoting education, since its missionaries' moral intent became clearly bound up with the Colonial Office's overall strategy of preparing the colony for modern politics. However, it was the antagonisms between coloreds and planters that served more directly to put education into this role. The planter, who, in 1926, advised the Legislative Council that education for the people would mean the collapse of the colonial economy, had his forerunner in the nineteenth century planter who despised education and with it the educated coloreds.[8] For this threatened dominant class, policies favoring education spelled disaster in two important ways: education would advance their lot through societal reconstruction and would also reduce the labor available for the plantations.

Clearly, race and color remained prominent factors in limiting access to economic power and education provided a measure of respite. Paths to development open to other minorities were simply closed. Take the Jewish element, for instance. Before and during the Crown Colony era, Jewish politicians displayed the opportunism expected of a social group vying for political and economic power. There were those who sought and acquired close relations with the planters (Andrade, 1941); on the other hand, others were quick to turn a favorable situation to their advantage. Late into the Crown Colony era and beyond, their new-found economic power made them full-fledged members of the capitalists. By the 1930s commentators could speak of their opposition to anti-establishment politics and their "elitist tradition of crown-colony politics" (Gannon, 1976, p. 47). In the 1940s, so entrenched were they within the capitalist camp that those among and close to them who supported Jamaican self-government were threatened with recriminations (Holzberg, 1987, p. 197).[9] The black educated class, in the meantime, still had to wage battles with the expatriate and colored elements for civil service positions.

Toward the end of Crown Colony government, the greatest benefits of education accrued naturally to the blacks, since the increasing rationalization of the political system, along with greater access to education, logically brought an incipient black middle class to the fore. At the time of the labor rebellion in 1938, the political and economic impetus of the coloreds was dulled. They had not breached the ranks of the planter class as a force; that distinction belongs to the Jews. Although some of them played a role in important expatriate organizations like the Jamaica Progressive League, on the whole the coloreds found themselves, by dint of their ideology and common perception of the black population, aligned to planter class culture. They continued to hold important positions in the government and within the professions but the prospect, as a group, for a direct, active role in elective politics, rapidly evaporated.

The patterns the coloreds had inaugurated, however, continued. The black middle class also saw its future linked irretrievably to education, as the testimony of J.A.G. Smith before the Moyne Commission reveals. Now conscious of its value, the class deplored the fact that the local civil service was being seen as "a system of outdoor relief" for the rejects of British society:

> We seek to bring the problem to the attention of the Commission that while it does not affect the workers and the peasants, who have no aspiration to enter the Colonial Civil Service, we feel that it is a question of vital importance to large sections of the West Indian communities, especially the intellectuals and middle classes whose future so largely depends upon finding employment in the Government Services.[10]

As was true earlier of the coloreds, J. A. G. Smith and his peers were forced, in a sense, to take maximum advantage of their invaluable role in the political arena, given a history of systematic denial of avenues for economic consolidation. Black and brown businessmen bent on developing the tourist industry were not allowed to purchase Constant Spring Hotel in 1941. J.A.G. Smith took their case to the Legislative Council to plead for black entrepreneurship (Gannon, 1976, p. 117). His efforts were unavailing.

At the close of the Crown Colony period in 1938, the middle class was firmly established as the seed bed for the would-be political functionaries within the government. As F. R. Augier (1962, p. 36) observes, the Crown Colony government bequeathed a legacy of rationalization to political institutions. While the practice of assigning British personnel to prominent government posts continued for a while, the stage was set for the gradual rise of the educated elements of the black middle class. At one level, it was now obvious that events favored the enfranchisement of the bulk of the black population. It was unlikely, also, that the newly empowered would easily forgive and forget the role played by planters and their retinue. The mediation of educated blacks was now imperative, a structural necessity. Their rise was also aided by the coloreds, whose attachment to British values was predictable. Like Harold Allan and other black political figures who opposed adult suffrage for educational reasons, there were prominent coloreds whose opposition to the advance of the subordinate classes espoused identical principles.

The importance of the educated elements of the black middle class grew, nonetheless. In the 1920s, for example, they formed powerful Responsible Government Associations, directed to furthering their own narrow interests and those of democratic government. Initially, leadership was provided by servicemen returning from the First World War who pressed for greater representation in the legislatures and for federation. The main tactic was to hold public meetings and prepare petitions to the Colonial Office (Proctor, 1962, p. 34; Brereton, 1980).

In the mid-1930s, just before the labor rebellion of 1938, the black educated middle class used such organizations as the Jamaica Progressive League to insist on their rights to lead and to press for independence. W. Adolphe Roberts, for example, insisted that on the issue of "autonomy there could be no compromise."[11] These organizations were not unique. There were similar organizations in all parts of the West Indies. In Trinidad, the future of these elements was intimately bound up with political movements led by Captain A. A. Cipriani and Tubal Uriah "Buzz" Butler (Williams, 1962; Ryan, 1974). In Barbados, Grantley Adams, a brilliant, black barrister, formed the Barbados Progressive League. The League featured middle class nationalists (prominently represented by the black intelligentsia) who were committed to overthrowing the "dictatorship of landowners" (Gooding, 1981, p. 76). In St. Kitts, the middle class-based Labour League was engaged in similar activities at this time (Gooding, 1981).

Modern Period (1945–1970s)

The constitutional reforms instituted from 1944 to independence in 1962 favored the ascendancy of the educated middle class. During the 1940s there were 20,000 professionals in Jamaica nearly all of whom were from the middle class (Roberts, 1957, p. 87). Constitutionally speaking, the increasing political power of the people went hand in hand with the increasing power of these strategic elements within the state apparatuses. As we indicated in Chapter 5, the capitalists could not hope to arrogate power through electoral politics. In fact, it can be argued that the foundation of the strategic middle class was laid by those elementary school teachers who figured so prominently as early members of the PNP and the JLP. Their children and others influenced by them would become the mainstays of state bureaucracies. Indeed it was the Jamaica (Constitution) Order in Council of 1962 that formalized the "Jamaicanization" of these state apparatuses, as the more influential positions once occupied by expatriates were now open to native Jamaicans. The organs of government had finally become potential instruments for the exercise of real power.

Economic policies in this same period (1945–1960s) also favored the growth of a vibrant middle class (and assisted in consolidating the position of the merchant/commercial fraction of the capitalists, as distribution took precedence over agriculture and manufacturing). Because the economy lacked the capacity to absorb the growing numbers of middle class representatives—there being a far too small and much too undifferentiated capital base—the state provided employment. This responsibility fell to the civil service, which concentrated most of its activities and employment opportunities in Kingston, the capital city. The number of civil servants (excluding teachers and police officers) increased from 8,700 to 15,570 between 1963–1975 (Bell and Gibson, 1978, p. 10).

The rise of a middle class ethos, eagerly championed by Norman Manley and the more organized elements of the class, had now become a reality. Strong emphases on industrialization and education together served to promote this end, although in a manner that makes the attempt to establish causality somewhat difficult. Combined, though, they affected the patterns of consumption especially of the urban areas. For instance, middle class housing started appearing conspicuously on the outskirts of key urban areas—Mona Heights (St. Andrew), Paradise Acres (Montego Bay, St. James), and Harbour View (Kingston). The total capital investment in private housing represented 16.6 percent and 15.9 percent of the total investment in 1960 and 1961 respectively.[12]

Without question, education had rendered its most effective service in the sphere of ideology. Here mimesis became canonized, whether or not there had been a monetary basis for this posture. Consumerism had taken on that enviable North American twist in which the motto was "buy now, pay later." Also in vogue was the itinerant shopper whose discriminating tastes would often cause him to forsake the sparsely stocked shopping centers of Kingston or Montego Bay for the glitter of Miami Beach.[13]

Data on installment credit tell us much about a consumerism on the rampage. In 1969, for example, the Installment Credit Outstanding (January 1965 = 100) jumped from 231.4 points at the end of 1968 to 296.5 points; this represented an increase of 65.1 points, compared with an increase of 48.2 points between 1967 and 1968. At the end of 1969, commercial banks, rather than finance houses, and dealers—the main sources of instalment credit—claimed 53.4 percent of the outstanding credit; this figure stood at 47.2 percent for the preceding year. Household goods and appliances together with automobiles accounted for most of the business (*Economic Survey of Jamaica, 1969,* p. 106). The middle class purchased and the merchant/commercial fraction of the capitalists reaped the profits.

Up to the mid-1960s, while their material position had demonstrably improved and their political power was on the rise, real economic power was elusive. The educated black middle class, as Munroe (1972) noted, held appreciable authority but no power. This state of affairs was not fixed, however. While the odds in favor of a substantial rise of the fraction were not great, they nonetheless existed.

A Fraction on the Move

A representative number of interview responses on the changing contours of the fraction depict a fraction on the move during the 1970s. The first comment is by a black attorney whose law firm was amalgamated with one of the old established, white dominated firms. His is the satisfied voice of upward professional mobility: "We were invited in and now we own the firm. My target is to produce [so many thousand dollars] worth of business. I have already met the target and it's only July."[14] The agonies of missed opportunities: "They used to refer to me as Jamaica's most educated tailor. [This individual holds a Ph.D.] Just a bad break and I would be gone through. I started playing golf." The resourceful are now admitted to the table: "Things different now [sic]. Now the banks treat us like Cuffie. Always finding ways to lend us money."[15] The old order changes "Yes. Many of my personal clients are now the bright businessmen who did not have the opportunity before."[16] The voice of the perplexed:

"They have bought all the big houses in Ironshore and Torado Heights. It's ganja money. My husband [a black attorney] always says, 'The niggers are coming.'"[17]

Certain elements within the middle class were now preparing to shine. The professional, administrative and technical groups to which they belonged had increased by 3.4 percent since 1943. In terms of racial composition, in 1960 members of African descent dominated the group (43 percent) followed by Afro-Europeans (22 percent), Europeans (14 percent), Chinese and Afro-Chinese (8 percent) and "Others" (12 percent) (Clarke, 1975, p. 153). Between 1970 and 1982 these elements increased from 7 percent to 11 percent of the labor force, ranking with service workers as the fastest growing occupations. Although statistics were not readily available, it is probable that members of African descent increased in these subsequent years, partly as the result of the supportive policies of the Manley regimes of the 1970s.

The mindset exhibited above has a public sector analogue that we have termed the "fringe bureaucratic" mentality. The growth in public expenditure under the PNP increased the numbers of ministries and parastatals. The Bank of Jamaica, the Jamaica National Development Bank, the Jamaica National Export Corporation, all these and others came under the control of middle class meritocrats, most of whom were black. Even supporters of the JLP comment favorably on this development. A high official of the Private Sector Organization of Jamaica volunteered that this was easily among the PNP's most noteworthy achievements.[18]

The fringe bureaucracy was created partly by political circumstance. The early ministers of government were largely unlettered men who had difficulties with the civil service, especially since its well-educated and well-placed members often created obstacles for them. To avoid these additional delays in an already cumbersome system, politicians resorted to an increasing number of statutory bodies that provided direct technical expertise without the impartiality, delay, and occasional ineptitude of the civil service.

Members of these statutory bodies were often in exclusive service to ministers—a role enabling the bureaucrat to mediate or act at the most crucial interfaces between the public and private sector. Many were employed under lucrative service contracts, and a kind of independence developed, enabling the exercise of real power. But ministers could not always rely on their loyalties, since often the exercise of independence and a sizeable severance payment could give rise to competing brokerage arrangements. This was the case with a very prominent official who delayed passing vital information on to the minister by whose portfolio his functions were controlled: "They were projects with excellent prospects. When I told him [the prime minister] about them, he told me not to tell him [the minister] about them."[19]

These are some of the factors that combined to transform a loose grouping from the middle class (the black and brown educated professionals) into a bona fide fraction of the class, which we have termed bureaucratic/entrepreneurial. What distinguishes this fraction is its ability and willingness to move in and out of government and to use state access to support entrepreneurial activities. Its members are familiar with the workings and personnel of government and its agencies, usually through participation in electoral politics or appointments to the government bureaucracy. They may, either concurrently with or subsequent to government service, engage in entrepreneurial activities, often linked to governmental initiatives or somehow dependent on the government and its agencies for their success. It should be noted that, unlike the coloreds, they were probably more favored. They had access to financing institutions such as the JDB.

The bureaucratic/entrepreneurial fraction, then, should be distinguished from the traditional civil servants, who are interested "only in safe careers," as Manfred Halpern (1968, p. 196) found in his study of selected middle class elements in North Africa. The security of tenure in state employment, with its well-known "perks," holds little attraction for our group. Instead, they display an engaging individualism. Quite often, employment in the state bureaucracy had less to do with salaries and stable employment than with the opportunity to be part of, or close to influence, valuable information, and powerful networks. As one of our sources put it, members of the fraction are sorely disappointed if they are unable to leave their substantive positions after a few years. This period of time allows them to do what is necessary for "opening their own Chinaman shop." Parenthetically, the notion of "the Chinaman shop" embodies two important changes. First, the idea of "shop" signifies the impatience with education as a "means of production" and a new willingness to engage in previously scorned economic activities. Second, it highlights new values, that is, to go from the status of servant to master. (Heretofore these educated citizens could at best look forward to important jobs in the economy.) These changes, occurring by the mid-1970s, contrast vividly with the experience of some of their African counterparts (Ghanaian, Tanzanian, and Ugandan) who found that their future was clearly bound up with "a technocratic upper-middle class of organization men [rather than with] members of a presumptive ruling elite" (Barkan, 1975, pp. 187–189).

It goes without saying, then, that this fraction would support governmental intervention in the economy, such as there was in the period of PNP rule, which would increase their access to resources and help consolidate economic as well as political power. Of course, economic success would also spell full entry into the ranks of the capitalists, lending a transitional quality to the fraction itself, and, potentially, to its politics.

There is an additional component to the fraction: people who do not participate officially in government but work closely with its members. A growing phenomenon in underdeveloped countries like Jamaica, this component includes, in the language of organized crime, "bag men," who handle the "deals" that the public service code denounces. In other words, the unscrupulous politician or bureaucrat needs and usually has an accomplice. Such practices can be observed in Zaire where Mobutu's clan, the "reigning brotherhood," works closely with accomplices in establishing a system of "connaissances" (personal connections) (Gould, 1980; MacGaffey, 1983).

Limited but suggestive data exist on the investment practices and peccadilloes of the fraction. Disclosures by the Jamaica Development Bank, the Agricultural Credit Board, and the Small Business Loans Board—all state agencies—indicate some of the advantages some members of the fraction and politicians obtain as a result of their privileged positions. Perhaps the most celebrated case, which has earned widespread coverage by the local media, is that of Dexter Rose, the Managing Director of the State Trading Company (STC), the powerful intermediary between local importers and foreign suppliers and buyers. Though an exact sum has not been ascertained, it has been reported that illegal transactions have resulted in millions of dollars being salted away in Swiss bank accounts.

On the whole, however, their financial adventures have not been mired down in corruption of this kind. More often there is a pattern of preferential treatment. Ministers of government have received red carpet treatment from leading lending institutions. The Jamaica Development Bank extended loans on the order of several million dollars to politicians; the *Jamaica Gazette* of February 12, 1976, cited seven such individuals who owed the bank some $2 million. The minister of health, Dr. Kenneth McNeil, owed the bank $1.8 million of this sum.

The Jamaica Development Bank's other lending practices provide an added dimension. Beginning with the bank's Annual Report of December 31, 1976, the pattern of lending started to cause alarm: "The problem did not stem from the usual cause of expenses overtaking income, since the normal expenses of $1.89 million were well below the income of $3.5 million. However, the provision for losses account was increased to a whopping $2.6 million for 1976, which completely wiped out any surplus. In the previous six years, the total reserves against losses had been $2.6 million but in 1976 alone the bank had to write off $2.6 million" (*Daily Gleaner,* May 7, 1978).

The interim auditor general's report, prepared in September 1978, uncovered related problems. Although the Jamaica Development Bank was forbidden to lend more than $500,000 to a single enterprise, many firms received loans in excess of that amount. The report also revealed that 98 percent of the bank's clients were in arrears on the order of $94.2 million (*Daily News,* September 6, 1978).

The spotlight fell on Noel Chin, the bank's managing director. Here is a case in which a parvenu entrepreneur used his strategic position to build an empire. In early 1978, Chin complained about government pressures on him to approve transactions that would expose the bank to excessive risk. By May of the same year, the minister of finance, Eric Bell, reported that Chin's own business practices warranted public scrutiny. Early investigations showed that Chin was also a director of two companies that were clients of the bank.

The paper trail led to other banks. The Victor Engineering Company, of which Chin was a director, negotiated a $200,000 loan from the Workers' Bank while he was acting chairman of that bank. The Workers' Bank, it is reported, lost $2.5 million in four years, leading to the assessment that it "had become a feeding tree for developers and other businessmen" (*Daily News,* May 31, 1978).

The circle takes us back to the Jamaica Development Bank and Noel Chin. The Bank negotiated a contract with the Clover Construction Company, a preferred customer for government awards, to add a new wing to the bank that generated cost overruns of $750,000. The principals of Clover were close relatives of a senior member of the bank (*Daily News,* February 7, 1979). Other members were also favored:

> The Auditor General found improper relationships between bank personnel and clients. For instance, in January 1976 an officer of the JDB bought into and became a director of a client company. The board was not informed. However, the officer participated in and influenced the processing and approving of a loan of $186,000 and guarantees totalling $286,363 to the company. On August 3, 1977, he was appointed chairman of the company (*Daily News,* February 7, 1979).

The performance of the fraction ranged between astuteness and ineptitude. Some were commended for their ability to find their way around the tough business world. We were told in an interview:

> Keith Roache [the managing director of the Workers' Bank] told me about this guy. They called him in to discuss his loan. He said, "Gentlemen, we all know that this is a development loan. No threats. You have to tell me how we can restructure it to meet your needs." As cool as a cucumber.

But an accountant with one of the leading accounting firms criticizes the business acumen of others:

> From what I have seen [his company worked for some members of the fraction] the money is there but it is not invested well. Take (X) [reputedly the private sector partner of a prominent minister]. He has invested a fortune in restaurants. He should have invested in manufacturing to form a solid base.

But some of the predictable behavior of this parvenu stratum began to force itself into the public awareness. Conspicuous consumption, an ostentatious lifestyle, and bravado were as much distinguishing features of the fraction as were the habits mentioned above. The fraction had distinctly moved away from the earlier perceptions many observers were fond of recalling. C.L.R. James (1970, p. 193) captures the definition quite well:

> For generations their sole aim in life was to be admitted to the positions to which their talents and education entitled them, and from which they were unjustly excluded. On rare occasions an unexpected and difficult situation opened a way for an exceptional individual, but for the most part they developed political skill only in crawling and worming their way into recognition by government circles or government itself, they either did their best to show that they could be as good servants of the Colonial Office as any, or when they rose to become elected members in the legislature, some of them maintained a loud (but safe) attack on the government. They actually did *little*. They were not responsible for anything, so they achieved a cheap popularity without any danger to themselves.

Others saw its members as "the nervous people sipping Scotch by the poinsettias on their patios . . . asking one another how long they will be safe in their beds" (Lowenthal, 1972, p. 299).

The evidence shows a qualitative break with this tradition. It was acknowledged even by the capitalists. The Chamber of Commerce referred to the politically influential fraction as "powerful and dictatorial"; the Chamber also dedicated itself to its destruction "before its is fully grown and ready to be born, thrive and grow."[20]

The Posture of the Fraction at the Time of Crisis

As a whole, the middle class lived up to expectations in 1972. It gave the PNP impressive support at the polls. The support must be seen against the backdrop of an enviable middle class ethos. Modern office complexes, attractive, affordable housing, and a North American consumerist ambiance yielded 70 percent support for the party.

The upper stratum, mainly its black members, continued to push their way into key positions in state bureaucracies. In the private sector, they would occupy highly visible positions in the insurance industry, though the power exercised may not have been proportional to the level of visibility. A few took positions in the stock market but, as we shall see, their passage was sedulously blocked by the capitalists.

The Subordinate Classes

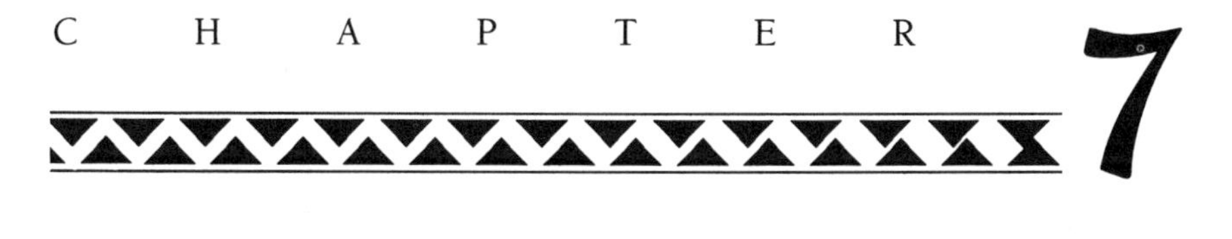

Thus far, we have devoted considerable time in this section to demonstrating two important components of our thesis: the relative weakness and fractionalized nature of the Jamaican capitalist class and the rise of intermediate sectors whose task it would be, among other things, to address the structural need for political mediation between the capitalists and the subordinate classes.

What remains to be done is to disentangle the various strands comprising the subordinate classes, thus revealing their individual contributions to the evolution of national popularism. These contributions, as we will see, were both positive, in the sense that they were supportive or active ingredients in the creation of the agenda (as was the case of Rastafarianism and the predominance of petty-bourgeois consciousness), and negative, by which we mean the turbulent or destructive class expressions that called attention to the urgent need for new mediations. Primary among the latter were the large numbers of the lumpen and subproletarians and the threat that their increasing violence posed for the continued existence of a liberal democratic state.

▶ Workers, Peasants, and the Subproletariat

We noted earlier that the working class was slow in its growth and failed to display a dynamic working class consciousness. The causes of this relative conservatism were structural and economic in origin. Structurally, the party/trade union symbiosis placed constraints upon the development of the working class. Up to 1972 over 90 percent of all unionized workers belonged either to the BITU or the NWU. These highly privileged workers had access to the best-paying jobs and the protection that a union-backed government must provide. This practice effectively disengaged the rest of the working class and large sectors of the sub-proletariat and placed them in an antagonistic relationship to these privileged, unionized workers; consequently, class solidarity and the development of class consciousness were seriously undermined. The equivalent of a caste system in the labor movement provided another structural impediment. It was and still is a rare rank and file member who attained influential positions within the unions. These slots were reserved for the educated middle class, whose ideological tendencies leaned more toward conservatism than radicalism. Accordingly, workers were divided, self-seeking, conservative, and lacked a progressive sense of their interests as a class.

These organizational arrangements encouraged vertical personalistic relationships between rank and file and party officials, union leaders, and politicians instead of horizontal ties uniting the respective membership in class terms. In the end, clientelistic compacts proliferated, while inter-union animosities were rampant (Stone, 1980; Gonsalves, 1976).

At the same time, since undercapitalization went hand in hand with small-scale operations, the factory system provided its own obstacles to the advance of the workers. The worker per factory ratio was stable over a thirty-year span—the period of most rapid industrialization—even though the absolute number of workers increased. For the communist cadre, high worker per factory ratios are crucial to the efficacy of working class organization. According to Leninist practice, they are needed to transform the worker into an "absolute proletarian." These ratios instill "collectivity," "the capacity for unity through interlocking, mutually supportive and concerted practices" (Therborn, 1983, p. 41).

These factors undoubtedly impeded the Jamaican working class movement and contributed to the low levels of consciousness it exhibited, as the WPJ repeatedly pointed out (Munroe, 1981, pp. 70–71). Numbers serve to underscore the position. In 1973 some 120,000 industrial workers were flanked by 130,000 peasants and 230,000 small business people and petty traders (Nelson, 1974). As late as 1976, the communists complained that the working class still held the capitalists in great esteem and looked to them for leadership and guidance.[1] This was however, an overstatement, prompted largely by the communists' impatience with the pace of proletarianization. Although few and far between, there were actions betraying disgruntlement and a measure of consciousness from the class. We cannot fully appreciate the significance of the preponderance of petty traders and peasants without referring to the traditional practice of the "casualization" of labor. Export agriculture still maintained the practice—dating back to pre-emancipation days—of seasonal employment, forcing most agrarian workers to spend the rest of the year scratching out a meager existence on rented land. With the passage of the "Squatters Laws" in the late nineteenth century, peasants who had acquired the right to possession of land by long usage were displaced. By the early 1900s, some 275,000 acres had been so repossessed and thousands of peasants became tenants of one kind or another. During the mid-1940s, the sugar and banana plantations bore evidence of the fruits of these highly dependent land tenures: 56 percent of all field workers occupied holdings that they worked "out of crop," (outside the growing season). At crop time, 39 percent of them obtained housing from the estates and 25 percent received small parcels of land to cultivate. Out of crop, 50 percent of these workers got free housing while 41 percent received allotments for cultivation (Munroe and Robotham, 1977, pp. 66–67). So this agrarian proletariat was really an odd mixture of petty bourgeois, peasant, and proletarian. It is no wonder that Jamaica does not boast a strong peasant tradition, which in other countries has engendered strong, politically active peasant associations, such as those in Venezuela (Powell, 1971) and Nicaragua (Bamat, 1988). Indeed, this made for a conflation of ideologies up to the time of the crisis in the 1970s and beyond. In Munroe's words (1981, p. 88), "The fact that so many workers are halfway peasants or other types of small proprietors and are also halfway wage workers [makes] the situation . . . very complicated—one consequence . . . is the fertile ground provided halfway ideologies."

As was the case for unionized workers, another factor hampering the development of a strong class consciousness was the political clientelism that pervaded the rural areas. Patronage was simply a way of life, with government supports, services and related favors being distributed on the basis of political connections rather than more rational calculations. Occasionally, as happened in 1969, old patronage bodies like the Parish Farm Boards were replaced by new and supposedly neutral entities such as the Parish Land Authorities (Smith, Sealy and Gordon, 1972). Yet patronage practices were remarkably stable, a sign of their deeper roots in the social structure.

These practices, coupled with the industrialization impetus of the 1950s, set in motion new waves of internal migration. Small peasant holdings (under five acres) decreased in number by 9.5 percent between 1930 and 1954 and by 18.5 percent between 1954 and 1961 (Marshall, 1968, p. 258). The rate of migration was staggering: between 1955 and 1972, some 339,000 persons from the rural areas relocated to Kingston and nearby St. Andrew (the Corporate Area) (Munroe and Robotham, 1977, p. 170).

Rural migrants flowing into the Corporate area did not readily blend into the capitalist economy. The occasional surveys of the areas receiving the majority of rural migrants (the slums of Kingston) tell the same devastating story: casual and intermittent participation in the labor force, one third or more of heads of the households admitting to unemployment, and deplorable living conditions (Nettleford, 1970, pp. 49–51). Another knowledgeable source (Eyre, 1983, p. 238) describes a situation that has not changed with time:

> The gainfully employed are mainly of two types: (1) the pitifully small proportion of the regular wage-earners who travel out of the area, sometimes many miles . . . to work; and (2) the self-employed, usually home-based workers. [There is a fascinating] variety of income sources in this informal employment sector: higgling (petty trading) of "pants lengths" and sundry other items; the collecting, sorting and reselling of glass bottles; tinsmithing; shoemaking; hustling for a fee to arrange transport for people leaving the city; raffia plaiting and embroidery; the cooking and selling of lunches; as well as the invariable activities of such areas—gambling, prostitution, and drug peddling. Many young men in the area emphasize the difficulties of finding gainful or "straight" employment if they try to leave the more lucrative role of paid gunman or racketeer.

The estimate is that perhaps well over 40 percent of the Kingstonians continue to be engaged in such occupations, as well as car cleaning, gardening, and car watching (C. Clarke, 1983, p. 230).

If these migrants and their offspring found themselves cut off from the promises of economic modernization, they were equally left out of the political system or any urban-led political movements that might seek and promise redress. They fanned out, instead, into an assortment of groups—Rastafarians, Rudie Boys (rebellious, anti-establishment youths), and Revival yards (revivalist-type churches). To the extent that they became politically involved, it was these institutions that provided them an indirect and unorthodox introduction into formal politics. They worked through the system of patronage, which included serving as paid gunmen for politicians. And when the largesse or spoils were slow in coming, they were not above turning to violence. Indeed, warnings were becoming more insistent by the late 1950s that the contrast between their deprivation and the conspicuous consumption of the middle and capitalist classes was creating an explosive situation. The severity of need and lack of alternatives are evidenced by the fact that in 1972 70 percent of this constituency sought patronage (Stone, 1973, p. 81).

As Eyre notes, street crime has always kept close company with the activities of the more intrusive members of the subordinate classes. Crime statistics are notoriously unreliable, since increases may simply reflect increased police activity which, in turn, is sensitive to the politics of law and order. This is particularly so when, for instance, criminal activities leave the confines of the ghetto and begin to intrude on the lives of the more prosperous members of society. Yet the large increases in crime reported for the 1960s (Senior, 1972; Lacey, 1977) and the shift toward younger and more violent criminals hardly seem the result of statistical manipulation. As Lacey noted (1977, pp. 67–69), fully 71 percent of the serious crime incidents reported in the decade of the 1960s occurred between 1966 and 1969. In 1974, robberies rose to 418 cases per thousand population from 43 cases per thousand in 1960–61. Over a similar period, violent crimes increased by 50 percent.[2] To counteract the perceived crisis, expenditures for police and military services kept creeping upward. One can hardly speak here in terms of "social banditry," as for the most part, these offenses were not politically progressive. From the 1960s the ghettoes of Kingston bore steadily increasing evidence of destructive energies having little association with conscious agendas for change.

Close to these activities lurked the machinations of the political parties, as political patronage became entwined with the destructive and intimidating activities of the growing numbers of lumpen and subproletarians who accounted for some 15 percent of the population, and, together with the unemployed, constituted 37 percent of the total population in 1973. When not in the immediate service of the politicians, these elements simply engaged in wave after wave of wanton physical destruction, leaving huge sections of Kingston in ruins.

If these lumpen elements were not of the ilk to engage in constructive social change, they were nonetheless a significant negative factor in the crisis of the late 1960s and 1970s: more and more they posed a threat to the very viability of the two-party political system, drawing attention to the need for new patterns of mediation by the state. Political violence was both the cause and the effect of changing politics. Politicians on both sides voiced their concern. Manley charged that the pattern of political victimization was such as to constitute violations of human rights. Prior to the 1972 elections, with religious leaders intervening as well, Manley and Seaga signed a "peace pact," that was designed to put a stop to political violence and the cycle of victimization and patronage. State resources were no longer adequate for the purposes of containment, as unemployment levels, crime statistics, and a growing threat, especially to the capitalists, dramatized the impotence of traditional forms of state mediation. In fact, what appears to have been occurring was a growth in the education and organization of criminal elements. Guns were introduced into the ghetto to secure politicians' careers in the 1960s. The economic crisis and the fiscal crisis of the state caused the gunmen to turn their entrepreneurial acumen to other uses reminiscent of the activities of organized crime, including the drug trade and preying on capitalist businesses. Indeed, this threat and the corresponding paucity of solutions offered by the JLP would appear to bear some responsibility for the shift to the PNP among the capitalist class: 75 percent of them (included in the category of "upper class") voted for the PNP in 1972, as opposed to 48 percent in 1967 (*Daily Gleaner,* June 8, 1980, p. 7).

As can be seen in other Third World settings, the ideological expressions of these lumpen elements are varied and quite unpredictable, tending to swing sometimes widely from right to left. (Taylor, 1979, p. 231; Markarkis and Ayele, 1986, p. 151). Where party buccaneering is involved, where these elements assist in maintaining a politician's majority or influence, these compacts are contractual and are not necessarily blessed by the morality of a common ideological stance. Indeed, the loyalty of one's thugs is usually only as enduring as one's cash. There was, nonetheless, an ideological tendency that enjoyed significant success among many of the dispossessed, much to the dismay of their "betters": Rastafarianism was on the ascendant.

▶ The Rastafarians

By the late 1960s the Rastafarians had shed some of their outcast image and had begun to carve out a place for themselves in Jamaica's sociopolitical life. The brethren's path had been rocky. Having trekked into Kingston in the early 1940s, the movement had found followers among the urban youth, its influence increasing particularly between 1955 and 1963 (Chevannes, 1981, pp. 392–393). Revivalism, with its unique stress on personal salvation, was declining at that time and furnished Rastafarianism with new members, attracted by the movement's quest for a new African identity as well as its more earthly concerns with economic hardship and discrimination (Simpson, 1986). The fact that the movement grew to embrace, either as members or as sympathizers, most of the residents of the Kingston slums (Nettleford, 1970, p. 49) speaks of the extent to which its emphases struck at the core of felt oppression.

Success was due partly to internal changes, since the movement was becoming more heterogeneous. Especially in the 1960s and with the advent of the Black Power Movement, there was a marked shift from retreatism, with its emphasis on repatriation to Ethiopia, toward Pan-Africanism and humanism. In the words of an adherent, "Ethiopia means Africa; it was the name for the whole continent, and the capital of Africa was Abyssinia, now Addis Ababa." While the emphasis on African identity remained strong, there was also clear recognition that brotherhood went deeper than one's skin color—a precept that would open the door for Jamaican whites and Chinese to become part of the movement. As it shed some of its retreatist image and attracted new adherents, Rastafarianism also evolved as a philosophy of life, a way to reconnect with nature and one's nature, a way to live in love and peace. This did not mean that the movement was poised to attempt what the working class would and could not do—assume a leading role in an attempt at radical transformation of the society. Rather, its contribution was to provide a new "system of narration."

In the classical Marxist sense, the central theoretical themes of Rastafarianism betrayed (and continue to do so to this writing) crippling contradictions. For one thing, the movement's legendary contempt of politics removed the brethren almost completely from an activist role in the revolutionary agenda. For the brethren, "politics was not the black man's lot but the white man's plot."[3] Politics means "many parasites," because the prefix "poli" in Greek means "many" while "tics" (equated with ticks) connotes "parasites." Even the most politically disposed—the Dreadlocks—scorn "the subservience to party politics" (Chevannes, 1981, p. 394). The Twelve Tribes, a splinter group that counted reggae star Bob Marley among its members, remains anchored to tradition; this cleavage operates from within an internally constituted government—the Jah Rastafari Holy Theocratic Government, which mediates between the movement and the wider society. And while there are marked tendencies toward a secular approach to politics, the Twelve Tribes still favors repatriation to Ethiopia and maintains a quasi-religious character to its political exercises.[4]

Another contradiction surfaces at the economic level. The principal economic activities of the brethren include fishing, making handicrafts, selling ganja, and the like. These occupations are marginal and occur mostly within the informal sector of the local economy. They come into contact with proletarian exploitation only in a restricted sense—a sense that does not suggest a major leadership role for the movement. For the brethren to have a more global and uniform impact on the oppressed classes, their political philosophy must become more generally acceptable.

There are indications that this level of generality did begin to emerge in the late 1960s and 1970s, growing particularly in the latter part of the 1970s. Coexisting with the separatist strands mentioned above, there was a growing recognition among influential adherents that narrow claims and perspectives must give way to involvement in the initiatives of the wider society. As one of them put it, the pool of social architects must include Rastafarians as well as others, giving rise to the infiltration "of the Pride of Lions (Rastafari) with wolves (criminals), goats (hypocrites), foxes (tricksters), and jackasses (fad followers)" (Tafari, 1980, p. 3).

With regard to events of immediate interest, however, the movement's impact on the course of democratic socialism was more pronounced in the realms of culture and ideology than in the articulation of a collective consciousness for the oppressed classes. In fact, the movement achieved new prominence across the class structure for its reformulations in such areas as art, religion, social theory, and linguistics. More than becoming a vehicle for revolution, it became "national property." We trace below the twisting path leading to this end.

Success with the dispossessed and discontented was not likely to be immediately translated into more accepting attitudes among upholders of the status quo. With independence in 1962, the philosophy of the common good that the PNP had introduced into Jamaican political life in the 1930s was translated into the national motto, "Out of Many, One People." Like its predecessor, the motto was laudable in intent, but in fact spoke of a continuing aversion to according proper weight to the African roots of some 90 percent of the population. Eurocentric sentiments were strong: one must look to England for help for the young nation, for, what could Africa possibly offer? Further, these naysayers threatened the very future of the new nation which, it was felt, depended on the productive cooperation of all its citizens. Rastafarians were vilified: they were "criminal, lazy, irrationally emotional, hopelessly mad and dangerously violent" (Nettleford, 1970, p. 55). The bearded, unkempt outcasts were hardly the personification of a young society preparing to take its place in the comity of nations.

Antipathy shaded into harassment. Since as early as 1945, governments, particularly those of the JLP, had taken to destroying the brethren's dwellings, ostensibly to make way for urban redevelopment (Phillips,

1981). The highwater mark of harassment was perhaps reached in 1963 following an incident at Coral Gardens in Montego Bay. On this fateful Thursday (now remembered by the brethren as "Holy Thursday"), a beleaguered group of Rastafarians bent on asserting their right to walk across the Rose Hall golf course became involved in a struggle with security, as a result of which a gas station was burned and a policeman was attacked. Predictably, the government reacted repressively to this incident, and thereafter continued to violate the brethren's civil and political rights almost as a matter of public policy.

Yet this outrage yielded unexpected returns. In 1960 a study team from the University of the West Indies had begun to dispel the cloud that surrounded Rastafarians by exploding the myth that the brethren were lunatics and criminals (Smith, Augier, and Nettleford, 1960). Their report was at the center of much controversy, but was followed by other studies, government missions to Ethiopia to explore migration to that country, and visits by African dignitaries. All these activities began to confer a measure of legitimacy upon the Rastafarians and the events following the Coral Gardens incident placed the brethren's plight at the center of the debating stage, resulting in more sympathizers and increased membership. Certain members of the movement contend that perhaps the swelling of their ranks was the most important consequence of the incident.

The brethren's political value gradually became evident. The more astute politicians quite early on recognized the movement's importance in competitive politics. As a young JLP politician, Edward Seaga (a future prime minister), blended the brethren into a significant part of his Tivoli Gardens constituency in Kingston. In the elections of 1962, he was able to outmaneuver his opponent, Dudley Thompson, partly through the adept use of the Rastafarian ethos. Along similar lines, his colleague, Wilton Hill, attempted to put the movement on the front page. In 1966, a month after a wildly successful visit to Jamaica by Ethiopian Emperor Haile Selassie, the Godhead of Rastafarianism, the then Senator Wilton Hill moved a constitutional amendment to replace Queen Elizabeth with Emperor Haile Selassie (Nettleford, 1971, p. 64). Not long after this abortive, but highly publicized ruse, another colleague, Dr. Herbert Eldemire, a former minister of health and a candidate for a constituency in St. James, took his turn. The popular Eldemire (Herbie) grew a beard and called himself "David." While name and affected visage are of patently Rastafarian influence, it should be noted that Dr. Eldemire is a Jamaican white!

The PNP had its turn. In the 1960s, the Rastafarians, enjoying a modest upswing but still suspicious of the intent of "Babylon's" newly sympathetic policies (missions to Ethiopia, for instance) dubbed Norman

Manley, "Man-lie." Yet by late 1971, Michael Manley was able to adopt successfully a political rhetoric that was unashamedly Rastafarian and carried distinct lower class cultural nuances.[5] His meetings would be punctuated with "reggaeisms" and begin and end with the Rastafarian exhortation: "The Word is Love." He had most of the popular reggae composers and singers at his beck and call—an alliance that produced some of the more memorable political songs in the country's history. "Beat Down Babylon," "The Rod of Correction," and "Let the Power Fall on I" are a few examples of songs directly influenced by the brethren. "The Rod of Correction" was especially significant, because it alluded to a walking cane the younger Manley was supposed to have received from Haile Selassie on his visit to Ethiopia in 1969. Thus PNP politics and Rastafarianism became even more closely joined.

Although the PNP had scorned the brethren in the 1940s, in 1971, the band of the Peacemaker's Church—the church of the Reverend Claudius Henry—supplied the fanfare for the PNP's announcement of candidates. The Reverend Claudius Henry, we must note, was imprisoned for sedition in 1959. At the time of his imprisonment, his following numbered in the thousands. Since then Henry had avoided active politics, but he still commanded a large following through teachings of marked Rastafarian provenance (Morrish, 1982).[6]

Let us leave this brief account of Rastafarian-influenced politics on a note of levity. In 1971 when the prime minister, Hugh Shearer, was pressured to announce a date for elections, he declared: "There is only one man who sets the election date. And that man is I-man." To this Michael Manley, the leader of the Opposition, replied that he was happy with the choice, as the date in question "turned out to be the birthdate of I-Mother" (Waters, 1985, p. 123). The Rastafarian language had become the currency of politics.

At the societal level, the impact of the movement escalated. During the mid-1960s, the middle class was easily the most virulently opposed to the brethren. The brethren saw its black members as "Judases" who had been selling out their own to a largely alien way of life and aping "Babylon" (that is, the oppressors) in the pursuit of middle-class values. Since the brown members of the class were referred to as "mongrels," one can easily understand their fury. How could anyone dare insinuate that they were of inferior breeding? Women did not escape unscathed. Black middle class women were chided for "frying" their hair (straightening it with a hot comb) and "reddening their mouths with lipstick."

In spite of this, the movement extended its reach. By the late 1960s and the early 1970s, it had among its adherents an impressive number of professionals (Tafari, 1980, p. 3) and had made inroads into the secondary schools and other institutions of higher learning. By 1978, over 50 percent of the advanced students (fifth formers) at one of Jamaica's leading high schools (Cornwall College in Montego Bay) "espoused the cause of the self-styled Rasta" (*Daily Gleaner,* Sunday Magazine, March 9, 1978). At the University of the West Indies, the movement was vigorous, with members of the faculty embracing the faith. As Dennis Forsythe (1980, p. 62) has correctly insisted, "Rastafarianism is the first mass movement among West Indians preoccupied with the task of looking into themselves and asking the fundamental question, Who Am I? or What Am I?" It seems clear also that the movement's philosophy now drew substantial attention, since it was at this time that many, especially the young people from all classes, really began to question the relevance of the traditional order. Further, whereas earlier political movements—Bedwardism, Garveyism, and the PNP under Norman Manley—were unable or unwilling to insert race and color into politics, Rastafarianism did. The conjuncture was at last favorable.

In the realm of art, the impact of reggae music perhaps most pervasively speaks to ideas in transition. In the visual arts, the bulk of the population could now experience themselves through lenses crafted by Rastafarianism. One notable proselytizer is Stafford Harrison, himself a Rastafarian, who has received critical acclaim for his works as a playwright. Kumina and Ettu, traditional folk dances that were gradually slipping into oblivion have been revived and now form a part of the entertainment package sponsored by the Jamaica Tourist Board. Of course, such acclaim will legitimize these neglected cultural forms, providing wider dissemination in the process and might even strengthen the nativistic turn of local politics introduced by Rastafarianism.

Rastafarianism became the focal point for all the more important class expressions because it crystallized certain crucial forces within the class struggle. Barry Floyd (1979, p. 140) provides useful insights on the process:

> Despite or even because of the persecution, new recruits were drawn into the Rastafarian movement, the majority from the ranks of the underprivileged and unemployed, but some from middle- and upper-class homes, rebel teenagers or university students—often very light in colour—with an excess of social guilt and reforming zeal. A true understanding of the "Ras" was not helped by this "boom in plastic Rastas" . . . brought about by the children of the Jamaican "bourgeoisie," seeking a new image for themselves to contrast with the "Afro-Saxon" establishment ways of their parents.

The despised son was being invited home.

▶ The Subordinate Classes and National Popularism

It has been our contention throughout this book that the agenda for change the PNP advanced for Jamaica in the 1970s was based on the creation of a national coalition dedicated to the pursuit of a newly redefined common good. The success of this project would require the state to attend to the respective interests of all participating classes as well as stimulate and inspire them with an ideology that could hold these disparate interests together. The ideology that became the vehicle for the PNP project had at its core the Rastafarian message. This, then, was the major contribution from the subordinate classes.

An explanation is needed. While the fact that Rastafarian language and symbols entered into common political usage is beyond contention, the real issue is one of interpretation. Questions as to how this very contagious ideology, surfacing from oppressed blacks, achieved such lofty status have been answered along two lines. First, many have seen the ideas and practices of the movement merely as a valuable political resource to be coopted by opportunistic politicians. Here we have the PNP coopting the Rastafarian political saws, extracting their pregnant socialist and humanistic messages, and transforming them into the ideological ballast for its agendas. At one extreme of the interpretative schema, these are the practices of the paternalistic, "hero and crowd" tradition, in which such practices are simply part of a pantomime designed to please and appease the masses, leaving the political and economic elites reasonably free to carry on without impediments from below. The name of national unity may be invoked, but the reality is the pursuit of narrower interests.

The second line of reasoning, espoused by the local communists, asked about the movement's potential contribution to the revolutionary struggle and saw these developments as secondary. Closely toeing the Leninist line, it ascribed the evolution of progressive ideologies to the interaction between a working class vanguard and a revolutionary intelligentsia. It was conceded that the Rastafarian movement contributed to revolutionary consciousness but only at the level of "individual expression," rather than serving directly the cause of "revolutionary mass struggles" (Munroe and Robotham, 1977, p. 170).

Both interpretations fall short of appreciating the ideological shift that occurred in the 1970s and the real contribution of Rastafarianism to it. The process, as we see it, was one of fusion rather than cooptation. This point requires grounding the movement's ideas and practices in a dynamic theory of ideology. At the center is the concept of interpellation. Briefly, interpellations are the mediated expressions of the social structure, providing the individual with an identity and a way of understanding the world. Interpellations are grounded in one's experiences, which may include race, religion, kinship, and nationality, as well as class. These mediated experiences also articulate, however, with interpellations from other sources, crossing class boundaries. So, depending on the state of the class struggle, subordinate class interpellations can and do conflate with those of dominant classes (Laclau, 1977).

As previous chapters have shown, the 1970s saw the emergence of a "strategic" fraction of the middle class that was preparing to breach the ranks of the capitalists at the time of the crisis. A key issue was how to harmonize capitalist ideology (to which most of the fraction was and remains wedded) with the negative inflections of Rastafarian and subordinate class interpellations. It should be remembered that for the brethren the existing socio-economic order was Babylon! The task required a reordering of the commonplace ideological strands and discourses that had been hovering in the air since the late 1930s.

In the first place, Michael Manley and the PNP paid close attention to socialism and pandered to the traditions of liberal democracy; they were acutely aware of the race and color issue but did not tie it to class relations until forced to do so. In other words, they did not invent any new discourses but merely renovated existing ones, carefully blending them with themes of the common good from the PNP philosophical tradition. Indeed the renovation conformed to common sense—meaning an agenda in the best interests of all classes. The Rastafarian creed, with its themes of love, brotherhood and justice, was well suited to the task. In the second place, however, given the temper for change suggested by the times, this formula could prepare the way for new "systems of narration." And with most of the ideological expressions not necessarily having an a priori class determination (as we saw earlier, even the communists conceded the proletariat's dependence on the capitalists for leadership, while the loyalties of the lumpen and sub-proletariats were unpredictable at best) the process of reordering enjoyed a considerable degree of latitude. This was especially so because the capitalists' momentary loss of power and control was counterbalanced by an insistent, though brittle, coalition of class forces from below.

That strategic fractions of the middle class will attempt to align economic power to their political control over state apparatuses is nothing new. The success of these fractions seems to depend on the level of class conflict and the structure of the state. Powerful working class movements supported by active and well organized elements from the other subordinate classes are usually enough to block the designs of the middle class.

On the other hand, an authoritarian state is often the middle class' best ally, as it expropriates and nationalizes assets from foreign investors or outlaws trade union activity in ways favoring the middle class (Ogbuagu, 1983, p. 266). While some form of ideological consensus makes the process a bit more palatable, it is not necessary because the regime has the force of compulsion (the army, for instance) on its side.

It is very different with a parliamentary democracy, such as Jamaica, where these modes of compulsion are absent. Here consensus must surface from a broad meeting of ideological minds. The Rastafarian movement provided the support for the strategic fraction. Admittedly, the activities of the PNP regime regarding democratic socialism (unilaterally reorganizing the bauxite industry and land reform, for example) made middle class political leadership more believable. But there was nothing about the tenor of ideological discourses up to the early 1970s that singled out the fraction for direct support. Indeed, as we have pointed out, the subordinate classes always distrusted the middle class, though they were closely bound to it by shared aspirations. The force that momentarily superseded these age-old animosities was cultural in nature.

Race and color, like nationalism at certain junctures, refused to take a back seat to social class. If Black Power, dependency theory, Marxism, Rastafarianism, and Third Worldism—if all these converging forces, which hugely affected local ideas and action—had a specific message, it was that fundamental changes in self-perception should be undertaken with dispatch. While on the global plane the need for control over economic and political institutions of the Third World surfaced as an important issue, at home the focus was on the narrower issue of the ideational transformation needed to adequately accomplish this change. To tackle related problems in Jamaican society, one must launch an offensive against the traditional bulwarks impeding the desired measure of institutional control. At their fundamental levels, these problems issued from the body of ideas that kept Jamaicans attached to an outmoded system of other-directed values. This system, Eurocentric to its core, fostered and maintained social relations based largely upon the infelicities of colonialism, with their propensity to degrade, minimize, and ridicule the black race, its achievements, and aspirations.

Indeed the script now demanded an inner-directed system of values, with the heritage of the bulk of the population (African) at the forefront. To avoid remaining "a dirty version of white," Jamaicans must now dedicate themselves to the creation of a black, Jamaican ethos. Furthermore, while commanded by black leadership, this ethos should not court or defend racism. As Black Power advocates and Rastafarians insisted, the new perspective would be black-centric and foster genuine leadership that the black race could provide. The movement did not preach black racism, only its prophetic role to lead in the interest of all races (Owens, 1976, pp. 60–62).

Black Power themes and Third Worldism were relatively new proponents of this new *Weltanschauung,* their offerings being mainly in the nature of prescriptions. Rastafarianism, with its infusion of subordinate-class culture, had been crusading for such changes since the 1930s and had created strategies for cultural change. By the late 1960s when the cultural renaissance was demanded, the brethren had already begun to affect the perceptions and cognitive styles of a significant portion of the population through language, artistic expressions, and social and political consciousness (Pollard, 1980; Chevannes, 1977; Nicholas, 1979; Faristzaddi, 1982).

The rebirth gave rise to a system of priorities from which the bureaucratic/entrepreneurial fraction immediately benefited. It is true that the productive and social relations with which it was enmeshed were heavily criticized by the brethren and the subordinate classes, yet cultural issues prevailed. They formed the basis for a new "narration" with pressing immediacy. Because of a shared history, the new discourse allowed the fraction to effectively join cause with the people in creating a momentous sense of unity, in plaiting a whip with which to flog the largely white capitalists for their own transgressions and those of their forebears.

In spite of the convergence, we still need to address the apparent contradiction between the subordinate classes and the middle class at the level of capitalist ideology. Rastafarian and lower-class interpellations were at once hostile to and supportive of capitalist ideology: was not the black middle class—the "Black Man Judas"—the brethren's favorite target? The explanation for unity lies in the interplay of cultural factors and shared, unmet expectations. Like the Rastafarians and the majority of Jamaicans, the strategic fraction of the black middle class had its own grievances with the status quo.[7] While this form of oppression was clearly different from the popular vintage, it had a common source against which to grieve and the proper political climate in which to do so. So the process promoted a closing of the ranks that was simultaneously hostile to and supportive of capitalism.

Hostility arose from the practices concomitant to capitalist ideology, which throughout Jamaica's history has never measured up to the deeply held democratic ideal. Instead, it can be readily perceived that capitalism brought with it racism, color discrimination, and economic exploitation. Norman Manley, the founding father of democracy in Jamaica, was by general philosophical bent not unlike the Rastafarians, the Bedwardites, the Garvey disciples or the bulk of the intelligentsia who so effectively defended the cause of democracy beginning in the mid-1960s. Recall that the Black Power advocates of the late 1960s embraced the principles of dependency theory whose main thesis was equality within the existing system; that while there were radical elements within Rastafarianism, the bulk of the movement, though overtly apolitical, preferred, through humanitarian conditioning, a political philosophy in the accustomed vein of liberal democracy;[8] that the bulk of the subordinate classes continued to depend on liberal middle class leadership; and that Marxism-Leninism was then and still remains a minor force in local politics, a point apparently conceded by the communists, even well after the initial crisis (Munroe, 1987).

Hostility, then, surfaced as a deep disappointment with the unfulfilled liberal democratic promises. The largely black bureaucratic/entrepreneurial fraction of the middle class which, like the bulk of the population, had not yet lost faith in liberal democracy, benefited on two fronts. First, it was seen as a viable alternative to the self-perpetuating power brokers. Second, from the van of a shared sense of hostility to the white-dominated capitalists, the fraction was aided by the anti-imperialism that pervaded the political air toward the end of the 1960s and that predisposed the people to support new contours of leadership. As a member of the intelligentsia remarked during the course of interviews, "It isn't that I am preaching racism, but I would prefer to see P. J. [Patterson, a black man who was the second most powerful PNP leader] or Pearnel Charles [Patterson's opposite number in the JLP and himself unmistakably black] become prime minister. I would not feel so bad even if they hardly improved things." Race was indeed mediating class relations. In the good old days, choice would be directed, in line with the widespread dictum etched in the memory of all school children, "not by the color of the skin but the true heart that beat within!" That heart, of course, belonged most often to the white- or brown-skinned.

The cultural component of Rastafarianism fit into this framework through a process of fusion and accommodation. Overall, the two tendencies (fusion and accommodation) collapsed in favor of continued capitalism. The cultural dimension of Rastafarianism was immediately at severe odds with the dominant capitalist ideology, which of course abhorred the mere mention of race and color. It related dynamically to that ideology at the level of political structures where the strategic bureaucratic/entrepreneurial fraction held enormous influence. In a real sense, then, the interpellations of Rastafarian ideology became articulated with capitalist ideology. However, this was an extraordinary process, since there was not a natural coincidence of interests between the two.

On the other hand, the articulation of Rastafarian interpellations with those of the strategic fraction of the middle class gave rise to opposing tendencies: one supported the capitalist perspective favored by the fraction while the other, subordinated for the moment, still bore distrust. The strategic fraction of the middle class benefited directly from the conjunctural importance of commonalities (race, color, culture, and anti-European ethos) that momentarily suspended its class antagonism with the brethren (and the subordinate classes). It is largely this feature of Rastafarianism that gave the movement its autonomy and freedom from cooptation. At the same time, we should be struck by the intrinsically fragile nature of this alliance, which eventually fell apart, as we discuss in Chapters 9 and 10.

Also promoting the compenetration of middle class, capitalist, and Rastafarian interpellations were the dynamics of the change process itself. When fundamentally new ideological forms are required to promote capital accumulation, a dissociation from the old order is often pursued. In the Third World, this usually takes the outward form of a break with a colonial or imperialist past—a tack that the strategic fraction of the middle class in fact took, as it inveighed against the capitalists at home and abroad.

Underlying appearances of self-serving cooptation—as when both parties utilized the Rastafarian ethos for good political measure—there was, then a deeper process of symbolic transformation and cognitive change. A good example is provided by the kareba-wearing professional or politician, the unrepentant "me a Jamaican" figure wedded to "rootsness" who flourished after 1972. Of considerable note was the changed profile of the legal profession that is famous for wearing conservative, formal dark suits in all kinds of weather. Now the informal kareba (bush jacket) was the new fashion on the street and in chambers. The molders of public opinion, the Jamaica Broadcasting Corporation's personnel, were also to symbolize this new Jamaican person.

We recognize in these exemplary citizens only lingering "Afro-Saxon" traits for which their parents were well known. It would hardly be a disservice to begin with Michael Manley himself. From early in his political career, he displayed the fairly novel practice of "going down and rubbing shoulders" with the "small man." In a well-recounted Jamaica Broadcasting Corporation strike, he engaged in civil disobedience with striking workers and was arrested.[9] Alexander Bustamante was arrested on behalf of the people but the new trend was different. The practice of "going down and rubbing shoulders" meant actively participating in the earthy culture of the "sufferer": playing dominoes, organizing "curry goat feeds," playing card games like poker and the local "pit-a-pat," and attending dances. Sir Alexander's internments had an aura of martyrdom; acts of the strategic middle class bespeak the "one o' wi" conviviality of intimate relationships.

This pattern of behavior is to be contrasted with that of the "Afro-Saxon" mindset. Such distinguished members of the "Old School," men like the late Norman Manley and B. B. Coke, Howard F. Cooke, and the governor general, Sir Florizel Glasspole, are examples of black Englishmen the people were taught to emulate. The prescribed behavior here is a kind of paternalistic distancing: the "Afro-Saxon" is still embedded in a culture that counsels the respect of the "small man" but frowns upon the practice of rubbing shoulders with him. There are several Jamaican sayings discouraging these associations: "Play with puppy, puppy lick yuh tongue," and "Play with dog, yuh rise with fleas," are two well-known examples. Falling out of line might earn one accusations of emulating "the dress of people in some uncivilized state, . . . no collar, no tie, no jacket?" And, worse still, of overall slackness: "One feels that slack dress means slack mentality" (*Daily Gleaner,* May 9, 1961).

As the transformation of the persona continued, the sounds of the well rounded English language yielded ground to the "roots" language heavily influenced by Rastafarianism. Listen to Michael Manley translate some of the prominent songs of the early 1970s (*Daily Gleaner,* February 7, 1972, p. 1):

> When a man sings "Better Must Come" he means no sedition, he means no violence. He only means that he is suffering and looking forward to a better day. "Let the Power Fall on I" means every man who can't find a job and goes and sees others with opportunity and privilege, and who says if there is a God, "Let the Power Fall on I." It means every woman who says "me find some way to send my child to school in a way that I would like." "The Rod of Correction" says that a man looks around and realizes that graft and corruption abound and henchmen grow rich, and that if it takes a rod of correction to bring justice, then justice must come. When Junior Byles sings "Beat Down Babylon" he is not talking about the police. The police are honest people with a job to do. "Beat Down Babylon" says: remove oppression, . . . and let justice rise in the land. Oppression and corruption are rampant in Jamaica and I am going to beat down oppression.

The new style of the fraction was not a conspiracy hatched to get closer to "Quashie." Rather, the fraction needed to identify with the broadly attractive ideology now being sketched, one that legitimately allowed it to point collectively with the people at the capitalists as the forces of oppression, one that allowed it to come to terms with its blackness as a source of value and pride, and a strong foundation on which to build. What Garvey was singularly unable to do, the brethren had undertaken with some success.

Hobnobbing with the "sufferer" (a practice that earlier would have been done at one's peril) was an approach to self-definition that exposed crucial points of contact with the subordinate classes as well as providing valuable materials for political strategizing. Examples of the new affiliation exist at many levels. A most glaring accommodation was color. After 1974, the fraction became particularly conscious of blackness, led without doubt, unintentionally, but certainly forthrightly, by Michael Manley's friendship with and marriage to Beverley Anderson, a very bright, attractive television personality. The high political profile of a "young, gifted, and black" woman was, according to a fair complexioned Jamaican woman, the reversal of an old habit: the traditional practice of Jamaican men "marrying up" in color. In her words, "after a time, light skinned women could not find a husband or a suitor. All the men, black or brown were getting attached to black women."[10]

The sound education of the fraction served both to debunk and elevate it in the eyes of the people. Anthony Spaulding, the minister of housing in the PNP government and Wilton Hill, himself a minister of housing in the JLP government, were well educated by any objective standard. Both were university graduates and prominent attorneys. But they were among the most effective in consorting with the "sufferer," for which they received significant popular acclaim. Their educational and professional accomplishments earned them the title "big brains." Ever so sensitive to the exploits of the educated, the people were equally influenced by what the fraction accomplished as professional politicians, social critics (as contributors to *Abeng*) or businessmen and were largely guided by patterns of behavior in other spheres. As a corollary, the fraction's credibility quotient increased as it led, collaborated with, or followed initiatives directed at effecting change within the social structure. Education still had magic but it was now assisted by another factor. Members of the fraction did not suffer from the alienating social distance that placed the capitalists out of reach. Along with a shared blackness, the fraction could often point to modest origins not far removed in time and place from the subordinate classes.

It is this wider, defining principle that created a special kind of accommodation between the people and the strategic fraction of the middle class. Here a new "system of narration" was introduced to redress an ideological imbalance in terms of the requirements of class relations. Tinkering with language and art forms, and rearranging commonly held notions of history and religion greatly undermined the Eurocentric basis of the dominant ideology. Of no small moment was the conversion to the movement of many bright middle class individuals, resulting in the capture of their minds "from Marx, Hegel, and even Jesus Christ himself," as an interviewee put it. Anyone who can so decisively rearrange a perception of the self that had lingered at the very outer edges of consciousness should be taken very seriously.

To understand this process of cognitive transformation, we must pay close attention to the nature of relations between classes. Cognitive change was not an imposition or manipulation from above but was, in fact, itself a creature of the changing consciousness of both classes, in dialectical harmony. These were subordinate class counterparts of the "Afro-Saxon," members whose consciousness was focused on the consciousness of the dominant class: "if Teacher Cooke [Howard Cooke is one of Jamaica's finest elementary school principals] talk patois [Jamaican creole] to mi . . . ah wonda if him tink mi no understan' English?" On the other hand, the counterparts of the incipient black capitalists (part of the strategic middle class) were those members of the subordinate classes whose consciousness, though undoubtedly freighted with the dominant ideology as well as other reactionary forms, was nonetheless more firmly located in their own experience and had a more positive valuation of it. In part, this was the contribution of Rastafarianism.

▶ Concluding Note

There are very specific things that we can say about Rastafarianism at the time of the crisis. While it had revolutionary potential, as its branch in Grenada demonstrated, such impulses remained undeveloped in Jamaica. Nonetheless, it was a movement of enormous political and ideological transformative power that during this conjuncture served to further the process of capital accumulation. The chief beneficiaries of these historical circumstances were members of the strategic middle class and capitalism, in the sense that the immediately influential Rastafarian elements stemmed from the principles of liberal democracy. If socialism were to be installed at a future time, it seems clear that much would have to come to pass in the way of ideological and political education and a decaying of the socio-economic infrastructure before Rastafarianism could be directly enlisted in the service of that agenda. At present, however, class alliances, such as they were, betrayed characteristic brittleness resulting from fundamental differences revolving around basic class practices. While the bureaucratic/entrepreneurial fraction of the middle class could join cause with the brethren against the capitalists, the alliance could not prosper because of other ideological differences.

The Rise of a New Politics

C H A P T E R **8**

The previous chapters in Part III examined the process of capital accumulation in economic terms. This chapter will analyze the accompanying politics, changes in the political leadership, and acts of distancing and redefinition made by the PNP to bring Manleyism into phase with prevailing conditions. It would not be inaccurate to say that the radicalism and discontent that surfaced during the late 1960s had much in common with the rebellion of 1938. The political sentiments and the ideologies of the subordinate classes were still cast in a liberal democratic mold; the middle class was still pursuing power, only now the strategic fraction enviously eyed the economic spoils of the capitalists; and the PNP, still built on the foundation of Manleyism, would play its role in containing radicalism, largely because of structural continuities.

A missing dimension in analyses of the period is an evaluation of some of the empirical expressions of the special relationship the PNP bore to the ongoing radical tradition of Jamaican politics. We have selected two PNP slogans that appropriately captured the crucial elements of this new politics.[1] These are "Better Must Come" and "Power for the People." Both slogans have their origins in a climate of discontent to which the JLP was unable to respond in a direct way. Apart from the momentary rise of a leftist tendency, the politics of the party remained hidebound and attached to "a deep respect for the folk traditions and methods of community and individual problem solving." (Stone, 1980, p. 114): if it isn't completely broken, don't fix it. For the JLP the upsurge of radical political economics amounted to political sacrilege. This was so mainly because it could not support or initiate policies that discouraged foreign capital and promoted state ownership.

The failure to incorporate even the mildest of these radical stirrings, combined with factors of a moral nature—frequent charges of corruption and open quarrels between prominent JLP politicians, for example—served to erode the party's political base and its authority to rule. While most of the ideas captured by the slogans were anathema to the JLP, the party did not remain immune to the movements within the social structure. During the late 1960s, fissures began to appear, as a leftist tendency made its entrance. Among the causes were the growing threat posed by unemployment and skewed distribution of resources.

Nonetheless, the JLP leftist tendency could not circumvent or overwhelm the main principles of the party. The faction's leader was Edward Seaga, who, as minister of finance between 1969 and 1971, introduced several pieces of legislation aimed at resource redistribution and a modified form of state interventionism. These were an income tax bill, a company tax bill, and transfer and Jamaicanization policies. These initiatives did not get very far because powerful segments of the business community termed them communist in nature (Senior, 1972, pp. 28–29). In 1972 they registered their disapproval by abstaining from voting and by shifting their support to the PNP (Stone, 1973, p. 17).

So these political slogans were rooted in a climate of change beyond the immediate capacity of the JLP. They garnered much of their ideological and ethical content from the Black Power Movement, Marxism, and other radical expressions energizing local discontent. At the same time, the political leadership was also changing. The era of Bustamante and Norman Manley was drawing to a close, and its replacement was being fashioned from the conditions reflected in these slogans. The politics of paternalism was shading into the politics of social class.

▶ The Changing Panorama

Norman Manley's farewell speech at the Annual Conference of the PNP in 1969 fittingly spoke of unfinished business and the need for new commitments. His generation had fulfilled its task in securing independence. The task at hand, he offered, was to redress economic inequities; political independence must be followed by economic betterment (Nettleford, 1971).

It was quite an admission, since the policies of industrialization by his first two governments produced a skewed form of distribution favoring the middle class and the capitalists. In part, the Marxists were expelled from the PNP in 1952 for advocating economic development without foreign capital. But in 1964 economic nationalism reasserted certain features of the common good that had appeared in the party's agenda of the late

1930s, calling for (1) an absolute limit of 500 acres on the ownership of land; (2) the right of government to acquire land compulsorily from holdings of 100 acres that were not in productive use; (3) restrictions on foreign ownership of land; (4) nationalization of the Jamaica Public Service Company to begin the transfer of the vital sectors of the economy from private to public ownership; and (5) the rejection of private capital from overseas as a means of development (*Daily Gleaner*, November 7, 1964, pp. 1, 10; November 8, 1964, p. 6; November 9, 1964, p. 10). At the same time, these radical ideas were being supported by new, progressive groups such as the Young Socialist League (YSL), a group of young socialist lawyers and other intellectuals who advocated economic nationalism and trade union reform.

Manley's change of heart was not totally voluntary. He did speak of imminent political changes as when he said, "I sense profound changes ahead of us. I am convinced that Jamaica is going to change" (*Daily Gleaner*, September 18, 1967). But he was rushed into agreement by the threat of revolt, as when, for instance in 1964 ten members of Parliament from the PNP threatened to desert the party (Gannon, 1976, p. 188). A political party that favored radical political views from its inception had to take stock of growing unemployment and the grossly unequal distribution of income. After all, these disparities were not the quid pro quo the subordinate classes expected for meeting the radicalism that appeared before and during the labor rebellion. If Manleyism was to maintain its place as the guiding philosophy of the PNP, assisting the party to either continue its role in capitalism or become an agent of social transformation, it must be in the context of a reassessment of prevailing social conditions and a redefinition of principles.

We will recreate the pivotal circumstances that served at once to activate radicalism and transcend the old political style. Our class analysis in Chapter 2 gave a sense of the patterns of class transformation occurring between 1953 and 1969. Our analysis here will treat the slogans separately, even though in reality they interpenetrate. "Better Must Come" will be analyzed relative to material needs and the lack of economic opportunities; "Power for the People" addresses the issue of political power and its distribution and will be analyzed in that context.

► "Better Must Come"

The slogan, "Better Must Come," was taken from a popular reggae song that the PNP used effectively in its campaign of 1972. Today, its message is routine and might even be considered tame in comparison with the excoriating lyrics of international reggae artists. However, in the early

1970s, this was the "message music" of the people, not yet the national treasure that reggae is today. Before such songs and their messages gained currency, they were widely ridiculed for their grammar, spelling, and odd vocabulary.[2] To the degree that this slogan communicated a sense of social and political urgency, it did so in a way that reflected the trepidation of the subordinate classes. The slogan in its entirety, "Massa Day Done, Better Must Come," referred to a mix of class-inspired expectations.

For the people, the song spoke of the wishes, grief, and aspirations of the Rastafarian, the pimp, the "scuffler" and other members of the subordinate classes in ways that the JLP, seen largely as the rich man's party, could not. Middle class aspirations were of a markedly different kind, though they also placed emphasis on material gains and the need for power. From the perspective of these two class configurations, the policies of the 1940s had failed. Industrialization by invitation and education, the main policies employed between the late 1940s and the early 1960s, were a mixed blessing. Unemployment was pervasive, but the rich were getting richer.

These policies exerted pressure on the class structure. Apart from the more discrete class configurations to emerge—the middle class and the merchant/commercial fraction of the capitalists—the rural areas were being visibly transformed. Many former peasants were now flocking to the city of Kingston in search of non-existent employment; between 1962 and 1972, about 50 percent of the peasantry migrated to the Corporate Area (Kingston and St. Andrew). In time many would join the ranks of the lumpen proletariat.

The PNP's educational policies provided an exit from powerlessness and economic deprivation, but only for some. Successive governments headed by both political parties stayed the course but the results were palpably of middle class provenance. In a real sense, this was the partial fulfillment of the democratic ideal that Jamaicans consciously embraced after the labor rebellions of the 1930s. These educational policies mirrored the economic dependency of Jamaica. As Malcolm Cross (1979, pp. 122–124) notes, education became relevant as an instrument of social mobility only when emerging intermediate classes were able to lay claim to professional and commercial occupations. These circumstances put their stamp on the system, especially in the Caribbean and other Third World countries, with regard to two interrelated needs. The first was a need focused on the middle class ethos, which was seen as the legitimate "pathway to affluence and status"; second, much that was required psychologically to establish this status was perforce tied to values conducive to a materialist culture.

But all that glittered was not gold. The lot of the majority had not materially improved. The phenomenon of dependency (openness of the economy, a consumerist orientation, and preponderance of primary product exports) triggered countervailing forces against economic development. Agriculture, the main employment outlet, steadily declined. During the years 1950–1955, the sector contributed 4.9 percent to the Gross Domestic Product (GDP); during 1965–1970, that figure dropped to 1.6 percent. This triggered unemployment, which climbed from 13.5 percent in 1960 to 23 percent in 1969. Urban areas were markedly affected. Between 1960 and 1972 urban populations grew from 32 to 42 percent; the Corporate Area grew by 13 percent over the same period.

The drive toward industrialization proved unequal to the task. For example, up to the late 1960s the labor force increased by 20,000 persons annually, but manufacturing could absorb only 5,000 of this number annually (Jefferson, 1972, p. 35). Wages also made a pathetic show; fully 60 percent of the working labor force earned only $10 per week.[3]

If rising expectations exposed a yawning gap between ideology and performance from the point of view of the subordinate classes, disillusionment ran even deeper for certain segments of the middle class. The developmental script, which relied heavily upon an exalted role for the middle class, was partially confounded. Indeed, the class could not be exempted from the disease of scarcity as the disparities in the distribution of surplus and available jobs applied equally to it. Unfulfilled expectations generated political and ideological responses.

Discontent and political activism tended to be largely urban-based, arising from three segments of the population—the subordinate classes, the middle class, and the Rastafarians. Although the Rastafarians belong to the subordinate classes, we have singled them out for special comment because of their crucial role in the area of ideology.

Among the subordinate classes the role of the working class was not especially significant. While its numbers increased appreciably and, with them, the support of unionism—a study for 1972–1973 showed as much as 82 percent of the working class favoring unionism (Gordon, 1978, p. 316)—overall, the class was politically undeveloped. Only one quarter to one third of the industrial proletariat opposed "imperialist and capitalist ownership of the economy" (Gordon, 1978, p. 318). In the words of a former high-ranking official in the National Workers' Union, "the working class were not converts to democratic socialism. They were the beneficiaries of social legislation. They never led any marches for it and they refused to have it shoved down their throats."

Even up to the late 1970s and the elections of October 1980, the class had not developed an engaging proletarian tradition. They were criticized for this lack by the communist Workers' Party of Jamaica, according to whom the class must master working class ideology, combat the false hopes held by workers regarding capitalism, and systematically teach working class discipline (Workers' Party of Jamaica, 1981).

Strikes represented by far the highest level of activity by the class, yet, in the context of a limited and reactionary trade union movement, it is hard to interpret these as anything more than self-seeking. Between 1965 and 1970, 15 percent of the organized working class were regularly engaged in strikes (Stephens and Stephens, 1986, p. 50) that were caused as frequently by grievances about wages and working conditions as by complaints about management attitudes. As the workers put it, they were offended by management's displays of "a superior attitude."

The linkage between trade unions and political parties facilitated the growth of a practice by the class that was fundamentally opportunistic. Immediately after the 1972 elections, 5,739 members of the BITU, the trade union adjunct of the JLP, defected to the NWU, the trade union adjunct of the victorious PNP (Gonsalves, 1976, p. 250). In the case of a JLP victory, the pattern would be reversed, since workers utilized this strategy to hedge their bets in a system where employment was bound up with patronage and political parties with a pronounced and growing employment function. As the private sector became more dependent on the state, politicians exercised increasing leverage over job placement. It is no surprise, then, that in 1974, 66 percent of all union members had fully paid their dues although unemployment hovered around 25 percent. This pattern was the norm because unionism provided protection and appreciable levels of comfort. It should be noted that the state employed over 15 percent of the work force.

We should not leave the impression that the working class was totally devoid of class consciousness. The urban segment of the class occasionally expressed themselves tellingly during this period. In May 1966, for example, disgruntled workers in Kingston vented their frustration on government property. With the assistance of members of the lumpen and subproletariats, they attacked and threatened the property and employees of the Ministry of Labour, the Ministry of Trade and Industry, the Ministry of Finance, and the offices of the BITU. Also, certain sections of the class did form alternative unions in the late 1960s to promote full trade unionism and the democratization of the movement. These unions united under the umbrella of the Independent Trade Union Action Council (ITAC), with the Municipal and Parish Council Workers' Union and the

University and Allied Workers' Union (UAWU) among their more prominent members. And, although strike figures can be variously interpreted (as we have noted above), it is nonetheless of some significance that during 1969 and 1970—the years immediately prior to the PNP's victory in 1972—there were more workers on strike (42,174 and 39,901) than for the entire thirty-year period between 1943 and 1974 (Gonsalves, 1976, 265).[4] Many of these were wildcat strikes. These were limited actions, however, that do not significantly detract from our argument that the working class did not constitute a major force for change. Thus the class does not figure as prominently in our analysis as do other segments of the subordinate classes.

We now turn our attention particularly to elements from the lumpen and subproletariats. By 1962, the year of independence, these classes had been making pointed references to unfulfilled promises. According to Wendell Bell (1964, p. 51),

> many lower-class persons were saying that their interests were not being served. They were raising questions about the civil service ("a middle class preserve") the new office buildings for the Ministries ("Why not more housing? The brown men take care of themselves!"); they were seeing new factories built and a new air terminal, but they saw their own circumstances improved less than they expected.

In the meantime, their dismal condition sent clear signals of an impending social explosion. At the time of independence, Ruth Glass (1962, pp. 26–27) speaks of the "dreadful shack dumps of West Kingston" that form a part of "Kingston jungles [that were] exceptional both in the extent and the degree of abandonment they expose." By the later 1960s, only the description of these physical conditions remained stable: the people were no longer obscured oddities, as they were now forcing themselves into the political reckoning.

But the accompanying ideologies were a pastiche of distinct strains. First, they were greatly influenced by the middle class ethos. For the handcartman and the "scuffler," betterment nearly always meant raising enough money to start their own little business. Such individuals swelled the ranks of Kingston's growing informal sector.[5]

Another distinct expression was militant by disposition. It was untouched by either intellectual perceptions of politics, or such notions as national development. These militants' understanding of political economy was basic to a fault; it rose and fell with naked bread and butter issues. This is termed in some quarters the "mi a look two, bawsie" mentality. The phrase means, "I am prepared to make a living by any means necessary."

Another significant tendency was as much a creation of scarcity and deprivation as it was the result of conjunctural political practice. Inevitably, the fierce competition for scarce resources led to a political patronage system, which rapidly developed under the new leadership of both political parties. These ruffians often worked closely with politicians to maintain "discipline," which in most cases meant intimidation and strong-arm tactics in the quest of political "support."

Then there were the politicized elements. They often followed in the tradition of radicalism kept buoyant by chronic scarcity. Overall, their political vision and sense of economic justice were tempered by the impact of the Black Power Movement and the resultant upsurge of radical ideologies, such as Marxism, that reasserted themselves, thanks in large part to the radicalism promoted by the Black Power Movement and dependency theory. In an important sense, these elements were the counterparts of the buccaneers employed by the politicians; they were attached to radical middle class intellectuals, mainly from the University of the West Indies, who began to enjoy increasing grass-roots support. It should be noted that this in itself represented a noteworthy break with tradition. Intellectuals in the past had little time for the popular classes, let alone "going down and rubbing shoulders" with them in these socially and politically productive ways.

The middle class produced three main grieving segments: those in civil service and managerial positions, the unemployed educated, and the educated radicals. For the first segment, satisfaction with status started giving way to the urgencies of materialism. Up to the late 1950s, the British standards of public service prevailed (Hamilton, 1964, pp. 134–135). It is true that this segment was among the main beneficiaries of industrialization, but, like Oliver Twist of Dickensian fame, it wanted more. Civil servants seemingly monopolize the national "feeding tree." The people claimed, "fruits, leaves, branches, tree trunks, everything went into their pockets. Nothing comes to us, the poor people" (Lowenthal, 1972, p. 310).

For these "Oliver Twists," the demand for more bread was prompted by two major causes: a comparison of their lifestyle with that of their counterparts in North America and structural changes within the state. The mining corporations were the first to contribute to a heightening of expectations: the much more attractive employment packages they provided dwarfed the traditional wage structure.[6] The multinational plants and operations that followed bauxite had similar effects and more. With the rapid Americanization of the economy came the related business practices of providing housing allowances, company cars to drive, trips to overseas headquarters, entertainment accounts at hotels and restaurants—"perks," nearly all of which were denied civil servants.

The structural changes within the state revolved around the execution of policies. The British civil service model proved inflexible. Politicians found its practices restrictive of innovation and efficiency, and far too many obstacles were put in the way of expediting policies needed to address urgent socio-economic change. To circumvent these obstacles governments created parastatal entities. The first of these were the Jamaica Industrial Development Corporation (JIDC) and the Jamaica Agricultural Development Corporation (JADC), both created in the 1950s. These organs were manned by educated personnel whose salaries and conditions of work bespoke material betterment and the exercise of real power. These developments led to the increasing demand for more autonomous bodies, which would at once serve the ends of the politicians as well as the upward mobility of middle class elements. This trend meshed with the forces giving rise to the bureaucratic/entrepreneurial fraction of the class.

In a real sense, economic nationalism, which surfaced around this period, was directly in phase. Longing eyes were cast at the bauxite/alumina industry. Of course, the PNP's adoption of economic nationalism would partly serve these ends. Yet it should never be thought that the JLP was without support for the doctrine. In 1966, the minister of trade and industry, Robert Lightbourne volunteered that bauxite and alumina ought to be transported by Jamaican ships (*Daily Gleaner,* September 13, 1966); in 1970, Hugh Shearer, then prime minister, suggested that the industry ought to exercise good corporate citizenship by redressing the imbalance in the proportion of bauxite revenues that were retained in the Jamaican economy (*Daily Gleaner,* April 23, 1970).

It is inevitable that the disparities between rising expectations and available resources would impact negatively upon sections of the middle class. It should be remembered that, beginning with the first governments of the PNP (1955–1962), the class stood alone as the carrier of fundamental change. The strategy was the concretization of Norman Manley's declaration in 1939. For him, major societal change was synonymous with education (Nettleford, 1971, p. 102).

The economic exclusion of the unemployed educated exposed a major contradiction in the educational policies of the times. For while the social institutions and practices inculcated the middle class ethos, the economic forces entrenched stagnation and downward social mobility for these class elements. For many of them, unfulfilled expectations led to the reexamination of prevailing ideologies. The reliance upon radicalism was pronounced. Many drifted toward Rastafarianism, giving rise to the seeming incongruities of black Rasta, brown and "red nigger" Rasta, and "Chinie Royal" Rasta. The movement transcended class, color, and economic status. Others were drawn to Marxism. Still others courted a radicalism that often verged on what might be termed "Robin Hoodism": taking from the rich to give to the poor. In this respect, we recall the famous bank robbery that took place late in 1971. Here a group, including a former student from the University of the West Indies, robbed a bank in Runaway Bay (one of the many tourist centers on the North Coast) and took the manager hostage. After a long police chase, the robbers were apprehended, but not before they distributed the proceeds of their daring adventure to several poor and needy on the way.

To these segments should be added the host of sympathizers, those smitten by what Rex Nettleford dubbed "brown man guilt." There also emerged a radical, educated left, having strong connections with the University of the West Indies. But because these expressions fall more appropriately within the framework of our other slogan, "Power for the People," we will take up the analysis later.

For their part, the Rastafarian brethren were obviously concerned with material betterment, but their immediate preoccupation appeared to be freedom from persecution:

Yuh can tek wey mi yam an' mi dumplin'
I don't care a kick about dat
Take me back to Ethiopia
Mek mi mark out mi burial spot

These pregnant lines from one of their famous songs underscore the cardinal importance of freedom to the Rastafarian vision.

In summary, "Better Must Come" meant different things to different people. It even had significance for sections of the capitalists. It is clear that many of them, the industrial fraction mostly, supported the regime in the expectation of increased gains. At the same time, it should be noted that as a whole these reactions were animated largely by the politics of the day. "Power for the People" is a slogan that turns more on the political variables associated with the crisis.

► "Power for the People"

It emerged from the Black Power Movement of the 1960s and represented the reappearance of the radicalism of the 1930s that had been held in abeyance by the inherent promises of self-government and independence politics coupled with the economics of industrialization. The social conditions breeding this slogan had not suddenly descended upon the political scene—they had been festering for quite some time.

Economic and political change of the kind induced by post-rebellion tinkering had placed a temporary check at best on radicalism. Indeed, new political expressions indicative of impatience with the abject conditions of the subordinate classes were emerging. We will cite three significant instances of these expressions, two of which departed from the tradition of modern (post-labor rebellion) politics in that there seemed to be a preparedness to engage in calculated, political violence.

The first instance of these belligerent postures was the so-called Rosalie Avenue Movement under the Reverend Claudius Henry, the founder and pastor of the African Reform Church of 78 Rosalie Avenue in Kingston. This movement, which reached its flashpoint in 1959, had religion, once again, as its principal foundation. Although reliable numbers are not available, the movement attracted "thousands" (Nettleford, 1970, p. 83). Claudius Henry, the "Repairer of the Breach," had as his divine mission the return of the Rastafarians and other oppressed Jamaicans to Africa.

A movement consisting of thousands of poor, disgruntled Jamaicans under the leadership of a radical-sounding preacher espousing a "Back to Africa" theme was bound to catch the attention of the status quo. Allegations were made that the movement had been building an arsenal, threatening bloodshed, and consorting with Cuban communists. Predictably, these allegations were more than enough to set the authorities in hot pursuit. The Reverend Claudius Henry was convicted under the provisions of the laws governing treason, firearms, and drugs. Hitherto unapplied laws were employed to secure his conviction.

The second expression was quasi-revolutionary and derived largely from the Rosalie Avenue Movement. The leader on this occasion was Ronald Henry, the son of Claudius Henry. The younger Henry was the leader of a gang, including a number of Rastafarians, who confronted a police and military detachment in the Red Hills section of Kingston on June 21, 1960. In the ensuing melee, two members of this detachment were killed. The incident, quickly put down by the authorities, was surrounded by melodramatic tales of revolution. For instance, the authorities supposedly discovered a cache of military weapons stockpiled for revolutionary purposes and a letter to Fidel Castro, requesting his assistance in overthrowing the Jamaican government (*Daily Gleaner,* October 13, 1960). There was also some suggestion of the draconian discipline one comes to associate with revolutions, such as recalcitrants ignominiously shot in the back of the head and buried in shallow graves. Ronald Henry was tried, convicted, and hanged.[7] The extent to which the evidence may have been exaggerated or even fabricated to justify an authoritarian government response is unclear. The incident does suggest a refocusing of political practice by some Rastafarians from repatriation to social change.

The third expression was far more sedate, though no less critical of color and racism. This was the founding of the People's Political Party by Millard Johnson on April 16, 1961. Johnson, a member of a prominent, well-to-do black middle class family from the St. Andrew area, was trained as a barrister. A student of Garveyism and an avid devotee of Africa, Johnson aligned the PPP closely to Garveyism, especially with regard to his mentor's emphasis on racial pride. The PPP appealed to the poorer segments of the society through a manifesto that enshrined such timely issues as equitable distribution of wealth, the eradication of racial and color discrimination, and political independence.

The PPP's showing in the 1962 elections was lamentable. In spite of a strong appeal to the electorate's African heritage, the party could only muster a little over 2 percent of the votes in the constituencies it contested. Of the sixteen constituencies contested, seven were situated in the Corporate Area; Johnson himself could muster only a shade over 4 percent of the votes polled in his constituency of west central St. Andrew. Millard Johnson's appeal to race and black dignity was conspicuous by its lack of success.

What should we make of these political offensives? Why were they so ineffective? No single explanation will suffice. We agree with Rupert Lewis (1976, pp. 170–171) that the political status quo represented here by both the JLP and PNP, deliberately emasculated such political efforts. The subtle constitutional principles embodied in the law and order imperative were an effective agent. The predominant tendency is for a citizenry to stand behind the state in its guard against disruptive or subversive political machinations. This is a sovereign duty that has little to do with partisan political interests; it arises, as Antonio Gramsci argues, from the hegemonic tendencies of capitalism. Additionally, whether the charges are true or not, the state's control over, and definition of matters of public interest is such as to appeal rather quickly to citizens. Very active here is people's sense of responsibility to their government that is to be attributed, in no small way, to the gradual installation of political society.

The Rosalie Avenue Movement as well as the Red Hills incident implicitly invited the intervention of the state. In theory and in practice, both postures offended notions of the public interest that the state interpreted with the concurrence of the Jamaican people. Rex Nettleford (1970, p. 84) observes that these incidents "merely confirmed in the minds of the wider society that people who preached Back-to-Africa were violent." It also put to the test the "passionate commitment to the rule of law principle" some scholars see as an essential trait of West Indians as a whole (Forsythe, 1974b, p. 404). Also, and perhaps not least of all, there simply was not the political consciousness and class support for their success. While many from the subordinate classes joined or subscribed to these political movements, engaging in political action of this kind is another matter. Ronald Henry was led to expect an outpouring of active, revolutionary support but this did not materialize; the movement of his father, Claudius Henry was at bottom wedded to Rastafarian precepts, having little to do with violent change as a defining principle. One can see distinct parallels between these expressions (the Red Hills incident excepted) and Bedwardism, Garveyism, and Rastafarianism.

The failure of these attempts at social change has much in common with the failure of Bedwardism and Garveyism. Another factor may also be important, however. There is some evidence that until the mid-1960s or so, disenchantment, though growing, had not reached critical levels. Notions of the common good crafted by Manleyism during and after the late 1930s had not totally lost their appeal. Education, though directed mainly at the middle class, still held out promise. The University College of the West Indies (a precursor of the full-fledged university) had been established in Jamaica in 1949 and children of the subordinate classes began to take their places there. Education was always seen as an avenue for social mobility, if not for oneself, certainly for one's children. Similarly, until the mid-1960s industrialization, which also embodied in principle many facets of the common good, had not yet lost its appeal.

These offensives were also undermined by the special interests of the JLP and the PNP. It should be remembered that, for all their differences, these political parties were essentially apparatuses of a capitalist state. That being so, they were constrained to ensure that all the ideological and political expressions within the society accord to acceptable pro-capitalist notions of give and take.

The Henry offensives could not be so sanitized. Both operated outside the ambit of constitutional politics. The Rosalie Avenue Movement was, for all practical purposes, impatient with institutionalized politics, except to the degree that Henry found Manley politically congenial. For their part, the "revolutionary" forays of Ronald Henry spoke cogently of growing impatience with the prevailing political order.

The two-party system seized this opportunity to entrench itself as the only viable option. Supported somewhat by the universal appeal to the theme of law and order and, by extension, reason and the notion of "properly constituted authority," the system proceeded to safeguard its special interests through thinly disguised psychological intimidation and outright repression.

The Reverend Claudius Henry and his followers were subjected to the now standard Bedwardian treatment: demoralize and scatter the flock by irreparably destroying its leader. In the very same way that Alexander Bedward was portrayed as a demented idiot, caught inescapably in the web of wild religious fantasies, so was Claudius Henry. Briefly, Henry made messianic claims that provided justification for a societal reaction based on the precedent of Bedward, who, it should be recalled, claimed to be the reincarnation of the Prophet Elijah. For all but the most intrepid and committed, the attack upon Claudius Henry was as devastating as it was demoralizing. Imprisonment is often a precondition to political martyrdom, as Alexander Bustamante's own case confirms, but madness was quite another matter. Followers do not easily rebound from the implicit accusation that they lack even the elementary perspicacity to distinguish between madness and sanity in those who lead them.

At another level, color and race incurred the collective wrath of both the JLP and the PNP. As these issues were related in a distinctive way to the Rastafarians, both Henry movements, and the PPP under Millard Johnson, the response of the status quo was uniformly negative. But the PPP presents an interesting case. Millard Johnson did not opt for the unconventional approach that had earlier disqualified Bedwardism and, to some extent, the Garvey movement; he followed the canons of organized political behavior—parliamentary politics.

In 1959, Johnson joined the JLP, thinking that he could introduce his brand of militant Africanism to a population ready for his message. He was wrong. For one, the JLP was impatient with what it construed as the preaching of race hatred. Johnson's exhortations on black racial pride and dignity were like the proverbial red flag before a bull. The similarities between criticisms of Johnson and those leveled at Garvey in the 1930s are uncanny.

The PPP was neutralized by the combined forces of the JLP (because of the race issue and the threat of a third party), the middle class (because the notion of race offends its cosmopolitan vision), and the PNP (working in tandem with the JLP against the rise of third parties). On this last point, Norman Manley was emphatic: it is, he noted, in the interest of our "peace and security and freedom that we should have two parties in this country" (Nettleford, 1971, p. 47). In later years, he volunteered that the two-party system was ideal for countries like Jamaica "young in nationhood, without established traditions" (Nettleford, 1971, p. 49). It should also be noted that Norman Manley was intolerant of anyone using race and color as political vehicles (*Daily Gleaner,* October 31, 1960)

Finally, failure was related to the opportunism of the electorate. The PPP's effort had been frustrated by the absence of patronage. Political loyalties were secured less by adherence to ideology than by the availability of political goods and services. This thesis is amply supported by many examples illustrating the positive correlation between the supply of these goods and services and political party alignment. The sagas of Ken Hill and A.G.S. Coombs strongly support the case.[8]

Ruthless prevention now became a key feature of the political agenda. Between 1955 and 1962, the PNP issued some 200 prohibitions against the entry into the country of well-known socialists and revolutionary figures, a practice that continued unabated into the subsequent JLP governments between 1962 and 1972. In the later years, the dragnet ensnared even non-communists and other intellectuals whose ideas offended government policies (Gonsalves, 1979, p. 15; Munroe and Robotham, 1977, p. 153).[9] This intensification of repressive action by both parties must be seen in the context of events that were not fully within the control of the political leadership: the disgruntled popular classes had begun to link their struggles to those of other oppressed people. The civil rights march on Washington, D.C., which took place in 1963, was one such example. While the solidarity with African Americans was impressive, the placards and banners on display dramatized the new and, for the political leadership, troublesome aspects of local politics. "Equal Justice for All," "Time for Socialism," "Babylon Beware"—these slogans cast long shadows back to the labor rebellion, with discontent drawing sustenance from a wider field of definition. These expressions certainly foreshadowed things to come.

The politics in vogue were reminiscent of the old days, with radicalism and discontent rising and surging, being kept under rigid discipline by both political parties. By the late 1960s, political patterns changed perceptibly.

The case of Walter Rodney—one of those excluded from the country because of his political beliefs and activities—illustrates the new political pattern. This was characterized by a tug of war between the products of middle class education and the turning of mass sentiments against the JLP. The slogan, "Power for the People"—and our analysis—should be understood against this relief. Rodney, a young Guyanese historian from the University of the West Indies at the Mona Campus in Kingston, was at the cutting edge of politics springing from the malaise of the JLP and the outmodedness of the Manleyist common good. In 1968 Rodney attended a Black Power conference in Canada and was denied a re-entry visa into Jamaica. The prime minister, Hugh Shearer, was specific in his condemnation: Rodney was a communist who had made three visits to communist countries in 1962 alone. At home, he continued, Rodney was closely associated with the Reverend Claudius Henry. In Shearer's words, "in recent months [during 1968] Rodney stepped up the pace of his activities and was actively engaged in organising groups of semi-illiterates and unemployed for avowed revolutionary purposes" (*Jamaica Hansard,* 5, 1968, p. 392). Further, like Bedward, Garvey, Henry, and Johnson before him, Rodney had incurred the state's displeasure by discussing race: he had "openly declared his belief that as Jamaica was predominantly a black country, all brown-skinned, mulatto people and their assets should be destroyed" (*Jamaica Hansard,* 5, 1968, p. 394). For the record, this is a gross misrepresentation of Rodney's position. Forthwith, the JLP government banned the importation of any literature associated with the Black Power Movement and Black Muslims.

This incident provoked a student demonstration in Kingston, that was joined by faculty from the university campus and many members of the middle and subordinate classes. A confrontation erupted between the police and the marchers in which three persons were killed. Property damage from the ensuing riot amounted to $2 million, mainly affecting commercial and financial enterprises.

What should we make of these events? First, that the new politics had little to do with revolution. The events lacked the "circumstances" and "currents" needed to "fuse" into a "ruptural unity" (Althusser, 1971, p. 99). The structures and relations needed to precipitate antagonistic contradictions—a rapidly deteriorating politico-economic infrastructure, a well-organized mass movement, and the collective consciousness—were absent or not fully developed.

At the same time, while the pure elements of revolution were not present in 1968, evidence of the class struggle was marked. Writers like Stone (1974) and Lacey (1977) state rightly that the Rodney Affair was neither revolutionary nor insurrectionary. Others, such as the contributors to *Socialism!* have taken the opposite position: the people were ready to lead a revolutionary offensive but were pushed back by the forces of repression.[10] In many ways both positions are extremist and merely station themselves at opposing ends of the great political debate. It is misleading to suggest that the lumpen proletariat and the urban poor merely used the demonstrations to loot and plunder (Lacey, 1977, p. 97); nor were they thugs on the rampage, setting fire to buildings, and menacing city life in a haphazard, grossly senseless manner (*Daily Gleaner,* October 17, 1968). Similarly, one is hard-pressed to find patterns of organization and compatible expressions of class consciousness to warrant the claim of a revolutionary agenda.

The correct interpretation, we suggest, lies somewhere between these extremes. The Rodney Affair really provided a reliable view of class relations. The dominant class clearly had the state on its side. Between the *Daily Gleaner*—the chief media outlet for the class—and Parliament, enough power was exerted to put down the Affair. The minister of home affairs, Roy McNeil, and the police force exposed strength and weakness in complementary ways. The police acted with dispatch (and extreme brutality) to restore order and neutralize suspected and potential leaders. The Jamaica Defence Force later admitted that it was at best "a half-cock urban insurrection" without any leadership (Lacey, 1977, p. 99).

There is little to be further said about the leadership. Faculty and students at the Mona campus of the University of the West Indies had begun to form what appeared to be a vanguard for change. In the ensuing debates and confrontations with politicians and the press, much of the apparent revolutionary zeal visibly disappeared. Not only was the structure of a well-formulated political leadership absent, but the politics embraced by the more visible and vocal of the vanguard could best be described as social democracy (Gonsalves, 1979). One putative leader could only speak of these events "becom[ing] a part of the permanent political experience" of those involved and concerned (Girvan, 1968, p. 67). It would, nonetheless, be misleading to leave the impression that class relations were not being urged on to a higher plane of development.

Research conducted in 1967 indicates that the urban youth, many of whom participated in the riots, were certainly more than looters and robbers responding to the basic instincts of survival. As Garth White (1967) has shown, the ideological stance of its most active faction—the "Rudie boys"—had been undergoing qualitative change. The "Rudies" were urban-based youths who adopted an aggressive political posture toward the latter part of the 1960s. Often, they were the partly educated unemployed who could not be absorbed into the work force. The age group from which they were drawn (fourteen to nineteen year-olds) was among the most savaged by unemployment. In 1960 unemployment among this group was 39 percent—a figure that climbed to 68 percent by 1970.

Rudie had a distinct culture. It amply served notice to the rest of Jamaican society, expressing itself in distinctly anti-establishment dress, music, and dance. Many aspects of the culture surfaced in popular music, which reflected underlying political and ideological currents. Songs such as "Poor Me Israelite" and "Shanty Town" spoke respectively to class-based deprivation and bouts of ideological redefinition engaged in by the Rudie boys and the forces of law and order.[11]

Rudie often went to jail and was repeatedly "gone 'pon probation" but nearly always his actions challenged the paternalist order and the political leadership to meet the urgent call for power to the black people.

It was a call not anchored to racism, which the old leadership (Bustamante and Norman Manley) and part of the new (Shearer, especially during his prime ministership in 1969) was unable to fathom. Of the top leadership of the JLP, only Edward Seaga seemed to have sufficiently grasped the new political realities, although the degree to which his initiatives went beyond improving his own political fortunes is unclear. Among the main new themes embraced were developing a thesis of the haves and the have nots; the organization of youth and Rastafarians of his constituency; and the promotion of local folk culture. First, he skillfully characterized the PNP's performance from 1955 to 1962 as one perpetuating the perilous condition of the class structure. He suggested that 93 percent of the population should be categorized underprivileged. This won him tremendous support. Second, he identified closely with the brethren of his constituency and was made an honorary member (Waters, 1985).

Another development of some note was the impact of Rudie culture on the youthful middle class. Popular radio programs often carried many requests for Rudie and Rastafarian music. In a sense, the scions of the middle class were reacting to new cultural and ideological impetuses that were at odds with the European tradition to which the class owed much of its gestalt. A former high school teacher, a mistress at Calabar High School, remembered her students' addiction to Rudie's music and dance: "I would be driving up to school, and there they were. The students. They were walking up to school, fists clenched, chests tucked forward, and shoulders rocking side to side to imaginary music. It was a sight."

This exemplifies an important facet of cultural change that has been confused with revolutionary potential. Here the Rudies were engaged in replacing the "tenk yuh, Buckie Massa" postures of their parents with "a new sense of class confidence" (Stone, 1974, p. 16). If nothing else, the Black Power Movement especially counseled them to genuflect no longer to the erstwhile colonial masters or their North American replacements nor to submit uncritically to their ongoing poverty, powerlessness and deprivation. Massa day done!

However, the culture's anger and venom were seldom vented on the capitalists or the middle class. The main victims were the distressed and the wretched living in the depressed urban areas. The picture began to change in the mid-1960s. As White put it, the gangs were "gaining some sense of purpose other than self-interest." While the political patronage system was deeply entrenched at this time, "the greater end was still recognized" (White, 1967, p. 43). The Rudie boys now embraced as part of their agenda, albeit one without organization and ideological cohesiveness, the liberation of the "sufferers" from the tyranny of privileged society.

Could a political liaison between these elements and the Rastafarian movement have prepared the ground for a flowering of revolutionary consciousness and purposive political action? The brethren, especially in the 1960s, had greatly influenced a significant section of the urban youth. Up to the mid-1960s, however, their retreatist features were marked, with petty theft sustaining the demands of survival and "Robin Hoodism" very much a part of community relations (Chevannes, 1981, Munroe, 1972). To the extent, therefore, that this retreatist posture remained dominant, it actually exerted a measure of political constraint upon Rudie and other lower-class elements.

But there are qualifications. During the latter half of the 1960s these retreatist tenets were undermined. Rastafarianism could not remain immune to socialism and leftist liberal thought that intruded upon the scene through Marxism and the Black Power Movement. Both spoke the language of political activism, thrusting upon the movement a reckoning of the political here-and-now. The impact on Rastafarianism was visible: the hammer and sickle, prominently painted on the walls of the brethren's houses, became a common sight (Chevannes, 1981, p. 393). The movement also gave birth to the Dreadlocks, politically aggressive members "who would thunder their sounds of blood and fire at non-dreadlock members" and the rest of society (Chevannes, 1981, p. 399). Out of keeping with the fundamental principles of the movement as a whole, this splinter group did not rule out activist, violent politics. In Grenada, similar pressures figured in the participation of Rastafarians in the overthrow of the Gairy regime in 1979. There, activist politics found its way to the core of the brethren's message:

> It is of great significance that the progressive movement included Rastas and have even placed brethren in key governmental positions in the security forces and other agencies and to show no prejudice to natty dread. It is with this in mind that we deliver this message to you as a reminder that the founding principles of the Rastafarian faith directed that we should at all costs serve the people in the truth and the right, so it behoves us to state emphatically and clearly the position of Rastafari on issues of temporal realities such as socialism, and communism which is opposed to capitalism, underdevelopment and imperialism. Rastafari must take their proper place in the Third World Revolution struggle against dictatorship and oppression. Rastas cannot and must not become

> the pawns of reactionary capitalists in their attempt to maintain imperialism (*Free West Indian*, July 5, 1980).

The Grenadian experience suggests that the brethren's love of peace and brotherhood—the base upon which the movement is built—could be transformed into revolutionary fervor. In Jamaica, however, during and beyond the period of democratic socialism, the movement seized public attention but did not align itself with a revolutionary agenda. Many differences in the social structure can account for the differences in behavior. Undoubtedly, of major importance is the fact that in Jamaica the movement received recognition by the state. In the 1970s, the Ethiopian Federation Movement, the parent body of the brethren, was placed on the same footing as other religious denominations such as Anglicanism and Catholicism. With this recognition came subtle and direct injunctions with regard to the movement's duty in the preservation of the rule of constituted authority. Of course, this apparent constraint could not prove lasting: the cultural resistance embodied in Rastafarianism is but a short and natural step to political resistance essentially germane to it. Thus the movement's philosophy managed to promote political education in subtle, indirect ways. The energy displayed here was perhaps more militant than revolutionary and transformative.

The delicate interface comes with cultural and, especially, religious forms. Many members of the subordinate classes were exposed to a stern Christian upbringing that made it difficult at times to reconcile Christian precepts with social inequality and political oppression. The brethren were helpful here. Through its patented mode of religious demystification, the movement transformed political ethics to accommodate differing strains of radical thought and action. Very popular songs like "Burnin' and Lootin" and "Get Up, Stand Up" present clear examples of the mode of translation. These were popularized by the late Bob Marley and his group, the Wailers.

As scholars like Nettleford (1978, p. 21) have correctly stated, the movement gave God "a human face." A form of secular theology stressing the alleviation of oppression and injustice replaced the pre-existing colonial theology that remained shackled to an emphasis on individual salvation (Erskine, 1981, pp. 116–120). Perhaps there is no better description of the nuances of this new ideological form than Joseph Owens's perceptive observation (1976, pp. 254–255):

> The brethren have consistently and thoroughly shattered the imprisoning images and conceptions of traditional Christianity and have elaborated a new vision of the universe which strives to re-

> capture the revolutionary message of these ancient dreadlocks, the kings, priests and prophets of biblical times. . . . By announcing that God is fully a man living among men, the brethren have not debased divinity, but have enhanced a whole people's awareness of their own humanity.

Today the Rastafarian is priest, prophet, social commentator, and the molder of emancipatory ideologies. These ideologies no longer suffer from the charge that, as ideas originating in pure thought, they are "not easily embodied in the class organs" (Post, 1978, p. 186). As we discussed in Chapter 7, the movement has become more systematically inserted into the Jamaican social structure, with Marxism and other radical ideas imparting their influence. Politically influenced lyrics, such as those popularized by the late Bob Marley, a member of the Twelve Tribes, suggest a distinct shift away from retreatism to the politics of participation, from individual salvation to class relations. Such messages have been relayed by members of the movement in other parts of the world. Recall the experience of Grenada.

So there was support for radicalism in its different manifestations, but tension was apparent. While the movement spoke about and encouraged radical and revolutionary activity, it did so as a warning to the broad principles of democracy in which Jamaican politics is embedded. The message was ambivalent and proved itself attractive to the radical elements of the young activists and intellectuals, some of whom took their agenda well beyond the act of joining cause with Rastafarianism and the subordinate classes to the prospect of coordinating political action.

While Terry Lacey's analysis hardly supports the contention, the Rodney Affair did indeed create "tentative links between basically dissimilar groups of people" (Lacey, 1977, p. 98). In his troubles with the Shearer government, Rodney received support from Black Nationalists, Rastafarians, Garveyites, Marxists, Black Muslims, and other radicals involved in the human rights movement (Gonsalves, 1979, p. 21). This was a rare display of unity; indeed a radical horse of a completely different color. One should remember that during the radical era of the 1930s, Garvey pilloried the Rastafarians; the Marxists disagreed with and distanced themselves from Garvey; the Rastafarians rejected and ridiculed the "poco-ism" (Pocomania) of the Bedwardites, and so on.

We mentioned earlier that education contributed to the expansion of the middle class, which in turn led to a more complex and vibrant class structure. In a related way, it operated to keep radicalism on the boil by taking advantage of the openness of the class structure and the relative absence of a commanding set of dominant ideas in the Jamaican society. This statement requires us to engage in some theoretical clarification concerning class ideology before proceeding with our analysis. As is well known, Marx and Engels (see the *German Ideology*) took pains to explain the thesis that the ideas of the ruling class are in every epoch the ruling ideas of a society. In other words, the ruling class in every epoch controls the means of material production as well as the means of mental production, which are geared to maintaining domination by that class. For instance, the middle class in a capitalist society would be beholden almost entirely to the ruling class, demonstrating most of this allegiance through its involvement in functions that reproduce capitalist relations.

This thesis does not properly fit the Jamaican case, because effective domination depends largely on simultaneously strong political and ideological apparatuses. Given an underdeveloped economy, the middle class has far more autonomy, a fact that does not allow us to assume the automatic cooptability and an intrinsic servility of the class.

The role of the radical elements of the young intellectuals at the University of the West Indies is partially defined by this historical datum. These individuals, some of them very gifted, engaged the politicians in debate, opened effective dialogue with the public, and employed such media as the *New World Quarterly, Abeng,* and *Bongo Man,* among others to spread a new message throughout the Caribbean as a whole. Others became more activist. Perhaps the better known members of this group are Walter Rodney and Trevor Munroe. Rodney, as we discussed above, identified with the new ethics of the Rastafarians, much of it finding its way into his political primer, *The Groundings with My Brothers.* He attempted to organize the brethren and the urban youth into a political force and is still remembered by many in the ghettos of Kingston for his enlightening lectures on African civilization. On the other hand, Munroe threw in his lot with the unorganized working class. His intention was to break the monopoly of the major unions, the BITU, the NWU, and the TUC. He spearheaded the University and Allied Workers' Union (UAWU), which later formed the base for the Workers' Party of Jamaica (WPJ).

Many of these intellectuals and their educated supporters were not the scions of an old, well-established middle class with a long tradition of dependence on the local ruling class. They were unlike their counterparts in the advanced capitalist economies or in some oligarchies and dictatorships where their functions are fixed and predictable.

Our young intellectuals were not to the manor born. They emerged from fluid class relations. The twin forces of industrialization and education had, in spite of their shortcomings, opened the way for the sons and daughters of civil servants, teachers, nurses, policemen, and small businessmen to enter the university. In a culture where large families were often the rule and scarcity of resources prompted selectivity in their distribution, the social mobility linked to education could not easily elude the realities and rigidities of the family structure. The lucky son or daughter who entered the university remained very much under the influence of the traditional family setting.

Of course, the ensuing expressions of support took on many forms, but we will confine our discussion here to the details of the newly evolving political forms. As we mentioned above, the aroma of socialism pervaded the air but its main strain had neither of the two political parties in its future. On the other hand, a reformist version, one that could accommodate the socialism of the PNP, was on parade. Among its leading lights were Norman Girvan (see also Chapter 1) and Arnold Bertram, a student activist who figured prominently in the Rodney Affair. Bertram's part in the debate between university activists and the government helped to point activism in a specific direction:

> Jamaicans need no prompting to recognize . . . arbitrary and inhuman conduct, neither does someone need to incite us to act on our convictions. As such, Jamaicans not only played heroes in organising, but persuaded many of their peaceful counterparts to join them . . . In a country that has produced a Bogle, a Garvey, a Henry, does he think we lack revolutionary fervour? So Mr. Prime Minister, to set the record straight, we are one on this campus. There is complete solidarity on this issue. (Gonsalves, 1979, pp. 13–14)

This mild rebuke embodied a significant political tactic. Placing the name of Henry alongside those of national heroes was a calculated act, one of partial psychological refocusing. And who is this "Henry?" He is the Ronald Henry who made an abortive attempt to lead the Rastafarians and the urban youth in armed struggle against the government. If revolution and the threat thereof are actions justified on the basis of "our convictions," where are the references to the Marxists, Buchanan, Hart, and the like?

Actually, this crucial omission stemmed from the politically acceptable and the preponderance of the democratic ideal. The temper of the times did not favor Marxism and the basic political philosophy of the newly educated and would-be leaders could more easily countenance and identify with a moral revolution, such as that led by Bogle, than a communist revolution. Though both involve the loss of life and the destruction of property, their fundamental premises are different. Bogle's moral revolution did not so much fault political institutions as it questioned and deplored the morality of politicians and political leaders. A communist revolution threatens these political institutions and with them the moral priorities of a Bogle. In a crucial sense, the political morality of Bertram was nothing more than an upgraded, articulate version of "Bogle-ism." Its purpose was twofold: to bring political outcasts and recalcitrants into more socially and politically inclusive arrangements by way of consensus building and to align the heretofore aloof University of the West Indies closer to the political arena. As Munroe (1981, pp. 73–74) correctly notes, the university was not structured for the involvement of students in active political life. The disciplines germane to politics were not taught until the 1960s.

Other young radicals approached fundamental political questions in a roundabout way. They did not directly favor socialism and were unimpressed by the prospect of professional politics. A politician was someone like a used car salesman—pedestrian and sly. These radicals approached their political duties as custodians of the public good. Their quarrels with government, as a whole, tended to revolve around questions of equity and political morality. Interestingly, it was in these areas that the new vogue Rastafarian ethics either stimulated or became integrated into revitalized formulas. As beneficiaries of the social and political products of the labor rebellion, they could identify with the plight and aspirations of the people. Some of them would "make safaris to the yards of deprived blacks as a kind of ritual of re-unification with the source—a source many of them might never have known" (Nettleford, 1970, p. 127). Others, during the course of interviews, spoke of their familial ties with these tenement yards.

In the main, the political predispositions of the young intellectuals surfaced unequivocally in their writings, especially their many offerings in *Abeng*. Our research shows that the interpretation of the slogan "Power for the People," though sometimes hopelessly impeded by the eclectic turn to their discourse, really betrayed two central tendencies. The first was the unsparing condemnation of government and its unfulfilled promises to the society. Both political parties came in for harsh criticism. These young commentators took advantage of their independence and education to offer opinions and ideas ranging from the polemical to objective research.

The second point is, however, the more interesting. The young intellectuals did not hold a brief for the PNP but they did have a strong affinity for its philosophical principles. At every opportunity they noised abroad the party's embrace of social justice. The JLP's handling of the

Rodney Affair and the differences with the university strengthened their identification with the PNP. Rodney, they claimed (quite correctly), advocated the cultural reconstruction of Jamaican society in the image of the blacks. His advocacy of Black Power was not tantamount to support for racism, as he called for the power to control one's destiny. In fact, he supported the principle of a multiracial society. The JLP government, they claimed, avoided making crucial verbal distinctions out of sheer bad faith, political opportunism, and reaction. Unlike the PNP, the JLP threatened academic freedom because the government attributed the Rodney Affair to elements within the University, although such political expressions are the inviolable right of the university as part of the wider polity. Government action was part of a program of repression that later banned certain political songs from the air, such as "Pharoah House Crash" in 1969 and "Beat Down Babylon" in 1971.

The JLP fell further out of favor when it invoked law and order and rehabilitated the old-style paternalism. Here the Rudie boys' continuous attack on the old patterns of noblesse oblige conflated with the young radicals' own dissatisfactions. State directed violence increased: for an eight-month period during 1967 and 1968, the press reported thirty-one incidents in which sixteen persons were immediately killed and seven others seriously or critically injured. And as the young radical railed against these activities in *Abeng* and other outlets, Rudie expressed his intransigence in the lyrics of popular songs and acts of defiance against law and order.

A sizeable section of the new educated elite argued that the old-style leadership contributed to "the politics of non-mass-participation" that characterized politics up to that time (Singham, 1968; Munroe, 1972). Norman Manley was severely criticized for implying that the electorate should follow him wherever he led. As we shall see, Michael Manley later steered the new leadership away from this minefield.

For his part, Hugh Shearer, the JLP prime minister, was sticking close to a suspect tradition. His statement that Jamaicans should have "a passionate belief and confidence in the stability of [their country]" (*Daily Gleaner,* November 25, 1968, p. 1) should have at least included the requirement that government also discharge its own responsibilities to the citizenry. Instead, the government practised gerrymandering and failed to update the electoral lists prior to the 1967 elections. This resulted in the disenfranchisement of some 100,000 new voters—the majority of whom were apparently committed to the PNP. It banned public meetings of political groups opposed to its policies, even the PNP, the properly constituted opposition. Pouring "distrust [upon] ideas and ideology and

[upon] those who carry and harbour them" (Nettleford, 1970, p. 218), the JLP seemed bent on yoking the newly educated to the traditional structures of paternalism.

Traditional political and economic arrangements were already under attack by local dependency theorists, as we discussed in Chapter 1. In such a climate, the principle of self-determination also surfaced as a point of contention. It did so on two levels, one raising the issue of national independence, the other questioning the representative nature of the political system itself. On February 28, 1963, as a senator, Hugh Shearer had launched the JLP's anti-communist, pro-United States policy. The government had made arrangements with the United States and Canada for the defense of Jamaica in the case of communist attack. These pacts were now viewed as tantamount to the surrender of sovereignty and the voluntary return to imperial subjugation. It seemed conceivable that the JLP government could invite or accede to foreign intervention in order to contain political action of the type that accompanied the Rodney Affair, Black Power, and the iconoclasm of the brethren. Parenthetically, the PNP drew less suspicion, as its new leadership showed greater tolerance for diverse political thought; its embrace of economic nationalism was also reaffirmed (*Daily Gleaner,* October 13, 1969), making the party sufficiently anti-imperialist to be considered progressive. The fact that Norman Manley took sides with the JLP on the Rodney Affair after initially criticizing government action, did not substantially impede the new leadership. This was so because of the party's new policies of redefinition and distancing as we will see below.

At the other level, self-determination emerged as a question posed in relation to black autonomy: are the people independent? The young radicals and the Rastafarians disputed "Pharoah's" (Shearer's) contention that the presence of a black governor-general (Sir Clifford Campbell), a black prime minister and black ministers of government indicated black autonomy. The Rastafarians effectively used their condition to dramatize their disagreement. They likened the black politicians and other prominent elements of the middle class to "Black Man Judases" bent on betrayal. From the movement's point of view, this condemnation was amply justified. The news media records are replete with letters and editorials from members of the black middle class subscribing to Shearer's (and indeed Norman Manley's) view of racial equality (Palmer, 1989).

Color and race were debated, not as the focus of bigotry but more in the form of a neo-Garveyism. It is ironic that the JLP might have contributed to its own criticism here. Doubtless the upsurge of traditional folklore, including the National Festival instituted in the 1960s, fed into currents of racial pride. It was Edward Seaga who shouldered the responsibility for rehabilitating these forms. The tone of these demands was rooted in the spirit of the Black Power Movement: "In Jamaica true Black Power does not attack white as white, brown as brown. All men are equal. The attack is on white, brown or black as oppressing the Afro-Jamaican and as an oppressive economic and social class."[12] This position is extremely close to mainstream Rastafarianism.

In the aggregate, the platform of these radicals rested upon two realities. First, the old notions of the common good were out of date. Change was demanded in keeping with many of the formulas they articulated; the JLP still clung to intuition as the way to formulate national policies in preference to formal reason. Second, their choice of political party was the PNP, though the process was indirect. The fact is that their ideas on equity, social justice, the protection of civil liberties, and respect for the person all fit rather well into the philosophical ambit of Manleyism. It is true that the economists among them took issue with the PNP's developmental policies, but these were the handiwork of Norman Manley whose *mea culpa,* offered during his farewell speech in 1969, helped to restore some faith in the strong moral principles of the party. And his call for some form of economic nationalism to redress prevailing social conditions brought his parting shot within acceptable philosophical limits.

If the JLP found it impossible to react in time to the changing politics, the new PNP leadership was not so hampered. Michael Manley's social democratic figure cast an ample shadow under which many of the new political ideas could safely rest. Indeed the new policy of the leadership enjoined the twin practices of redefinition and distancing. It would redefine current demands in light of an appropriate version of the common good (that is, as modifications to Manleyism) and carefully distance itself from the old PNP and the JLP: economic nationalism, closer associations with Black Power and Rastafarianism (Manley visited Ethiopia in 1969 much to the delight of the brethren), thereby bridging the old gap between leadership and the "sufferer."

The new politics was also about national reconciliation: the churches, the JLP, and other powerful groups were invited to forge a program to eliminate corruption in government, outlaw the practice of political patronage and victimization, and remove political violence. These actions came in response to the "natural tendency" of Jamaicans to be democratic—a notion that was used to ground democratic socialism; a notion that is in direct descent from Manleyism.

So, while Michael Manley and the party remained within the tradition of Manleyism, these acts showed him modifying its principles to meet contemporary demands. These twin processes had other major consequences for the mediation of class conflict. By 1975, the ideological orbit was wide enough to accommodate the communists, whose earlier resurgence took place against a thorough condemnation of the PNP's betrayal of socialism. (See Chapter 2.)

▶ The New Politics in Perspective

There was enough to the character of the new politics to suggest particularity. The entire class structure showed distinct tensions. However, the tensions that figured most prominently in the crisis—those involving the capitalists, the middle class, and certain sections of the subordinate classes—were not consistent with the destruction or substantial undermining of the socio-economic order. These tensions reflected changing patterns of capital accumulation and carried with them distinct suggestions about future class mediations. Clearly, the politics around the Rodney Affair showed that the state would be expected to play a more direct role. The nature of race relations embraced by Rodney and many of the subordinate classes really attacked the traditional paternalistic notions in which the capitalists were seen as benevolent patrons.

Leadership was also undeveloped. The radical intellectuals who emerged as a result of the Rodney Affair in 1968 betrayed many weaknesses. First, they did not constitute a political organization, mainly because there was not enough impetus from below. The rural areas did not participate in the rising political tensions in Kingston, and there was no discernible movement of any ideological wave flowing perceptibly to those areas. This situation is to be contrasted with Ethiopia in 1974, where organized protest by the teachers, students, and taxi drivers of Addis Ababa quickly fanned out into the provinces. Second, the lack of connection between political constituencies made these potential leaders a force that could guide at best, not lead. Their brittleness in the face of status quo opposition, as we saw, said much for the kind of politics they were prepared to spearhead—our focus on Arnold Bertram should be recalled.

Race appeared as an issue, but hardly in a disruptive or threatening way. The Chinese businesses that were looted during the Rodney Affair attested more to the existence of an ethnic consciousness with a local history, than class consciousness. The Seaga/Thompson election contest for the Tivoli Gardens constituency in Kingston in 1962 further strengthens our point. Blackness did not help Dudley Thompson, because, as Adam Kuper (1976) showed, performance and sincerity with regard to the plight of the small man's cause took precedence over color. This view betrays in large measure the Black Power position declared by Rodney himself (1970, p. 64): "All white people are enemies until proved otherwise, and this applies to black intellectuals, all of us are enemies to the people until we prove otherwise."

Results from the 1972 elections provide support for these views on race. Carl Stone (1974) states that while race and color were important, they were not at the top of the voter's list of priorities. Questions in Stone's poll geared to determine black/white animosity revealed no threatening patterns of hostility. The feeling that whites have oppressed blacks in the society was shared mostly by "the wage labouring classes of skilled and unskilled labour" and less so by "small peasant farmers, middle farmers and white collar workers." Attitudes toward employment also tended to reinforce the general point. Workers from the major occupational strata registered mostly a note of indifference to the racial characteristics of employers (Stone, 1974, pp. 78–80).[13]

Our analysis has clearly illuminated a politics of ferment. However, it was not revolutionary, though it is clear that substantial changes occurred—changes in political leadership, in the level of class conflict assisted by global ideologies such as the Black Power Movement and Marxism, and changes in the level of expectations on the part of the Jamaican people.

Charting National Popularism

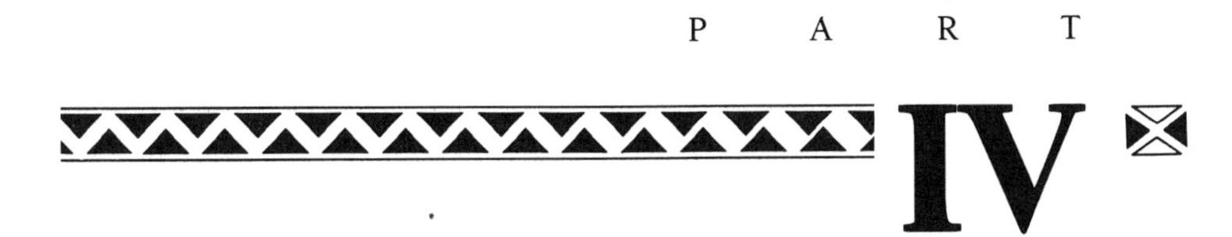

National Popularism and the State

CHAPTER 9

Our interview data support the interpretation that when democratic socialism was declared, most of its supporters saw it as the imposition of a more equitable distribution of resources on the existing structures. The reorganization of the bauxite industry, coupled with the huge levy extracted from it in 1974, served to deflect attention from the fact that equitable distribution had essentially to do with their interests. In a fit of myopia, many felt that new resources would be created and the question of redistribution would not arise.

The bulk of supporters of the PNP looked at the project opportunistically. The state promised new, productive opportunities, demonstrated its lack of fear of the big capitalists, and espoused an ideology of fairness to which the oft-maligned electorate could relate. It can be argued that supporters were actually encouraged in their opportunism and steadfastness to capitalism by the regime's definition of democratic socialism as an amalgam of capitalism and socialism—a reform of a known system, which would thus be more appealing (see Chapter 2).

In particular, a significant portion of the capitalist class quietly approved PNP policies designed to reduce North American influence in the Jamaican economy. In 1973, Stone (1973) informs us, 35 percent of the class abhorred the spread of the influence of the United States in the business world; our own interviews during the mid- to late 1970s found these sentiments very much alive.[1] While talk of state participation in the economy was suspect, the anticipated rise in government expenditures for economic and social projects was expected to provide new opportunities for involvement.[2]

Thus objective forces actually pushed in the direction of the restructuring of capitalism. Nonetheless, several factors contributed to ambiguous and contradictory ideology and practice by the regime: the motley composition of the PNP, the different class interests ensconced in various state apparatuses, the need to maintain the national popularist alliance, and the insertion of geopolitical issues in the final equation. Although we maintain that radical societal restructuring was not to be expected, the debate about socialism versus capitalism inside the state was quite real. It is also beyond doubt that even as stated by Bernal in the epigraph at the front of this book, the project of state monitoring of the capitalist mode of production would have altered significantly the relationship between the state and the capitalist class. We could expect, therefore, some resistance by social forces supporting the status quo. This chapter focuses on sources of pressure from outside Jamaica as well as from those in and around the state, while the next chapter discusses the regime's policies in the context of Jamaican social classes.

The project of restructuring capitalism required achieving a measure of control over economic processes and thus loosening economic control by the metropole (that is, the First World). The regime tried to accomplish this through what is termed "selective delinking" (Diaz-Alejandro, 1978). The strategy is associated with policies that diversify markets for goods and capital, promote commercial and financial links among Third World countries, socialist economies, and smaller advanced economies, break up import packages in order to foster economic autonomy and secure financial gains, and encourage entry into the marketplace at times favorable to the peculiar interests of the Jamaican economy. The tasks here are quite formidable.

Third World countries attempting to install such policies can expect fierce resistance from the capitalist world, especially when these policies are accompanied by a radical or socialist ideology. The examples of Mossadegh in Iran and Allende in Chile come to mind. We do not assume a priori, however, that the might of the United States and its hatred of socialism make the project impossible in the Western Hemisphere. Between 1974, the year the PNP declared democratic socialism, and 1980, the year the experiment failed, fourteen new radical governments were established. And while in many cases resistance from the capitalist world was fierce, including support for anti-government forces, these governments fared reasonably well. Our analysis, therefore, proceeds on the assumption that such regimes could be successfully installed.

We also assume, however, that Jamaica's dependence on North America would make delinking difficult. The metropole (of which North

America is a part) controls Jamaica's external financing and dominates trading arrangements. The attendant control over foreign exchange would give it ample opportunity to cut off the lifeblood of a new and competing set of economic arrangements. Acts of economic destabilization, of fomenting internal political violence and general socio-economic disruption, are common and have all been amply recorded in the case of Jamaica (Stephens and Stephens, 1986; Kaufman, 1985).

The question, however, is whether or not these external forces should bear the brunt of the blame for failure—as Michael Manley has continued to assert with regard to the activities of transnational corporations (Manley, 1987). We argue that the regime's success must be assessable in relation to policies inserted and nurtured in a milieu prepared for such resistance, in particular, one that boasts social relations such that support will not wane immediately in the face of the predictable metropolitan pressure. These supportive social relations, in turn, should include at a minimum a governing party that is solidly behind the policies and in control of key state apparatuses. That was not the case in Jamaica, as this chapter will demonstrate. We turn first to delinking and its difficulties.

▶ Economic Delinking

There can be little doubt that the attempted reorganization of the bauxite industry was the supreme act of delinking. The huge potential of the industry for generating foreign exchange was clearly the basis on which the new economy would be built. The process of delinking was organized mainly around the initiatives listed below.

Bauxite Initiatives and Their Intended Impact

Initiative	Intended Impact
State co-ownership of industry in partnership with North American transnational corporations operating in Jamaica	Reduce oligopolistic nature of industry, open it to local capital Increase and ensure greater availability of foreign exchange
State-sponsored joint ventures with regional neighbors: smelters, alumina processing, cement, other	Regionally/nationally controlled, vertically integrated aluminum industry Lessen dependence Change import patterns and utilize more local raw materials

International Bauxite Association (IBA) created	Wrest control of industry from transnational corporations Regulate income from industry and use it for internal developmental priorities Promote self-reliance
Diversification of credit: Hungary and the Soviet Union (1978); Nigeria, Iraq, Norway (1979)	Secure new lines of credit Foster bilateral trade, generate new business

Other state initiatives were also directed to delinking, such as state involvement in light manufacturing and an attempt to take over the formidable conglomerate, Grace Kennedy. The acquisition, which fell through when the necessary financial arrangements could not be made, would have meant significant control over key manufacturing and export/import operations. Additionally, the distribution of land to small farmers under Project Land Lease had as one of its main functions the reduction of imported foodstuff.

But the process of delinking became hopelessly strangled by several constraints. First, the joint ventures and country to country agreements aimed at transforming the bauxite industry had not been sufficiently sensitive to the internal politics of the participating countries. If the political and the economic climates were to change, so indeed would the fortunes of these ventures. Such was the case with Mexico, where internal political issues and changed circumstances caused the scuttling of carefully prepared plans (Manley, 1987, p. 248–250).

Second, the necessary trade realignments could not be easily achieved. On the surface, the data in Table 5 show a significant shift in trading partners. Undoubtedly, trade dependence seemed to have less-

TABLE 5

Diversification of Exports and Imports, 1970 and 1980 (in % of trade)

Type of Trade	U.S.		U.K.		Canada		Latin America		All Others	
	1970	1980	1970	1980	1970	1980	1970	1980	1970	1980
Exports	51.5	37.4	15.4	19.4	8.2	3.9	1.2	1.6	23.6	37.7
Imports	43.0	31.4	19.0	6.7	9.0	5.9	5.9	17.0	22.8	38.9

Source: Statistical Digest (Kingston: Bank of Jamaica, 1980), p. 108.

ened, as trading partners became more diversified. Yet, trade diversification could be considered a success only in terms of short-range political considerations. With regard to the broader economic picture, its potential long-term benefits could not be realized without immediate dislocations and bottlenecks in the economy. The difficulties involved in making the desired shift become obvious when we consider the economy's dependence on imports. Dependence had remained stable (at high levels) or increased during the period of industrialization, so that by 1978 the local manufacturing sector depended on imports, on the average, for 43.4 percent of its inputs. In some industries, such as tobacco and metal products, the dependence was closer to 60 percent (Ayub, 1981, pp. 60–61). Typically, the new trading partners receiving Jamaican exports would link them to imports from their countries. While barter arrangements—an interesting way to open up trade opportunities with soft currency countries—were hailed as successes, the goods that were exchanged were not always in line with pressing economic needs.[3] The conclusion is that trade diversification would require economic restructuring at a far faster rate than the economy was able to withstand.

The third difficulty with delinking involved the financial requirements of the economy. While loans from non-traditional sources were loudly acclaimed and widely publicized, the record shows an unmitigated reliance on Western (and capitalist) sources. Not for want of seeking elsewhere, the list of foreign lenders in the 1970s speaks only of continuity (Manley, 1987, pp. 280–281). The monopoly of funds available for foreign borrowing, in turn, limits one's options in the kinds of changes to be promoted: institutions such as the Bank of Nova Scotia, Chase Manhattan, Bank of America, Citicorp, and First Chicago Syndicate are unlikely to support delinking (with its attendant uncertainties) and, least of all, socialist change. While dealings by Western banks with socialist countries are not unusual, (Yugoslavia is a prime example) the temptation to dictate economic policies increases when the client is a small and dependent country, prone to destabilization, and in the shadow of the United States. The new practice of borrowing more from commercial banks than from traditional bilateral and multilateral sources meant shorter maturity periods, resulting in the regime's perpetual struggle to service these debts.[4]

Financial constraints also operated locally. Local banking and financial arrangements remained hemmed in by capitalist practices. This served to demonstrate that these state apparatuses could defeat socioeconomic transformation unless watched and monitored closely.

Finally, socialist solidarity was weak or missing altogether. As interviews revealed, the level of support received from socialist countries was a major source of disappointment: Guinea, for instance, appeared to have abandoned socialist cooperation for an increased bauxite quota and the Soviet Union's level of proposed assistance came up very short.[5]

▶ Ideology and Consciousness within the PNP

As was true at its founding, the PNP in the 1970s consisted of a coalition. It fielded a conservative wing, inhabited mostly by the 'old guard,' (Wills O. Isaacs, Allan Isaacs, Florizel Glasspole) and some younger blood (Vivian Blake and Eric Bell, for example); a moderate faction, by far the largest, which included such notables as P. J. Patterson, David Coore, Howard F. Cooke; and a left wing propped up by Dr. D. K. Duncan, Arnold Bertram, Michael Manley, Hugh Small, and Anthony Spaulding.

From the PNP's victory in 1972 until 1975, the relatively stable equilibrium of the factions within the party presented the mirror image of the class structure as a whole. The eight-fold increase in bauxite revenues (from $25 million to $200 million) in 1974 provided appreciable psychological and economic comfort. The business community was relatively satisfied with the removal of some restrictive policies. (For example, imports were temporarily relaxed). Many lower to middle income individuals faced the rosy prospect of owning a home through the National Housing Trust, whose enabling authorization was passed in March of 1976; worker participation was being studied for implementation; and the un- and underemployed continued to receive their subsidies, whether as Special Employment Programmes (SEP) or other forms of patronage.

The minister of finance, David Coore, painted a picture of continuing optimism during the budget debate on May 15, 1975. Education would receive an allocation of more than $40 million over the previous year to expand and improve the quality of facilities, training, and instruction. The budget for agriculture would also increase by some $20 million, to continue Project Land Lease and the Food Farms projects. Rural communities would benefit from an electrification project financed partly by a loan from the Inter-American Development Bank: the project would take light and power to some 170,000 persons at a cost of $75 million over four years. Clinics, health centers, and nutritional programs would receive an additional $4 million as well as new personnel with the hiring of community health aides. A cluster of projects in soil reclamation, land settlement, forest development, and water and river control, which would support special employment programs, was allocated $53.2 million, an increase of 53 percent over the previous year.[6]

These programs were expansionist and favored social equity but foreshadowed extensive overseas borrowing as well as an internal reallocation of resources. Because many of them were socialist in intent and non-productive, the class alliance would come under increasing strain: should matters of equity outweigh the necessity for growth? Economic and ideological factors came heavily into play.

On the first count, the economic problems went beyond the net outflow of foreign private capital that occurred in 1976, as international banks stopped extending credit to Jamaica: losses in exchange reserves amounted to $254 million for that year. Total nominal domestic savings, which stood at $175 million in 1970, fell to a negative $47 million by 1977 (Bourne, 1981, p. 64). As correctives, the government imposed new taxes on overseas travel, liquor, road traffic licenses and other preserves of the wealthy and middle class. Upper range salaries were frozen, as were rents until a more appropriate rent control system could be devised. Despite these corrective measures, the government was forced to seek the assistance of the International Monetary Fund. Before we deal with the role of the Fund, let us discuss the relationship of Michael Manley to Fidel Castro, which, along with the pursuit of equity economics, served to expose and ripen key contradictions within the PNP.

Even before the PNP's victory in 1972, Michael Manley had taken to providing ideological support for liberation movements and socialist regimes. For many PNP supporters, this was not given enormous credence, as in opposition a party is permitted certain excesses. In a sense, the role of the opposition is to be, politically speaking, attractively irresponsible! Michael Manley's visit to Cuba in 1975 really signaled the emergence of the party's contradictions. A power struggle surfaced within the party, as control shifted from the Central Executive to the National Executive Committee (NEC) and the Congress. The Central Executive, likened to a board of directors, had hitherto been the watchdog of the party, the seat of its conscience guarded by the conservative wing where founding members, the governor-general, for example, exercised enormous influence. The NEC/Congress entity was taken in by the left-wing ideological tendencies of the day. Up to this point, that is, 1975, it seemed clear that enough will had not been exercised to anchor the party ideologically to socialism. Apart from the confusion over its meaning, there was the obvious split within its ranks. Stephens and Stephens (1986, pp. 304–305) persuasively link this to the lateness at which the party's definitive document, *Principles and Objectives,* made the rounds—in 1979. Confusion reigned.

Innocuous statements made in opposition hardened into defiance, as Michael Manley courted Castro against the wishes of the Central Committee and its supporters. The Cuban visit had filled Manley with confidence and purpose. An NEC member, who did not support democratic socialism, volunteered that Manley returned from Cuba "sounding like Martin Luther King . . . been to the mountain top." That galvanizing zeal resulted in pronouncements of unequivocal support for Cuba and broadsides directed at the capitalists and the middle class.

Ultimately these events contributed to the disintegration of the class coalition that had emerged in 1974. For one, the middle class, which had always had a tendency to roam, started taking flight to North American cities of refuge: Miami, Fort Lauderdale, Atlanta, New York, and Toronto. This was largely the result of ill-considered pronouncements and tendentious and irresponsible journalism. The trip to Cuba enticed Manley to announce that under democratic socialism millionaires would be rare and that those who were not prepared to make their wealth the democratic socialist way should remember that "planes depart five times daily for Miami" (*Daily Gleaner,* July 16, 1975). The *Daily Gleaner* interpreted and pronounced this to mean the demise of capitalism.[7] This was particularly irksome to the bulk of the middle class, whose social mobility greatly depended on capitalism. Most of the powerful capitalists, on the other hand, especially those more severely affected by these policies—the merchant/commercial fraction—appear to have stood their ground. It was the ultra-conservatives from this class who invested less, exported capital, and even emigrated (Stephens and Stephens, 1986, pp. 94–95, and 100).

Before long, the alliance within the party began to break apart. Under severe pressure, the moderate and conservative factions helped to force a recantation, when Manley explained the party's Cuban policy in more acceptable terms, toned down attacks on imperialism, and distanced democratic socialism from communism (*Daily News,* October 4, 1975). The negative turn of events following the elections of 1976 increased the tensions, although they merely hastened what appeared to be inevitable.

Up to the 1976 elections (which the PNP won handily taking forty-seven seats to the JLP's thirteen) the regime was able to weather the storm. Election data showed that significant support came from the young voters, whose numbers increased by 267,000 between 1972 and 1976. A sizeable portion appeared to have gone to the PNP because of its identification with a positive African, pro-"roots" image. In terms of the economy, though, imaginative tinkering had run its course. As policies were introduced to push the economy further along a socialist path, the financing of these and other projects became more dependent on overseas capitalist sources, with the new pressures discussed above. The prime minister rallied the electorate with his celebrated "We Are Not for Sale" speech, delivered on the occasion of the 38th Annual Party Conference held on September 19, 1976, but the IMF was invited in to take corrective action.[8]

Manley continued nonetheless to pursue the socialist agenda. Along with the left wing, he was in control of most areas of government policy. Plans to acquire the Caribbean Cement Company and a number of commercial banks were announced in early 1977. The moderates and conservatives were apparently biding their time. The socialist economic alternative was embodied in the never released *People's Plan,* which was formulated between late 1976 and early 1977 and was modeled on the principles of "planning from below." Three essential elements were to be united: a cadre of radical economists mainly from the University of the West Indies; a Ministry of National Mobilization created after the elections of 1976; and the ideas and responses of the people.

The Plan failed in ways that weakened the coalition within the PNP and laid bare its fundamentally capitalist nature. First, the various apparatuses of the state, the Ministry of Finance and the Bank of Jamaica, for example, withheld total cooperation in ways that reduced the purpose and efficacy of the Plan. The reasons revolved around professional jealousy, as the radical economists and planners appeared to be "Michael's favorite boys;" the arrogance of "the boys" who were said to delight in pushing their weight around; and fundamental disagreements with either matters internal to the Plan or democratic socialism. Second, the capitalist and intermediate class elements began to withdraw support and fuel opposition, as the democratic portion of the democratic socialist label (meaning capitalist features) was being eroded by socialism. Third, widespread complaints were heard from below leading to the old charge of insincerity at the top, as it appeared that input solicited from the people, garnered largely by the new Ministry of National Mobilization, had not been treated with enough seriousness by planners and by Michael Manley himself.

Fourth, Michael Manley himself attacked the *People's Plan* for being unrealistic, with the observation that "Socialism alone cannot produce" (Stephens and Stephens, 1986, p. 166).[9] It is now clear that Manley felt the need to toe the line and support policies not of his own choosing. Capital formation was falling dramatically. Net fixed capital formation as a percentage of the GDP fell by 50 percent to single digit (7 percent) compared to 1975. In 1971, just before the PNP victory, that figure had been 18 percent (Boyd, 1988). There were ample signs that political support depended on economic performance. Now the capitalists as a whole began to restrict investment, exacerbating employment difficulties and threatening the political base of support of many PNP members. It takes little to realize that those who opposed democratic socialism were incensed by this destructive reality.

These dramatic developments emboldened the conservative and moderate wings of the party to go after the left. The faction had been reneging on its pledge of moderation on the issue of communism (*Daily News,* October 4, 1975). The visit of Castro in November 1977, combined with growing anti-communist sentiments at that time, placed the intra-party coalition under increasing strain. In spite of his appealing diplomacy, exemplified by his refusal to be drawn into local political issues, Castro's presence merely stirred passions. For the paranoid, he was there to ceremoniously drive the final nail into the coffin of Jamaican capitalism. It did not help matters that his visit had been preceded by that of Samora Machel, president of Mozambique, whose pronouncements were much more uncompromising and doctrinaire.

While such frenzy might not affect the more rational, there was sufficient grist for the mill in the actions of the influential PNP Youth Organization (PNPYO). The PNPYO advocated such immediate policies as the transformation of the judicial system into a socialist system of people's courts; the installation of a people's militia by restructuring the police force; the takeover of the media; and a general acceleration of communism (*Daily Gleaner,* September 2, 1977). The JLP had intensified its attack upon the PNP's socialist ideology and effectively utilized the PNPYO pronouncements to justify its accusations that the party had gone over to communism and was preparing to install a one-party state.[10]

Help soon arrived. With the sound of victory at the polls in December of 1976, the regime was able to temporarily resist the clauses of the earlier two-year Standby Agreement it had negotiated with the IMF. It argued that such IMF demands as a freeze on wages and appreciable devaluation of the currency (20–40 percent) were inconsistent with the people's mandate. By August 1977, however, the regime was forced to bend to the will of the IMF. The guidelines sought to undo the socialist agenda by imposing limitations on the line of bank credit to the public sector; placing ceilings on the domestic assets of the Bank of Jamaica (the nation's central bank); and placing limitations on the Bank of Jamaica's ability to contract new external debt for the public sector and insisting that debts contracted by the private sector be covered by government guarantees (*Economic and Social Survey, Jamaica,* 1981 6,6.1,6.2)

The powerful National Planning Agency, once tapped for ministerial status, was weakened. The economic package resulting from the IMF's intervention also reflected the directorate's diminished influence: actual state intervention into economic production appeared to be dropped, as "the intention of the Government," according to certain senior government spokesmen, "is to allow existing big businesses to continue to operate but . . . the thrust of Government policy would be to develop small and medium-scale enterprises, especially by providing financial assistance" (Chen-Young, 1977a, p. 6).

The IMF had set the stage for an intraparty showdown. The conservatives and moderates, who feared the combined power of the prime minister and the left wing, could now mount an offensive against the backdrop of IMF protection and declining popular support for the regime. Even the strongly pro-PNP capitalists registered their fear of the left wing (Stephens and Stephens, 1986, p. 161).

The IMF singled out Duncan, the most ardent socialist and the party's premier organizer, as the target. As minister of national mobilization, it was his responsibility to organize the people for socialism. He was relieved of his portfolio as well as of his post as general secretary of the PNP. The party did the rest. Catlike, the conservatives and moderates pounced upon their prey, while a confused prime minister was unable to do much about the carnage. On September 15, 1977, Duncan resigned, citing, inter alia, the lack of cooperation of fellow ministers in his task of mobilizing the electorate for socialism. In truth, this difficulty emanated equally from his unrepentant socialist posture and his perceived haughtiness. In the United States, the likes of Duncan are feared for their socialist revolutionary potential; in Jamaica this factor was compounded. Duncan, with his beard, dark glasses, revolutionary dress, and lack of respect for the traditions of the PNP, offended many. The governor-general, Sir Florizel Glasspole, one of the founders of the party falls into this category. Many ministers and backbenchers deplored his attempts to carve up their ministries in the name of national mobilization and could not tolerate his interference in their domain, as, for instance, when he countermanded their directions to organizers in constituencies or simply ignored the existing constituency infrastructure.

His eclipse also involved collusion. There is nothing like a palace coup, real, attempted, or rumored, to reverse political fortunes. As Stephens and Stephens tell the tale (1986, pp. 155–156), Duncan was accused of leading a plot by the left to overthrow Manley as leader of the PNP. "The rumor leaked to the rank-and-file activists and resulted in physical attacks on Duncan in several cases." Within the PNP, opinions of Duncan were divided. However, most of those with whom we spoke alluded to his superb organizational skills which, in the opinion of some, had gradually pitted him against his leader.

Duncan and the left wing were defeated at the PNP party conference in 1977.[11] During the ensuing party election of officers, the faction lost key positions within the NEC. To dramatize that capitalism had neutralized the challenge within the PNP, significant changes were made. The all-important Ministry of Industry and Commerce was given to R. Danny Williams, a very wealthy and successful businessman who figured prominently in the ensuing dialogue between the government and the business community. At his elbow was another prominent supporter of the capitalists, William Isaacs, who, as the new minister of labor, was obviously a compromise choice by the PNP to assuage the fears and misgivings of the capitalists.

The IMF kept up its support. In May 1978, the Extended Fund Facility Program underwrote funds of some US$250 million. The main conditions were an immediate 15 percent devaluation of the Jamaican dollar, with an additional 15 percent over the first twelve months of the program; a revised price and income policy calling for maximum wage increases of 15 percent over the first two years; increased taxes amounting to $180 million for 1978–1979; reduction of the net domestic assets of the Bank of Jamaica, which should increase at the rate of 16 percent for the end of the first year of the program; increased liquidity ratios of commercial banks (40 percent) and the imposition of a ceiling on commercial bank loans and advances to the private sector. The socialist agenda was further undermined, incentives such as wage increases to the working class, were removed, while the stock of the capitalists (measured in increased imports) improved.

Let us look at the inner coherence of the PNP as a political party. Most political parties are correctly seen as self-perpetuating oligarchies. The PNP and, for that matter, the JLP are no different (Bradley, 1960; Gannon, 1976). This characterization can, however, be misleading, as the element of self-perpetuation masks their intrinsic class base and the competitive quality endemic in its structure as an apparatus of a capitalist state. There should be little confusion here. Even non-Marxists recognize that the party feigns internal harmony based on a cause it is supposed to hold as an ideal (Miller, 1962, p. 80). Tocqueville (1945, I, p. 175) reminds us that for political parties "private interest, which always plays the chief part in political passions, is more studiously veiled under the pretext of the public good; and it may even be sometimes concealed from the eyes of the very persons whom it excites and impels."

The diverse composition of a party ensures a shifting equilibrium within it (Beckford and Witter, 1982, p. 158). That one faction or factions gain a hegemonic position at any given time is due less to consensus than competition that can at times be quite vicious. Factions within political parties struggle for leadership based, concretely, on strategic support (Poulantzas, 1974, p. 72). Further, political parties often present a comparatively united front in relation to their diverse compositions because of a variety of compulsions, expectations, and structural requirements. One pressing compulsion has, obviously, to do with gaining and/or retaining political office. The very existence of a party relies largely on feigned unity (except in the case of a mass party). If it is unable to resolve or contain its own contradictions and inner tensions, it can hardly begin to effectively mediate class conflict.

Let us look at these basic theorems empirically. When democratic socialism was announced, symbolic and, less so, rhetorical unity were evident. Old three-piece suits were proudly exchanged for the kareba (bush jacket) that was also in keeping with the burgeoning "roots" culture. But the wearers of the kareba had not abandoned the fundamental interests they represented: the hegemony of the left wing and a conjuncture favorable to radicalism were preemptive. In fact, Eric Bell, while attired in his kareba, volunteered that his training as a lawyer forbade the adoption of socialism; another leading socialist engaged in real estate dealings with a powerful capitalist that had more to do with profit maximization than with the directives in the party's *Principles and Objectives*. It is also doubtful that the rate of conversion to socialism among the PNP elite was impressive. Their economic activities since the fall of democratic socialism suggest that they had not acquired the necessary freedom from the taint of selfishness.[12] One is strongly persuaded that the same consciousness of opportunities for betterment exhibited by the bulk of the middle class and the subordinate classes is present here. The only difference is that scarcity pushed the latter into declaring their ongoing capitalist inclinations earlier.

From the very beginning, the fragmentation of the PNP, which surfaced boldly toward the end of 1976, was predictable. The hegemony of the left wing and the favorable conjuncture into which it was inserted were unstable and would be throttled by the constellation of forces ranged against them. The ouster of Duncan and the emasculation of the left wing resulted from economic, ideological, and structural causes, most of which will be introduced below. Within the structure of the party, limited though the analysis might be, Duncan's difficulties were linked to the internal leadership struggle, ideology, and the left wing's authorship of policies now affecting the constituencies of the other factions. Duncan's bedside manners were an issue quite inferior to the others. How could Eric Bell and Ken McNeil tolerate his left wing mobilizing in their constituencies when they were avowed supporters of capitalism? Should they not heed the complaints and threats of the middle class that felt that the responsibility for the have-nots was placed on their backs because of the new rash of taxation introduced by the government? A prominent city lawyer expressed these widely held sentiments in an interview:

> I had to work through many hardships to get where I am. Let them [the masses] go through those experiences. Handouts do not build character. . . . I have always supported the PNP; in fact, they wanted me to run for a seat. I cannot vote for them and I will not vote JLP. I have not given them any more contribution.

Just as leftists like Hugh Small and Anthony Spaulding saw the momentary economic dislocations and the struggle with the IMF as positive for the decline of imperialism and the fall of capitalism in Jamaica, some conservatives and moderates felt personally responsible for what they termed a tragedy (*Daily Gleaner,* November 6, 1977; June 6, 1977). Preeminent among the latter group was Sir Florizel Glasspole, the governor-general who, by 1979, took to openly opposing the PNP in public speeches. He was to openly defy the prime minister by rejecting the party's list of Privy Council nominees; in one case, he substituted a prominent founding member of the PNP like himself (Vernon Arnett) with a member of a very prominent capitalist family (Leslie Ashenheim). As if caught in the throes of a death wish, others began to leak Cabinet secrets to the JLP, which put the information to good use especially after 1978. For instance, the JLP was apprised of the nature of the negotiations between the IMF and the regime in 1977. It was able to indicate the level of devaluation—40 percent—insisted on by the IMF.

The emergence of the conservative and moderate factions from the hegemony of the left wing accelerated during the three-year period of IMF control of the economy (1977–1979). Concretely, the IMF's policy, elaborated below, reversed the policy of relief for the subordinate classes as a whole. Over this period, the government was compelled by these pressures to introduce policies to decrease the level of real wages; initially these losses to the worker were offset by subsidies, which increased from $44.1 million in nominal terms in 1976 to $157.9 million in 1977. But the IMF would not relent, so that by 1980 support was decreased dramatically (Boyd, 1988, pp. 103–104). The compound of declining wages and withdrawn subsidies threw leftist politics into a tailspin, as the displaced and the dislocated began to fit themselves into well-tried, if not often successful, capitalist practices: the informal sector (own account workers) grew at the rate of 20.1 percent over this period, much faster than that of the labor force (12.4 percent). In these trying times for democratic socialism, PNP advisers saw their desertion as distressing indeed

(Beckford and Witter, 1982, p. 159). What is more, desertion within the party was matched by withdrawal of middle class support in general.

▶ The Middle Class and the PNP

A Stone Poll (*Daily Gleaner,* August 6, 1980, p. 7) clearly shows that whereas 81.4 percent of the middle class voted for the PNP in 1972, only 40.2 percent did so in 1976. Our interviews confirm that loss of middle-class support was one of the consequences of the preferential treatment the PNP accorded to the subordinate classes. And while, after 1976, less was done to halt the erosion of labor's base, the middle class joined other classes in interpreting working class militancy as unreasonableness and lack of patriotism.

Providing for these social classes was acceptable until middle class interests were infringed upon or underrepresented. In 1974 the bauxite levy went substantially into a middle class pasttime—consumption—which accounted for over $100 million, while foreign borrowing provided the funds for the programs geared to the subordinate classes. When the trend in consumerism was halted and taxation imposed, partly to service the foreign debt, the middle class found it expedient to attack the government for its policies toward these classes, who were now seen as an enormous burden upon its unwilling shoulders. From this tension over resources surfaced sharp ideological differences, as the common denunciations of the class emerged from hiding: the subordinate classes suffer from a "freeness mentality," are shiftless and lazy, are parasitic, need to be taught the virtue of hard work and discipline, and so on. Hardly the stuff of which harmony is made; certainly the grist for increased instability and brittleness in class relations.

There was more to the loss of support than the issue of resource distribution and the fear of communism, however. Strong antagonisms swirled around the increased power—attempted and real—of the strategic fraction of the middle class that was promoting democratic socialism. In the heady days of 1977, before the IMF agreement, Manley declared openly that the business community should have a junior role in the economy (*Daily Gleaner,* February 8, 1977). This statement caused quite a stir, although this position was in keeping with PNP economic policy that, from the beginning, had given the state sector the leading role in the economy. The capitalist class, while divided on some issues which are discussed in Chapter 10, was of one mind with regard to allowing the state the power to determine the course of the economy. Winston Meeks, president of the Chamber of Commerce, made this an important portion of his address at the Chamber's semi-annual dinner on February 27, 1975:

> The other aspect that could also tickle the funny bone if it were a laughing matter, is that [the PNP government] ends up creating the same sort of structure that it purports to smash—elitism. That is, it replaces what is now called the business elite to which anyone can aspire . . . [with] a political elite which is by far much more powerful and dictatorial. . . . Recent history has shown the type of repressive elite that this situation has created. Fortunately, in our case we have the opportunity of aborting it before it is fully grown and ready to be born, thrive and grow. (Jamaican Chamber of Commerce, mimeo.)

Members of the bureaucratic/entrepreneurial fraction had always been prominently placed in the state bureaucracy; now they were thrust in the forefront of economic reconstruction. This role heretofore had been reserved for the expatriate consultant, aided by members or functionaries of the capitalists. It would now be filled by the National Planning Agency and the Technical Advisory Committee, which were among the first creations of the PNP government in 1972:

> [These put] together some of the key elements in the civil service like the Governor of the Bank and the Financial Secretary, the Head of the Planning Unit and put with them a couple of first class minds from the private sector and a couple of first class economists and so on and so forth, who have the business of generating ideas and plans and also translating into action the political input that comes from the official Economic Planning Council, a sub-committee of the Cabinet. (Hearne, 1976, pp. 114–15)

Among this elite group were former members of the New World Group and black, radical economists who had been engaged in the power struggle at the University of the West Indies during the 1960s.

This presence of the "university educated" brought to a head the tension between education as a "means of production" and the more traditional forms of power. Members of the strategic fraction of the middle class were seen as dictatorial. Education had allowed them to bypass the stringently guarded socialization rites of the capitalists: close social intercourse, productive business interaction, and a shared capitalist ideology that went beyond the principled support of capitalism to the embrace of specific ideological motifs of the class. The difficulties associated with acceptance did seem to vary, however, depending on whether the members of the fraction were engaged in business, members of the new political directorate, or members of the directorate and socialist in orientation.[13]

The new-found power of the fraction and its predisposition to wield it over society were alienating. This disaffection touched all classes, though the degree to which this occurred was not quantified. Even members of the educated middle class—the fraction's natural ally—were on occasion taken aback by its arrogance: "A friend of mine at the Bank [of Jamaica] told me that G. Beck [Dr. George Beckford, a member of the Technical Advisory Committee and the chief architect of the People's Plan] came down there demanding sensitive documents. Power is gone to the heads of these fellows." And while some of the traditional middle class celebrated the rise of the fraction, others saw its members as "the fellows from the University" with little "practical knowledge."[14]

Also put on show was the elitist conception of "government by the best brains." During 1978 and 1979, the educated middle class presented the controversial notion that the nation's current difficulties could be solved only with the help of the best educated located in government.[15] This provoked a range of responses from the other classes. In March of 1977, the Chamber of Commerce stated its belief that the fraction would find it "a source of gratification" were the private sector to collapse. These "Ph.D.'s with beards" were "a dangerous reactionary element" (*Weekly Gleaner,* July 10, 1977, p. 6). Merchants made good their promise to frustrate the political directorate of the bureaucratic/entrepreneurial fraction. Orchestrated shortages, business closings, and layoffs of workers in 1977 served to turn popular opinion against it. After all, the present plight of all social classes could be directly attributed to its failed policies! The initials IMF acquired new meaning: "Is Michael's Fault."

Working-class representatives registered their own condemnation:

> There are others, mostly among the intellectual middle class, who have not yet quite fitted into any of the power status left vacant after Independence, . . . This group constitutes a dangerous reactionary element. They cannot reconcile how Ph.D.'s can have less influence than trade union leaders. Hence, armed with all skills of statistics and attractive argument which are related to present social and economic condition and expressed in grossly over-simplified terms, they seek to organize the masses in order to take their self-conceived rightful place in the body politic.[16]

There was one more compelling reason to doubt the sincerity of the fraction. The occupational group from which it sprang was footloose. For the period 1973 to 1981, the group increased at an average annual rate of 5.7 percent. At the same time, the group experienced the highest rate of emigration—1,900 per annum or 32 per 1,000 workers (*Economic and Social Survey, Jamaica,* 1982, p. 17.5). Their commitment to democratic socialism rose and fell with government-sponsored opportunities. With characteristic verbal flourish, Jamaicans, reeling under the IMF austerity guidelines, blamed the regime for the perilous state of their economic condition and their decision to emigrate: IMF here translates into "I Must Flee."

▶ The State, the Party, and Change

Our analysis of intraparty conflict and desertions on the heels of the crisis orchestrated by the IMF should support the point made at the beginning of this chapter: that antagonistic metropolitan actions should be expected and that social relations should be such as to enable the state to withstand the inevitable political confrontation and concomitant economic destabilization while waiting for alternative economic structures to yield their first fruit. Clearly, social relations in Jamaica would not allow the required breathing space. There is, nonetheless, significant literature that would argue that but for the IMF, socialism could have been installed; this compels us to address the topic once more.

Let us, for the sake of argument, allow the point that the PNP was a revolutionary party. Ralph Miliband (1969, p. 53) correctly observes that even in the case where a revolutionary party has to function within a liberal democratic framework, we should expect it to play by rules "that are not of its own choosing." State apparatuses display "the characteristic features of rigidity and resistance" (Poulantzas, 1980, p. 232) that result from the interplay of bureaucracy and the dominant ideology. This holds true in all historical cases; Germany after World War I is a good example (Haffner, 1972).

There are several scenarios under which the state may pursue revolutionary aims. Closest to the Jamaican situation is one in which a party takes office after a substantial electoral victory in a quasi-revolutionary environment. Of course, as Miliband further reminds us, "electoral support of left-wing parties . . . [may] suggest a high degree of popular availabilty for extensive and even fundamental change" . . . but not necessarily "popular revolutionary fervor" (Miliband, 1969, p. 99). Even this model, however, is hardly appropriate for Jamaica, since the nearest we can get to a revolutionary situation—albeit with some charity—involves the events surrounding the elections of 1976. Popular support was certainly high but a concomitant requirement is the insertion of the party into conditions of institutional breakdown inherited from a prior regime.

In reality, regardless of whatever the IMF and other factors may have contributed, it was the PNP that was seen as the major architect of the disastrous downturn in economic conditions and precipitous decline in the standard and quality of life after 1976. These sentiments were amply reflected in the dramatic loss of broad-based support indicating that socialist fervor was wearing thin by the late 1970s. A poll conducted toward the end of 1978 showed that 40 percent of the electorate had no preference for either the PNP or the JLP. In October 1976, just after the prime minister's famous "We Are Not for Sale" speech, that figure had been 15 percent (*Weekly Gleaner*, November 20, 1978, p. 11). Likewise, by 1979 the PNP's party group structure had diminished from 2,000 groups to about 500 (Beckford and Witter, 1982, p. 99).

Miliband's "high degree of popular availability" for fundamental change appeared momentarily between 1976 and 1978 but was not sustained and its impact seemed marginal. Carl Stone (1980), George Beckford and Michael Witter (1982), and Evelyne Huber Stephens and John Stephens (1986) suggest that the requisite quality of popular support for the PNP existed, in spite of the IMF sanctions. Be that as it may, the general situation did not favor fundamental change. Table 6 on the performance of the economy under the regime is quite instructive. The state of near total economic breakdown conveyed by the figures for manufacture, construction, distribution, and (until 1977) mining might indeed be grist for revolutionary fervor—all other necessary variables being present.

TABLE 6

Real Growth Rates of Gross Domestic Product for Selected Industrial Sectors, 1975–1978 (% change)*

Industrial Sector	1975	1976	1977	1978
Agriculture, forestry, and fishing	1.8	1.1	3.0	9.7
Mining and quarrying	−20.2	−20.6	17.5	2.5
Manufacturing	2.4	−4.9	−11.6	−5.4
Construction and installation	1.3	−20.0	−20.8	3.6
Distribution trade	2.8	−18.4	−7.9	−1.4
Government services	5.5	15.9	6.8	4.8
Total growth of gross domestic product at constant prices	−0.4	−6.3	−2.4	0.3

*At constant prices

Source: National Income and Product, 1982 (Kingston: Government of Jamaica, Department of Statistics, 1982), pp. 26–27.

However, such an argument would beg the question of who should lead the revolutionary charge against this state of affairs, when the most likely candidate for leadership—the state itself—is seen as bearing the bulk of the responsibility for it.

In proposing that the party had revolutionary potential, both before and after the fiasco, politicians and analysts seem to anchor their position to a piece of Lockean rationalism: the left wing could, starting with the opposition within the PNP, reason the main segments of countervailing ideological and political thought into change. Lost sight of is Lenin's astute observation that revolutions must be made. These calculations also omitted his insight that a political party that seeks to pursue an unpopular or extremist agenda must be prepared to employ legal as well as illegal means to gain and/or retain power.

▶ State Apparatuses and Change

State power, as the PNP discovered, has peculiar constraints. At its source is that resistant cement produced by the dominant economic forms and ideology that binds state apparatuses to the imperatives of a particular mode of production. For example, civil servants, military personnel, and police are by subtle and direct ideological methods made to identify with and support the system. Some of these methods derive from status, privilege, power, and esprit de corps.

State power must be carefully distinguished from class power. It does not follow logically that the one shifts in marionette fashion to follow the dictates of the other. This occurs only if there is substantial compatibility between the two bases; even so, the process is not automatic, as state personnel can promote, inhibit, or curtail the execution of policies. Perhaps the best contemporary example of the class power/state power dissonance occurred during the Chilean revolution, where the class support Allende received from the coalition of political parties was no guarantee that all the state apparatuses would fall in line with the policies of the regime. In fact, certain key apparatuses of the state (banks, the judiciary, and law enforcement) sabotaged the policies at every turn (Debray, 1971, pp. 88–89). In Uganda, Milton Obote was forced to turn to Idi Amin and the military when state bureaucrats began to undermine the policies of his regime (Lofchie, 1974, pp. 489–92).

The great expansion of government activities that accompanied the installation of the PNP did much, of course, to promote the recruitment of like-minded personnel to perform key functions. The PNP also used to advantage the well-known ruse of creating parallel organizations when it was known that a particular state agency was likely to resist its policies. Such was the case, for instance, when the Trade Administrator's Office was bypassed in favor of the State Trading Corporation. Efforts to create alternative channels within the state were matched by similar efforts outside. The all-important task of molding public opinion was entrusted not to the *Gleaner* but to newly created or purchased organizations: a newspaper (*Jamaica Daily News,* founded in 1973), a radio Station (RJR)—which commanded 79 percent of the total radio audience, and a television station, the Jamaican Broadcasting Corporation (JBC). The increases in public expenditures tell the story: between 1973 and 1976 public expenditure as a percentage of the GDP rose from 21.2 to 35.2 percent (*Economic and Social Survey, Jamaica,* 1973, and 1976). Government subsidies increased; for instance, the subsidy to education increased from $6.7 million in 1969 to $26.0 million in 1974 and $43.5 million in 1976 (*National Income and Product, 1976,* Account 2).

State power does often operate as a transformer of wills. However, the implementation of programs to match attractive political pronouncements may sorely test the sense of professionalism and dedication of new recruits. There were recruits by the Manley regime who quickly became disenchanted when faced with such difficulties. Two such instances will illustrate the point. The first to tell of his experience is a current World Bank official: "I was in Canada at the time. They recruited me to help to work with the sugar cooperatives. I could not stay. What they failed to realize is that a business isn't based on just giving." A former director of finance, also recruited from Canada relates:

> The project [Cornwall Dairies at Montpelier, St. James] was to receive $900,000. When I left the position [for the United States] expenditures were over $2 million and still running. It was a mess. I remember asking . . . at the JDB [Jamaica Development Bank] how they determined the volume [of milk] to justify the price supports. . . . Do you know what he said? "Talking to the producers I got a gut feeling."

Our interviewees buckled under "the basic structures of domination which had been left intact" even in the thick of socialist transformation (Kaufman, 1988, p. 53). In all such cases, we detected a tension between the ethical commands of democratic socialism and its economic configuration (or lack thereof). For while the notion of social justice for the poor surfaced repeatedly, it was confounded by subtle and direct support for the intrinsic inequalities of social class and the ongoing legacy of scarcity. How far are bureaucrats prepared to go in support of these changes when they suggest attacks upon their vital interests? An interviewee talked to us about his experience in the Trade Administrator's Office:

> When I brought my Volvo down, a fellow there told me that I would have to pay 150 percent duty. There was another fellow standing close by listening. After the first one left, this other fellow told me to meet him in the parking lot. We talked and I got the car in for $500.

Our experience suggests that such practices were widespread. The soft underbelly of government service provides subsidies of a sort for salaries and wages ludicrously low compared to those on the outside.

State power also influences bureaucratic behavior. Those "fringe" bureaucrats who accumulate capital as capitalists are not immediately concerned with the maintenance and reproduction of the political and economic order. Not so the state bureaucrat who is socialized to maintain and reproduce the political and social order. Perhaps the commitment derives mostly from self-interest—a quality that compels concern with stability and feeds the peculiar conservatism of bureaucracies (Michels, 1915, pp. 185–86). Furthermore, it is doubtful that political neutrality ever plays as determinate a role as survival and the strengthening of the state apparatus that is one's preserve (Horowitz, 1982, p. 221). The point is brought home when we consider economic growth rates. Between 1950 and 1968 the Gross Domestic Product (GDP) grew annually at an average of 6.8 percent at constant prices. In the following period, 1968 to 1972, average growth was 7.5 percent. All this was reversed between 1973 and 1978, when the GDP fell to a negative figure, −2.1 percent. Any of the factors mentioned above could serve as incentives for state bureaucrats to undermine policies they associate with this negative performance. We detail below some cases of dissension internal to state apparatuses. The incidents are important enough to dispel any notion that they might be exceptional. We do not want to leave the impression that Manley was not aware of these potential dangers; he devoted thirty-nine pages in *The Politics of Change* to the issue (Manley, 1974, pp. 162–201). It is clear, nonetheless, that the necessary preparations for change were not made.

Central Banking

Predictably, strains surfaced early between the government's push for change and hidebound banking practices. As Adlith Brown (1981, p. 5) informs us:

> In February 1975, . . . while the Bank of Jamaica reduced the bank rate and relaxed monetary policy, as a continuation of the policy of gradual reflation, sections of the government and the ruling party inveighed against private capital and threatened non-wage earnings. While the actual measures introduced between mid-1974, after the bauxite levy, and early 1975 suggested a widening of the scope for economic activity, official political statements promised restriction and insecurity of earnings.

Here the banking bureaucracy was informing the government that its distributive policies were inconsistent with the needs of an economy requiring immediate private sector support.

These disagreements led to open defiance. Beckford and Witter (1982, p. 149), themselves prominent economic advisers to the regime, suggest sabotage and indirectly allude to party factions working destructively at cross purposes:

> The most crucial year of that period was September 1976 to September 1977. It was in September 1976 that Manley and Jamaican people declared unequivocally that WE ARE NOT FOR SALE. WE KNOW WHERE WE ARE GOING, at a mass meeting in the National Stadium. Unknown to the Jamaican people, the then Minister of Finance (David Coore) . . . had made a firm commitment to the [International Monetary] Fund at a September meeting in Manila, Philippines. And between September and December 1976, Bank of Jamaica officials held secret meetings with IMF officials deliberately "disguised as tourists" on the north coast of Jamaica.

Was Michael Manley speaking out of both sides of his mouth? No. At this time, while the left wing and Manley enjoyed political and ideological ascendancy, the moderates and conservatives controlled the institutions responsible for economic linkages with local and foreign capitalists. Intraparty rivalry had effectively disarticulated the politico/ideological aspect from the economic aspect. It was the power of the economic aspect that by late 1977 had reoriented large sections of the middle class and portions of the subordinate classes to capitalism; it was this that emboldened David Coore and his clique to make promises running counter to the political and ideological declarations from the left wing. Bureaucrats holding strategic positions within the Ministry of Finance, and at the same time staunch supporters of the JLP, had figured prominently in the task. For many from the left, this signaled a huge deficiency in the role of the party in maintaining commitment to integrity. The so-called "Pickersgill Commission" which was instituted to carefully screen role players for toeing the party line, had clearly broken down. A properly functioning commission would have weeded out such detractors.[17]

Coore delivered the package the IMF requested, resigned his portfolio as minister of finance, then took a position with the Inter-American Development Bank. The tradition of capitalist realignment continued under his successor, Eric Bell, another moderate conservative. He, too, left the government for a post with the World Bank after the PNP and the Cabinet had rejected the IMF's interim agreement in March 1980.

Banking and economic policies also suffered from predictable blows dealt them by the capitalists, who were informed by the government, for political reasons, about all major economic policies, but would regularly join cause with foreign capital in undermining the government's agenda. As we were informed by Richard Fletcher, an economic adviser to the prime minister, this meant, for instance, that the Private Sector Organization of Jamaica (PSOJ) was able to influence the conditionality of foreign debts.[18]

The regime was caught in a vicious circle. As long as its economic and banking policies were dependent on the existing capitalist framework, it was compelled to caucus with the capitalists, first for structural reasons, then because of its continuing commitment to convince the class that democratic socialism was also capitalist! The capitalists, especially the merchant/commercial fraction, were not bound by any rules of reciprocity.

Government Agencies

Bureaucrats in the various government ministries could not always be counted on to follow and uphold government policy. An investigation of the Trade Administrator's department—a crucial agency in the government agenda, since it issues licenses for imports—revealed that items imported under license as "unfinished goods" were really meant for direct sale to the public. Products imported "for export" were sold locally. The most notable scandal involved the denial of an import license to the State Trading Corporation and the subsequent grant of a license to two private-sector trading houses to import the same material at a higher price (*Weekly Gleaner,* June 18, 1979, p. 31; April 16, 1979, p. 2). There were cases of public enterprises that were overseen by persons with interests in private-sector businesses and who were not supportive of the mixed economy, such as Ariguanabo Mills, for instance (*Weekly Gleaner,* December 24, 1979, p. 11). The Jamaica Broadcasting Corporation (JBC), a government-owned entity, also displayed a degree of recalcitrance, balking at the suggestion that media presentations should be supportive of the regime. Its head, Dwight Whylie—relieved of his post in 1976—based his resistance on Manley's declared policy of a free press.[19]

The Armed Forces

As the enforcers of law and order and the legitimate possessors of instruments of repression, the police and the military are crucial to the maintenance of any state. The PNP had generated goodwill by providing substantial support for both the police and the military. Between 1972 and 1979 public expenditure for the armed forces increased from $7.220 million to $86 million (Bell and Gibson, 1978, p. 10). The increase was largely a function of a dramatic rise in the incidence of crime and violence but benefits came in the form of increased salaries, added personnel, and improved equipment. On the surface, steps were taken to pander to the armed forces at least in part because of their strategic importance.

But in a climate of mounting violence, where right-wing, centrist, and left-wing ideologies as well as sheer opportunism flourished, the element of mischief became more intense. The government could not maintain its grip on the levers of control, since the loyalty of the armed forces themselves was questionable. In the words of Keble Munn, the minister of national security, there appeared to be "a deliberate, organized, attempt to destabilize" the government from within this body (*Daily Gleaner*, December 7, 1976). To compound matters, the armed forces gradually took the view that the government had a direct hand in repeated attacks upon its personnel, in which many had lost their lives (Stephens and Stephens, 1986, p. 133). The PNP intensified the alienation by creating citizen groups to provide security at the community level and at political meetings—an action that could certainly be interpreted as designed to undermine the army and police. Thereafter, the armed forces threw in their lot with the JLP: "The PNP's own analysis of their defeat [in the October 1980 elections] identifies intimidation of their organizers and their voters by elements of the security forces as a principal factor" (Beckford and Witter, 1982, p. 128).

The alienation was compounded by an event with even deeper significance. On June 23, 1980, three civilians and twenty-six members of the Jamaica Defence Force were detained for an alleged plot to overthrow the government. The alleged plot was connected to fears of a special paramilitary unit created by members of the PNP, the "Brigadistas." According to a memo allegedly leaked by Anthony Spaulding (minister of housing) to the PNP National Executive and published in the *Daily Gleaner* on July 23, 1979, this unit had been formed in 1975 in order to develop a group of "conscious cadres." It was widely alleged that the unit had actively participated in quelling the gasoline strikes of January 1979 (*Weekly Gleaner,* January 22, 1979, p. 11). It was also widely known that Jamaicans were being sent to Cuba for training and that a people's militia existed under that regime. In the public's mind (and presumably those of the alleged plotters as well) this indicated creeping communism: were the PNP and the regime preparing to adopt Leninist political tactics and replace the police and the army with a people's militia?—asked the author of a letter signed "Thomas Paine," which the *Gleaner* published seven days before the arrests (*Daily Gleaner,* June 16, 1980). It was evident that the support or at least the neutrality of the armed forces, both of which were crucial to the regime, had been severely undermined. Yet the government tended to act in ways that suggest its ignorance of the perils on its path. As always, Manley's public reactions to breaches in support for the PNP agenda resided in voluntaristic appeals. Patriotism was used lavishly: in his broadcast to the nation on January 5, 1977, for instance, he had encouraged the building of a country "of self-reliance on ourselves and determined to win for all our people their rightful place in the sun." There was continuity rather than a break with his strong conviction, stressed earlier in his book (Manley, 1974, pp. 91–92), that a global patriotic response was logically part and parcel of the transformation of "states of mind" to be effected "from above."

Maintaining strategic control over the armed forces required either their effective subjugation or restraint by a broad and firm popular alliance. With regard to the first, the Brigadista Program (taken as a realistic response to the objective realities) came about too late. Indeed, it was sabotaged by elements of the party's left wing—attesting to the lack of solidarity within the vanguard. As for the latter, the relative ease with which the armed forces intimidated the people spoke less than eloquently of the existence of "a broad and firm popular alliance."

In fact, the absence of a unified revolutionary consciousness at this time was conceded by those having the greatest interest in the subject—the communists. As the Workers' Liberation League (WLL) put it, the proletariat was only "beginning to stir" and ask "dangerous questions." The communists went on to admit that the working class was "only beginning the long and difficult process of organizing in its own interests and the interests of a revolutionary democracy" (*Socialism!* 1, no. 6, [December 1974]: 14). At the same time, the lumpen and the sub-proletariats, whose reliability in the best of times was suspect, were opting more and more for self-service as the popularity of the government declined (Chapter 10).

▶ Concluding Note

Our analysis has begun to reveal the fundamental weaknesses of the state's agenda for change. It was not only that, as Clive Thomas (1988) put it, the regime's policies were largely ad hoc and rationalized later to match its ideology. The more fundamental difficulty was that the principal hub around which the transformative agenda revolved—delinking and state power—needed the support of a class structure strongly committed to transformation. In the case of a socialist agenda, this would mean, for instance, that popular forces were well situated in key state apparatuses or sufficiently strong to build viable alternatives—and prepared to make the necessary sacrifices. Instead, we witness a state divided and a national coalition steeped in opportunism rather than revolutionary fervor, whose brittleness was revealed precisely when support was most needed. At all crucial junctures the project was impeded by intact capitalist structures and relations.

Our analysis supports the thesis that changes within the Jamaican class structure were animated mainly by changes in the patterns of capital accumulation. Capitalist relations and structures were never profoundly threatened.

Safeguarding Class Alliances

Resource Distribution under National Popularism

CHAPTER 10

Previous chapters demonstrated that a socialist transformation was not possible for Jamaica in the 1970s. Rather, what emerges when we focus not on immediate utterances and possible intentions but on consequences at the deeper, structural level, is an attempt to restructure the capitalist economy. This chapter takes a close look at PNP policies in terms of specific classes and addresses the implications they carry for the restructuring of socio-political and economic relations. The following analysis will involve a discussion of the limits inherent in these policies that, in large part, explain the difficulties of the PNP agenda and the amorphous and brittle nature of national popularism.

▶ The PNP Regime and the Capitalists

We have stated that PNP policies were generally meant to mediate the crisis and further the capacity for capital accumulation of the economy. In this sense, then, policies favored the ongoing interests of the capitalists. This should not be confused with immediate and unmediated interests, however. If that were the case, we would have to explain why so many took to flight with their capital, or why they attempted, at all turns but especially after 1976, to undermine the regime. These contradictions are clarified if we understand that social relations operate simultaneously at two interrelated levels: the level of basic, everyday working and understandings (the social formation) and the level where these basic, everyday workings are objectively related to the systemic processes of production, distribution, and circulation (the mode of production).

People live their lives on the first level. For capitalists, this is where decisions affecting their interests are made against the backdrop of short-term statistics on trade, relations with trading partners, the hostile or congenial nature of ideologies, and the like. By habit, capitalists are mostly influenced by quantitative data—especially the profit and loss columns of their ledgers. Nonetheless, reference to capital *as a social relation* can provide an objective assessment of their actual conditions. This perspective involves an understanding of the important qualitative processes (economic, political, and ideological) that give rise to the creation of value from which the capitalist's profits are derived (Sweezy, 1968, p. 338). It is at this level that one speaks of the state's role in safeguarding capital accumulation. The capitalist's continued ability to mobilize the forces of production so as to create value and, in turn, extract profits, depends not just on narrow economic considerations but on broader social relations. When the government, for instance, introduces a policy to increase minimum wages, the potential consequence—regardless of immediate intent—is to increase the market for goods produced by the capitalist. Thus we see that the state's redistributive function often serves the fundamental interests of the capitalist class—although the class may not immediately appreciate this fact, and, indeed, not all its members may benefit.

While some PNP policies directly benefited the capitalist class overall—though some fractions stood to gain more than others—they were more in tune with the need to promote capital accumulation, operating therefore less at the level of the social formation than at that of the mode of production. This was a complex undertaking, involving not only the relationship of the state to the capitalist class but also, as we discuss below, such factors as the search for new ways to reduce the ever-present tensions with the lumpen proletariat and the working class.

Capitalism often expands through the workings of an ascending, dynamic fraction bent on eclipsing a once-dominant fraction. As we saw in earlier chapters, but for the interests and support of the British colonial government, the planter class would have suffered its demise much earlier than in the modern period. In turn, its continued dominance acted as a brake on the development of more dynamic avenues for capital accumulation, such as manufacturing. By the 1970s, the PNP government would set for itself the task the Colonial Office had carried out so well during the Crown Colony period: supporting one fraction of capital (the industrial fraction) at the expense of another (the merchant fraction). Promoting capital accumulation, then, was far from synonymous with safeguarding the interests of the merchant fraction.

As we have discussed, the economic control exercised by the merchants had become a major stumbling block. Merchant capital could not create "jobs that last long"—the need, expressed first in the 1938 rebellion, that for many had not yet been fulfilled through the modern economy. Agriculture had carried the burden of providing the bulk of employment (such as it was) until the 1960s but the costs of continuing this pattern were mounting.[1] The pull to industrialization was, therefore, strong. PNP policies were clearly directed at increasing the local manufacturing capacity, an agenda that would be realized through an alliance with industrial capital and the provision of support for small, productive, businesses. Henceforth, the merchant and the distributive sector would be relegated to the role of junior partner.

State-sponsored projects, the reorganization of the bauxite industry (with Mexico and Venezuela to invest in industrial operations) and projects revolving around infrastructural development were outside the direct organizational sphere of the merchants. They were within the ken of industrialists—as the prime minister declared in Parliament on July 26, 1972 (Hearne, 1976, pp. 114–115)—those whose roles "have a direct bearing on economic development," and whose organization, the JMA, would be involved in meetings with the appropriate ministers "on a regular basis to exchange and know of the ideas or the services supplied by the planning agency and its experts."

Merchants and the Chamber of Commerce were excluded from the PNP agenda. A parliamentary debate on November 9, 1972 left few doubts about the government's reasoning. The prominence of distributive activities in the economy had triggered excessive food imports, an "orgy of consumption of imported luxury goods," a pattern of personal savings inconsistent with the "development of the productive sectors of the economy," a "failure to develop an industrial base which was related to the processing of the basic raw materials [to provide manufactured goods]," excessive reliance upon inflows of imported capital, and failure to modernize export agriculture (Hearne, 1976, p. 126).

The regime had much more in store for the merchants. The banking system would be restricted in the provision of loans "for financing the distribution sector and personal loans." These policies were directed mainly at those "among the wealthier classes seeking only to maximize profits as quickly as possible." Salesmen were threatened with partial obsolescence: "there may be a reduced demand for certain categories of employees, for example, salesmen, but if you think of the shortage in other areas [it is] of the sort that a certain kind of salesman can fill." (Hearne, 1976, p. 125)

The message was clear. Early in his tenure, Manley had sent a signal that curbs would be placed on distribution. At that time, consumer imports were met largely through capital imports, particularly out of proceeds from the mining industry and overseas borrowing. Table 7 provides a picture of the foreign exchange situation that dramatizes the dilemma. We should note that the 50 percent increase in spending on imports between 1971 and 1973 was not the result of an expansion of the manufacturing base. The number of companies operating under incentives actually declined, from 192 in 1971 to 177 in 1973.

Such dramatic increases in the foreign exchange drain spelled an eventual disaster. The PNP imposed quantitative restrictions on imports in order to promote the development of local industry. There were 158 items on the restricted list in 1968, 201 in 1973—the increase reflecting, in part, the beginnings of world economic stagnation—and 334 in 1979 (Ayub, 1981, p. 33). Formulated in the same spirit were policies to protect the local industrialists against political and commercial risks of nonpayment as well as to give preference to exporters and so-called linkage industries.

Perhaps the first substantial move against the merchant/commercial fraction was the formation of the Jamaica Nutrition Holdings in 1973. This state-owned agency became responsible for the school nutrition program, which involved importing foodstuffs and other materials. This effectively reduced the scope of private-sector commercial activities. The merchant's loss became greater when this state agency was absorbed by the State Trading Company (STC) in 1977, which was granted exclusive license to import medical supplies, grains, and other staples.

The beneficiaries were the industrialists who, by 1970, were regarded by the communists as the "big national bourgeois" (*sic*) who exercised "a big influence over government policy" (Miller, 1976, p. 5). Indeed, Mayer Matalon, the architect of the bauxite levy, was known to be Manley's unofficial financial adviser. How the industrialists were expected to benefit from state policies will be discussed below.

There were limits, however, to the ability of the state to reduce the power of the merchants and at the same time propel the industrial frac-

TABLE 7
Foreign Exchange Drain, 1971–1973 (millions of $)

Foreign Exchange Flow	1971	1972	1973
Foreign exchange spent on imports	64.4	63.6	97.3
Foreign exchange earnings from exports	28.1	27.2	24.4
Drain on foreign exchange	36.3	35.9	72.9

Source: Daily News, June 7, 1978, p. 3.

tion forward. For one, the immediate result of PNP policies was a considerable increase in government spending, which led to solidified links between capitalists—especially the distributive sector—and the state. By the late 1970s roughly 32 percent of the total public expenditure went toward the purchase of goods and services. We discuss two aspects of these limits: insufficient differentiation between merchant and industrial capitals; and difficulties of the state sector in fulfilling its anticipated role of economic leadership. Other limitations resulting from the state's policies relative to the other classes, especially the lumpen and subproletariats, are discussed in later sections of the chapter that pertain to those classes. Suffice it to say for now that the allocation of scarce resources to the various classes became an increasingly intractable problem.

The Merchant and Industrial Fractions of Capital

The preferential treatment the government accorded industrial capital created tensions within the capitalist class but nothing of the rupture that signals the decline of the existing socio-economic order: while the two fractions compete when in pursuit of divergent interests, they will unite against a common enemy. During the 1940s, all three fractions of capital (planter, merchant, and industrial) formed the Jamaica Democratic Party to thwart Bustamante and Norman Manley. During the current crisis, merchants and industrialists joined in the Private Sector Organization of Jamaica (PSOJ), formed in 1976, and fabricated shortages and staged layoffs in a joint effort to topple the government in 1977.

Fractional antagonism alternated with class collaboration. The role of the Prices Commission to curb excess profiteering provides a good example of antagonism. Both fractions were distinctly alarmed at the import quotas. However, while the merchants expressed added indignation at proposals to phase out particular lines of imports (those ultimately taken over by the State Trading Company), the industrialists maintained a studied silence on this score. In fact, the more progressive industrialists even showed substantial antipathy toward the merchants. This attitude was marked in some of our interviews with JMA executives and members. Attacks on import substitution would be followed by another recurring theme: the need to remove manufacturing and management from widespread family control. There was particular distaste for the old patterns of taxation and incentives that impeded productive industrialization. Overall, the bias was evident, though it was clear that the organization entertained different shades of opinion and that at times private utterances and public positions would differ.[2] Thus it was that the JMA joined cause with the other major business organizations to oppose the STC but "STC officials on more than one occasion indicated that some members . . . had offered their support privately." (Kirton and Figueroa, 1981, p. 163)

Hegemony was also an issue. In July of 1977, the manufacturers negotiated special price margins in the absence of the merchants. "Manufacturers—countered the Chamber of Commerce—[should] . . . consult with the distributive sector when there was a need for increasing prices."[3] Further, the Chamber noted, "if members of either the Chamber or the Manufacturers' Association were suspected of arbitrarily increasing prices without consultation, the complainants should submit their complaints to the Chamber which should then intervene on their behalf."[4] The records indicate that the Chamber attempted to establish "closer collaboration" with the JMA but without significant success.[5]

Members of the industrial fraction were among the last elements of the class to desert the regime. The *Emergency Production Plan* of 1977 projected investment opportunities in garments, textiles, furniture, metal fabrication, construction materials, and food processing. New and expanded investments were earmarked for synthetics, leather manufacturing, plastics, sulfonic acid, and aluminum sulfate. Induced by these prospects, the fraction could continue to play a waiting game.

In fact, by 1977 the forces ranged against the PNP's socialist policies had not yet rendered the prospects of the industrialists completely unavailing. There were 143 small business projects in the pipeline, calculated to employ 831 workers.[6] Likewise, producers of construction materials were somewhat encouraged by the state's plans to construct forty-five housing projects valued at several million dollars. The IMF-sponsored *Emergency Production Plan* also provided industrialists with an indirect level of support, since it gave them some space in a way that the alternative *People's Plan,* prepared by University of the West Indies economists, did not. The *People's Plan* deemphasized medium to heavy industrialization because of its heavy foreign exchange input; it projected decreased economic activity and a loss of about 700 jobs; and it envisioned an industrial thrust directed principally at small manufacturers, who would utilize local raw materials.[7]

The affinity between the industrialists and the regime had its limits, of course, which were reached at the slightest hint that the government might be pursuing socialist relations. The Jamaica Manufacturers' Association (JMA) provides an excellent example. During the thick of foreign exchange shortages, which purportedly led to shortages of consumer goods and layoffs of workers during the weeks before the *Emergency Production Plan* of April 22, 1977, Hungary had offered a line of credit for

$8 million. It was not accepted. The president of the JMA gave his reasons (*Daily News*, April 10, 1977):

> We have tried to deal with them but found it difficult. There is the question of delivery time. You have to order way in advance. There is the matter of communications and language, and quality is a serious problem. To import equipment from them would be a disaster as we have always bought from Western sources.

There would indeed be difficulties in dealing with the Hungarians but they were certainly minimized by the economic crisis and, especially, the level of misery and inconvenience precipitated by foreign exchange shortages. The truth is that while the fraction favored rearranging the formulas for the generation and distribution of the surplus, it was not prepared to go so far as to establish relations with socialist governments in pursuit of its aims.[8]

In sum, there was no question of a threat to the interests of the capitalist class as a whole. The issue of intraclass distribution of resources resulted from the dynamics of non-antagonistic contradictions—contradictions that did not threaten the overall domination exercised by the class. Nonetheless, when dominance rather than continued existence is seen as the issue, it is clear that in the end the odds would favor the merchant/commercial fraction. The fact was that, short of the state succeeding in its project—which, as we have seen, was extremely unlikely—circumstances would favor established patterns. Thus it was the merchants that benefited from the "great absent partner's" support for capitalism writ large, as executed by the IMF and similar institutions. Equally, the state's increasing role in generating demand, extending subsidies, providing employment, and so on, shored up the fraction's fortunes. Between 1973 and 1976, 82,400 new jobs were created compared to 30,600 for the preceding twelve years, and real median income increased by 28 percent (Bernal, 1984, p. 63). Regardless of the government's stated intentions, much of the activity linked to this growth fed into existing structures of production, circulation, and distribution. Finally, given the lack of support of lending institutions, as discussed in Chapter 9, the stage was set for the undermining of the state's role in resource procurement and distribution. Particularly after 1976, with increased competition for resources, the state's relative autonomy developed visible fissures as the merchant/commercial fraction was increasingly able to reassert its dominance.

In practice, then, the economic stagnation and downturn that began to deepen in 1976 put to rest the possibility that the envisioned economic transformation would succeed. The efficiency and viability of the old structures were severely impaired but, far from dealing them the coup de grace, entrenched governmental demand management partly ensured that the old patterns were not completely forsaken. PNP policies highlighted the intrafractional antagonisms within the capitalist class but in the end did not succeed in transforming the basic class formation. Manufacturing actually declined by some 10 percent annually between 1974 and 1978 (Ayub, 1981, p. 46). What accounts in large measure for the unwanted conclusion of the contest between merchant and industrialist was the PNP's failure to transform government revenues into productive capital—and thus the failure of the state sector to become a dynamic force in the economy.

The State Sector and Economic Leadership

According to the PNP script, the rise of the industrial fraction was intimately connected to the rise of state economic power. For the state, as we have seen, this meant a much more interventionist role than had heretofore been acceptable. The role was clearly exceptional and, as the capitalist could well imagine, could easily stray across the boundary to the forbidden land of socialism. Perhaps most abhorrent was the specter of a state that was no longer the suitor of capital, forever standing at the door wooing its master with attractive gifts.

The experience of the merchants with the State Trading Corporation demonstrates that the state could, in spite of assurances to the contrary, remove resources traditionally controlled by the class. When the State Trading Company (STC) was created in 1977, its managing director, Dexter Rose, assured the business community that the corporation was a boon, since it had removed the most risk-ridden area of business from the merchant. In fact, the STC did remarkably well, succeeding in breaking the merchant's monopoly on trade. Between 1977 and 1978, it created three new subsidiaries—Jamaica Building Materials Ltd., Jamaica Pharmaceutical and Equipment Supplies Ltd., and Jamaica Textiles Imports Ltd.—which had threatened to eclipse the merchant. Their arrival signalled the obsolescence of the so-called "20/80 rule": initially, the STC controlled only 20 percent of the total value of discretionary imports while the remaining 80 percent went to the traditional business community. By the end of 1977, however, the STC was empowered to handle all imports. If these were the prescribed limits of the role of the "junior partner" under democratic socialism, as the regime defined it during that very year, then there was precious little for which the merchant/commercial fraction should be thankful.

One could argue that a substantial infrastructure remained capitalist and yet continued, as such, to receive PNP support. One could also suggest, as we have, that PNP policies deserved support, since they went a long way to promote capital accumulation. Yet, it went against the grain of capitalist institutions to support the local variety of state intervention, which clearly did not conform to standard capitalist ideology and practice.

These limitations were combined with the increasing difficulties the state experienced in obtaining resources for development purposes when, in the face of economic crisis, allocations had to go toward recurrent government expenditures. These were also linked to the reaction of the capitalist forces—internal and external—to the perceived threat posed by the *People's Plan*.

The economic activities of the government were thus bounded. The functionaries of the financial infrastructure would continue to insist that the government play by traditional capitalist rules. Let us examine the role of the Jamaica Development Bank (JDB) within the framework introduced by the Bauxite (Production Levy) Act and the *Green Paper on Industrial Development Programme*. The Bank was especially important since it controlled the Capital Development Fund (CDF), the repository of specially earmarked funds from the bauxite levy. As these were instruments of the new agenda, we might expect them to signal significant modifications in the Bank's operating procedures. Yet, the figures for 1974–1976, the period that witnessed democratic socialism at its most effective, show that the JDB played a limited role in development. This occurred when the Bank's notional accumulated funds for development purposes stood at $257.5 million—a relatively extensive source of investment.

It is no secret that the JDB operates mainly as a capitalist institution. Its major providers of funds—the World Bank, the IMF, and the Inter-American Development Bank—compel it to employ standard capitalist practices. Indeed, the *Green Paper on Industrial Development Programme* (p. 12) made it clear that the JDB would not provide funds for the state's industrial thrust. These biases were confirmed when, in early 1976, the Bank successfully resisted the government's attempts to use development funds to support state initiatives in agricultural development.[9]

Restraints imposed by the JDB were further augmented by the restrictive practices embedded in international borrowing. For instance, both the United States and Canada attach conditions to their loans that reinforce prevailing capitalist relations, as they require that 70 percent and 66 percent respectively of their loans be in the form of goods and services from their economies. This is an example of one of the many features enabling capitalists "to maintain the authority of the rules of the free market." (Robinson, 1979, p. 94)

And yet there was sufficient evidence that the PNP was not bent on totally destroying the capitalist foundations of the economy. The Industrial Incentives (Regional Harmonization) Act of 1974, for instance, was designed to develop interlocking links between the Jamaica Industrial Development Corporation (JIDC) and other parastatals. The legislation did enlarge the scope of the JIDC, but it did not alter its basic functions, which centered on administering incentives and inducements to foreign capital. The JIDC now worked closely with state-sponsored industrial projects and the Small Businesses Development Agency (a new entity created in 1977). The premises underlying this expanded role fell squarely within the law of value: to maximize profits through expanded and more efficient production. As we will discuss further, the state-sponsored joint ventures with foreign capital and foreign countries also operated substantially within well-tried capitalist principles, in spite of written declarations about local priorities. The key to the Harmonization Act was the offer of traditional incentives. Tax holidays were provided for ten years, with maximum income tax relief of 50 percent. In order to stimulate the use of local products and labor, incentives were tied to the local value-added. Foreign capital was still needed and policies were still designed to favor investors when they entered the economy.

The Urban Development Corporation (UDC) also played an important supportive role. Directed by a member of the Matalon family—one of the wealthiest families in the country and the most prominent representatives of the emerging industrial fraction—the UDC channeled significant amounts to the construction industry. At the end of 1976, it received $1.63 million. Planned activities fell within two broad categories: traditional infrastructural projects designed to stimulate investments, primarily in tourism and manufacturing, and socially motivated redistributive projects. The Portmore Housing Development in Kingston was one of the projects designated by the UDC and developed by the Matalons. The complex involved 2,000 acres of land with some 30,000 homes; it also boasted a network of factories and warehouses linked to the mainland by a 2.3 mile bridge.

The anchor to which these activities were attached was a reorganization of the banking industry, which, however, failed to materialize. The downturn of the economy and declining popularity of the government after 1976 were visible impediments, but structural limitations were the more significant barrier. As Reid (1977, pp. 22–23) argues, both options realistically available to the state—outright nationalization and a substantial takeover—are problematical. In addition, traditional business formulas and codes were too ingrained to be easily dislodged. Members of the business community welcomed government intervention only in the good old-fashioned way: to promote their own effectiveness and profitability.

At any rate, partial takeover of the industry (far more likely than nationalization, given the circumstances) would still leave intact the widespread banking policies that revolved around low-risk, short-term investments. The shift toward long-term investment was slow, due to the prevalent merchant mentality. A practice observed in the agricultural sector throughout the mid-1950s and the late 1960s exemplifies the problem. Government expenditures through such initiatives as the Farm Development Scheme (1955–1960), the Agricultural Development Programme (1960–1965), and the Farm Production Programme (1963–1968) found their way back into small manufacturing and distribution (Girvan, 1971, p. 110). While there was an objective misallocation of lending to distribution and personal consumption, the trader mentality of which Michael Manley spoke deprecatingly in his book *The Politics of Change* remained a large part of local investment calculations. Finally, the takeover of a limited number of commercial banks would not result in structural change. In fact, it was not even seen as high-handed or socialist. Many members of the business community felt these changes would be an improvement (Chen-Young, 1977b, p. 3).

Ultimately, the failed transformation of the economy must be linked to the difficulties the government experienced in turning a substantial portion of the bauxite levy into capital investments. First, the total sums earmarked for the Capital Development Fund (CDF) never materialized. Instead, transfers to the Consolidated Fund of the National Budget took an even greater portion of the funds, increasing from 50.3 percent to 73.1 percent from 1974 to 1976. Some 35 percent of the remainder of the bauxite revenues made its way into government revenue through investments in government securities. Only 18 percent of the combined funds from the levy was committed to the Jamaica Development Bank for potentially productive ventures (Reid, 1977, p. 99). Technically, nearly 70 percent of the CDF went into government securities, while only 18 percent was used for productive investment—a far cry from the original plan that called for allocations of 52 percent and 31.9 percent respectively (Capital Development Fund, *Annual Report*, 1976). The pattern continued: for the four years after creation of the CDF (FY 1975/76–1979/80), transfers to the Consolidated Fund of the National Budget increased at an annual average of $38 million or 53.8 percent.[10] This was a period during which government revenue from the industry had decreased, as the mining transnationals shifted production to their other subsidiaries to register their opposition to democratic socialism.[11]

Far from being the exceptional answer to unusual circumstances (at the time, the government's agenda was being steadily derailed), this escalated recourse to distributive (non-productive) policies was required to shore up the foundations on which the regime's legitimacy depended—providing sufficient palliatives for all classes in the national alliance. In this case, it was the popular classes that needed to be appeased. Indeed, we argue that here is a key factor enabling us to separate national popularism from democratic socialism—the degree to which the alliance around the state is maintained or threatened according to the state's ability to provide for all special interests, as opposed to a more solid alliance that is able to withstand the unavoidable hardships that accompany radical restructuring.

Resources formerly budgeted for productive ventures found their way into basic government operations. In turn, these fed into the existing patterns of circulation and distribution controlled by the merchant/commercial fraction of the capitalists. There are three principal reasons for this pattern of deployment: to comply with the ongoing nature of capitalism in the Jamaican economy (recall that the IMF intervention that spanned this period had unevenly favored the interests of this fraction); to mobilize popular support (for the fiscal year 1974/75 the regime created record levels of growth rates in current and capital expenditures that were tied to consumerist-driven incentives); and to maintain relatively viable levels of resources for the state's management functions.

Reinforced capitalist practices also came from two other sources. First, as the economic downturn gained in severity during 1976, the state was forced to rescue failing enterprises. By July 1976, for instance, it owned half of the tourist industry. The second set of factors was tied to the growth of the mass market. A growing mass market is essential for a wealthy and vibrant capitalist class. The PNP regime contributed to this pattern of growth in three distinct ways: by expanding state expenditures to meet increased employment in the state sector; by indirectly raising disposable income through enacting a national minimum wage and increasing National Insurance Scheme (NIS) pensions (both instituted in 1974); and by improving and expanding trade with other countries (Harris, 1976, pp. 282–305).

Capitalist interests are also served by certain state interventionist policies glibly termed socialist. It is well known that certain types of investment are not attractive either to foreign or local capital because they are too risky or unprofitable;[12] as Peter Evans (1979) states with regard to Brazil, these investments do not provide the investor with a "comparative advantage." The PNP regime took over many such businesses—sugar and public utilities, for instance—and contemplated others that the investor spurned. By mid-1979, via participation in industry, rescue operations, and the investments mentioned above, the state came to own just under 57 percent of the local productive capacity (*Weekly Gleaner*, June 11, 1979, p. 9) Such policies complement the capitalist in at least two major ways. First, they maintain the productive base of the economy, in that these investments are kept in an integrated system. Second, as public enterprises continued to operate according to capitalist practices (except for sugar, which was organized into cooperatives), the old profit-making linkages were not seriously affected.

Clearly, in the end, the state could not succeed in its drive to restructure the capitalist economy under its leadership. By the late 1970s the capitalist class seemed to be in control again and an appeased president of the Chamber of Commerce was able to state: "In all fairness, the government had now recognized the private sector as an integral part of the economy" (*Weekly Gleaner*, July 10, 1978, p. 8).

▶ The Working Class and the PNP Policies[13]

The PNP promised the subordinate classes more jobs, land, a decent living, and a democratized society in which they would be full participants. As Michael Manley put it to the PNP/NEC on September 18, 1975, "We believe that Parliament is the foundation of democracy but recognize that parliamentary democracy is only a beginning to the democratic process. To democratise a society you have to experiment in forms and institutions which bring decision-making to the broad mass of the people." (*Daily Gleaner*, Oct. 5, 1975, p. 8) The primary avenues for fulfilling the promise were the new industrialization policies, Project Land Lease, and worker participation. Cooperatives were envisioned for agricultural workers in the sugar sector that came under government ownership.

The *Green Paper on Industrial Development Programme* estimated that working class jobs would grow by 31,545 by the early 1980s (p. 84). The estimate seemed a reasonable improvement over the past, given that 30,600 new jobs overall had been created between 1961 and 1973. These new jobs would be developed from the broadened productive base created by the new agenda. Additionally, all jobs would be covered by the Minimum Wage Law (1975)

Complementary social programs included day care centers, nutrition subsidies, school feeding programs, free education, free school uniforms, free health care, special drug windows (to facilitate access to medical drugs), and subsidized purchase of foodstuffs from Agricultural Marketing Corporation outlets. The working class also benefited from the creation of the National Housing Trust (NHT). In March 1976, a compulsory national savings plan was instituted based on a 3 percent payroll tax added to a 2 percent share of gross earnings to provide housing principally for the working class. Improvements in the living conditions of the class were anticipated from plans such as "Build on Own Land" and "Home Improvement." By 1977 there were also plans for forty-five housing projects at a cost of $62 million. Of course, discrepancies appeared at the level of execution. For instance, even reasonably priced housing schemes was beyond the reach of much of the population. In one much publicized case, 1,000 government houses remained unsold, without any takers in sight (*Weekly Gleaner*, May 22, 1978, p. 3).

Support for the PNP by the class appeared to follow the promise and decline of such redistributive initiatives. The initial response was strongly favorable: in the 1976 elections, the PNP received 61.6 percent of the "lower class" vote, versus 43.3 percent in 1972 (*Daily Gleaner*, June 8, 1980, p. 7). However, as we recount below, much of the support had waned by the end of the decade.

Improving the lot of workers and achieving successful mediations between capital and labor, however, required an even more comprehensive approach. These interclass mediations had traditionally been achieved through the links between the political parties and labor unions, clientelistic relations of subordinate classes with the state, and the common tools of social control. As we saw, these channels of mediation were becoming less effective in the 1960s. At the deeper level of the mode of production PNP policies would address the need for new forms of mediation.

It is in this sense, then, that employment and redistributive policies may be considered secondary to two policies directed at the relationship between workers and capitalists: worker participation, unveiled and subjected to discussion between 1975 and 1976, and its kindred policy of creating sugar cooperatives, and the Labour Relations and Industrial Disputes Act of March 1975 (LRIDA). Worker participation never got off the ground, scuttled as it was by a combination of class resistance and, after

1977, IMF restrictions and economic pressures. Nonetheless, the debate around its implementation makes for an instructive study. The sugar cooperatives fared only a little better, as pilot projects were undertaken by the Frome Monymusk Land Company. The company's 70,000 acres produced some 28 percent of the industry's overall output. Taken together, these initiatives speak of the attempt to graft the politics of consensus onto a relationship that was replete with tensions. Both raised questions about the traditional role labor unions and political parties had played in mediating these tensions and set out alternatives intended, true to the "family model," to protect the interests of both capital and labor and curb any excesses on either side. Both, needless to say, were controversial.

Worker participation would curb excesses by muting the distinction between workers and owners. As Michael Manley put it, "people are equal . . . [but] an inferior status [for workers] . . . is the inevitable consequence of being a worker and not an owner."[14] The key provisions of worker participation were shared ownership, the involvement of workers in management decisions at all levels, including basic policy making at the level of boards of directors, and collective bargaining based on full disclosure of accounts and other relevant information.

The Labour Relations and Industrial Disputes Act (LRIDA) removed many of the archaisms that existed as throwbacks to master and servant enactments[15] and secured significant benefits for workers. It also, however, potentially limited union activities. The Act compelled employers to recognize unions democratically elected by workers, created a tribunal that would hear cases of dismissal and could order compensation and reinstatement in the case of wrongful dismissal, and provided for arbitration at the request of any of the parties to an industrial dispute. More damaging to workers were provisions that forbade work action before a dispute was sent to the Industrial Disputes Tribunal (IDT) and gave the minister of labor the power to declare illegal any strikes that were "likely to be gravely injurious to the national interest." Other provisions also had the potential of criminalizing normal union activity at the workplace and shifting the burden of proof from employer to worker.

These two initiatives must be seen against a wider social backdrop. There were mounting pressures from the capitalists to bring the unions to heel. As early as 1966, the Jamaica Employers' Federation (JEF) had lobbied for an industrial relations court to curb the activities of the unions. By 1969, the JLP yielded, creating the Permanent Arbitration Tribunal. An economic study conducted by the capitalists in the late 1970s called for establishing a South Korea–like economy in which organized labor would be neutralized in order to attract foreign capital in pursuit of cheap wages (*Weekly Gleaner*, December 4, 1978, p. 5). The JEF study on wages found overall increases of 6.5 percent for 1960–1967 and 10 percent for 1968–1972 (International Bank of Reconstruction and Development, 1974, p. 43). Over the same period, there was a rash of disputes ranging from dissatisfaction with wages, fringe benefits, and working conditions to strikes prompted by worker solidarity (Gonsalves, 1976, pp. 227–78). The clear implication was that curbs should be placed on labor's ability to command wage increases as well as on the supposed lack of discipline of Jamaican workers.

On the workers' side, the traditional base of labor unionism was becoming disaffected and there was appreciable unhappiness with the BITU/NWU oligarchy. In fact, research conducted in 1975 relative to worker participation showed substantial majorities of workers from different areas of labor preferring direct worker representation to trade union representation. In fact, strong preferences for direct representation (70 to 85 percent of workers polled) were recorded in all industries except private sector distribution, where substantial numbers (36 percent) did not record any preference (Stone, 1976c, p. 101). The Sugar Workers' Cooperative Council (SWCC), for instance, questioned a trade union presence in the movement. It felt that trade unions were not equipped to provide the necessary leadership toward ownership and control of land.

For the PNP government, these two pieces of legislation would remedy the situation on both sides. LRIDA brought a measure of control to traditional labor-management relations, while worker participation would fulfill a dual role required by the new politics of the common good: workplace democratization would promote the equality of workers and owners; at the same time it would educate the worker "so he can be more aware of his responsibilities toward national productivity and also to appreciate the circumstances that characterize his environment."[16]

The dilemma was that, while new mediations were needed, ongoing tensions between workers and unions and workers and capitalists militated against consensus building. Michael Manley might talk about equality, but the ground in which these seeds would be sown was replete with traditional and more recently cast antagonisms between capitalists and subordinate classes. A survey of workers' opinions undertaken by Stone in 1975 found a high level of dissatisfaction with working conditions, especially management-worker relations and facilities available to workers. Workers' feelings could be summed up by a respondent's comment that "we are not treated like human beings" (Stone, 1976c, p. 86).

The PNP's answer, conceding a little to either side through the LRIDA and changing antagonism to cooperation through worker participation, was appealing but bound to meet resistance. The history of worker participation—or workplace democracy—shows that workers disaffected from collective bargaining and established management practices may well support this alternative, but the equivalent motivation for management to share power may only be reached—and that with difficulty—as a last resort in the face of crisis. Thus, when at least 75 percent of the sugar workers, in a 1975 survey, expressed the wish to "have a cooperative," it was a simple matter for the government to decide to exercise this option with regard to the sugar factories it had salvaged.[17] But, while parallel research also confirmed high levels of enthusiasm for the cooperative concept as a whole (Stone, 1975), the pattern was not repeated. Management interest in worker participation centered around the desire to increase productivity and improve worker-management relations—insufficient motivations for power sharing. In addition, the concept was largely foreign to the North American milieu, which exerted a telling influence on Jamaican capitalists.

The capitalist class had little to say about LRIDA but was not silent on worker participation. The JEF took the position that the existing system of collective bargaining provided the working class with the necessary power "to influence the industrial system,"[18] further recommending that norms instead of laws be the basis of these policies, with legislation coming into play only when voluntary methods failed. It conceded worker participation at the shop-floor level and no higher. The Jamaica Chamber of Commerce felt that ordinary workers had just as many opportunities as the present managers and directors in the business community, many of whom had started as ordinary workers. The *Daily Gleaner* of July 14, 1975 averred that worker education had not yet reached the plane where workers "can go up the ladder and perhaps participate in management." The Private Sector Organization of Jamaica (PSOJ) invoked "the limits of good sense and practical efficiency in management" and lectured on the importance of "gradualism and pragmatism."[19]

For the working class, worker participation policies had to be put in the context of the antagonistic labor-management relations that had given rise to LRIDA. In spite of its high support for the concept of worker participation, the class was concerned about the level of protection it would provide. Creating a viable alternative to unions may have been devoutly wished, but if unions were replaced, how would workers be assured of their rights? Workers of the Jamaica Public Service Company, which was involved in worker participation, went on strike in the crisis-ridden days of 1978. The prime minister called their action an abuse of freedom (*Weekly Gleaner*, September 18, 1978, p. 7). In another incident occurring in 1978, forklift operators engaging in a "go slow" at Port Bustamante (Kingston) ignored the government's solicitations to tone down their demands in light of the socialist principle of worker participation. The job action only ended when the matter was sent to the Industrial Disputes Tribunal (*Daily Gleaner*, September 19, 1978).

Ironically, it had earlier fallen to the BITU, the trade union arm of the JLP, to remind the government that conflict, rather than voluntaristic urgings to consensus, was the rule in Jamaican society. As it stated to the Advisory Committee on Worker Participation, "the dualist nature of [Jamaican] society manifests attitudes which are more in conflict that in co-operation."[20]

The BITU and NWU supported worker participation in principle but both were concerned that implementation could undercut collective bargaining. The fear appeared justified. On at least one occasion, sugar workers negotiating workers' councils at the Monymusk sugar factory rejected trade union representation: "What they desired most was one united workers' organization which was essential to true worker democracy and unity" (Agency for Public Information, Release No. 197/77, January 28, 1977). On another occasion, the Jamaica Omnibus Service, which had become part of the public sector in 1974, attempted to set worker participation in place without union participation. The unions threatened to strike.

The government muddled through in its various attempts to clarify the issue, not providing satisfactory answers to anyone. While it became increasingly difficult to unravel actual consequences from original intent after 1977 and the restrictions that accompanied IMF intervention, these difficulties illustrate the limitations of consensus-based politics. As worker participation receded into the background, enveloped in a tissue of suspicion, LRIDA became increasingly a tool of worker control. The power of the minister of labor to declare strikes illegal was particularly troublesome, as the list of essential services serving the national interest could be lengthened arbitrarily. The unions saw this as particularly dangerous and sought to restrict the list to life-and-death services and to those attached to the protection of property.[21] The skepticism of organized labor continued into 1978. The Industrial Disputes Tribunal, a non-political body, was superseded by the Ministry of Labour. The Ministry assumed the monitoring of contracts, which meant that, in the national interest, it could reject agreements negotiated between workers and management, if the conditions listed above were not met (*Daily Gleaner*, May 22, 1978, p. 30).

Combined, these regulations greatly reduced the effectiveness of unions. Simultaneously, management's position improved vis-à-vis labor. Other legislation also came into play. The Employment (Termination and Redundancy Payments) Act of 1974, which was supposed to protect workers from wanton and unlawful dismissal, was largely ineffective. The matter was put to the test in January of 1977. Rumors of a planned layoff of workers by industry to protest government-imposed foreign exchange restrictions exacted from the government a promise of legislation to compel employers to give the Ministry of Industry thirty-days notice of any plans to reduce the work force. It was unclear what was to be gained by this: did it mean merely that the worker would receive advance notice or would the employer now have to show reasonable cause for these actions? These and other crucial questions were not addressed; in fact, this anti-layoff law was dropped. It transpired that the government, as the largest employer, would perhaps be victimized severely: government jobs based on patronage are temporary and any notice of their reduction would wreak unspeakable political havoc.

Together these events served to force labor and particularly the trade unions into retreat. Early in 1977, several unions of differing ideological persuasions attempted to consolidate in the face of the pronounced anti-labor posture. The attempt failed, however, as their ideological differences could not be resolved.

Trust in the regime and management was at a very low ebb. Unlike the capitalists, who declared after the 1977 *Emergency Production Plan* that "there is no room left now for any further lack of confidence,"[22] workers were both angry and demoralized, as established channels failed to provide any protection from the continuing attack on their wages and rights. In late November of 1979, 180 workers at C.M.P. Footwear in Kingston held the financial comptroller and the general manager hostage until their demands were addressed. Their prime grievance sprang from the Industrial Disputes Tribunal (IDT) award concerning their dispute with management. In fact, the Court Tribunal had almost no legitimacy left: "[The workers] brush it aside, even to the verdict of the Court they pay it no mind" (*Daily News*, January 21, 1980, p. 7). Numbers of the unemployed, for their part, took to storming government offices demanding jobs.

Political apathy was the other side of such desperate acts. In a survey conducted in 1978, fully 40 percent of those polled indicated no preference for either of the two parties. In 1976, that figure had been only 15 percent (*Weekly Gleaner*, November 20, 1978). The politics of creating consensus between capital and labor had reached its limits.

▶ The Agricultural Classes and Agrarian Policies

The crisis had again brought into the open the agricultural question. Unlike the 1930s and 1940s, however, social justice and the demands of workers and would-be small farmers were not as much at issue as the requirements of economic growth and political mediations. Internal and external migration had significantly relieved rural population pressure and redirected the aspirations of many among the rural subordinate classes. But vexing questions rooted in the state's management role, such as the size of the food import bill and the political pressures of the lumpen and subproletariats served to point to domestic agriculture as a possible solution. Agrarian support schemes, with the occasional infusion of community development components, would, it was hoped, not only help keep people in the rural areas productive but also reverse the trend of rural-to-urban migration, thus lessening the threat posed by the festering slums of Kingston and other major cities. Besides, tackling rural reform seemed feasible now that the economic role and political power of the big landowners had diminished considerably.

Some of the major concerns of the new regime were aired on November 9, 1972, during Michael Manley's first important statement as prime minister:

> We have failed to meet the domestic food requirements of our people, and . . . most of our staple foods are not those produced by ourselves. It is interesting that in 1962, our imports of food amounted to $30.4 million. . . . whilst the population increased by 12 percent during the period from then to now, our imports of food in 1971 amounted to $76.3 million, or an increase of 150 percent in the import food bill—an increase much greater than the population and price increase when we put them altogether. (Hearne, 1976, p. 117)

A revamped agricultural production, Manley declared, could also provide local raw materials for industry as well as new exports.

The *Green Paper on Agricultural Development Strategy* followed in 1973. The ramshackle agricultural credit system was to be restructured so as to promote the productivity of all farmers. Although the Agricultural Credit Board, the JDB, and commercial banks did, in theory, provide capital for agriculture, their services were largely outside the reach of the small and medium-sized farmer. The Agricultural Credit Board's records showed that it had extended loans to some 150,000 small farmers, with an annual provision of less than $20.00 per farmer. (Private Sector Organisation of Jamaica Task Force on Agriculture, 1977, p. 23).

A newly planned Agricultural Credit Bank would be more responsive to the farmers' needs. Certain modifications to the JDB's lending policies were also made and an additional $6.5 million was allocated in 1976. In 1975, price controls on domestic food items were removed. A system of higher guaranteed prices and subsidies to farmers coupled with sales through outlets of the state-run Agricultural Marketing Corporation (AMC) would ensure fair prices to the farmer and reasonable costs to the consumer. The Price Review Board was charged with maintaining the proper balance of agricultural prices: increased prices to farmers ranged between 100 and 400 percent by 1977 (*Daily Gleaner*, July 14, 1977).

However, the centerpiece was Project Land Lease, a reform program designed to affect different class configurations. Phase I was intended to assist the rural small farmers in bringing "their operations and production potential up to a level where they can earn enough to support a decent standard of living."[23] The government undertook to prepare the land and provided the required materials. In Phase II, lands acquired by the government were to be converted into 49-year leaseholds. The holdings would be allotted "in sizes to achieve the difference between what the farmer can earn from his own land, and a minimum income of $1,500."[24] It was evident that the government intended to support the creation of a class of small, independent farmers. Hereditary rights were attached to this form of tenure and compensation would be payable for capital improvements. Phase III was intended to be the core reform. State farms complete with community centers and other amenities would be created to provide basic training in farming techniques, primarily for young people. The trained farmers—many, according to the plan, from the urban areas—would then be placed in modern homes and on lands adequate to provide "a decent income."[25]

The first aspect of Project Land Lease was initially quite successful. A status report given in February of 1977 showed that the Project had placed 23,886 tenants and brought into cultivation some 48,000 acres with a projected crop value of some $20 million. However, the expectations the program raised were far greater than its ability to deliver. The government was faced with land "captures," (especially in the rural parishes of Westmoreland and Portland) in which the unauthorized occupiers of the land then turned around and sold or leased lots to others. By 1980 this most popular program was running far short of projections.

The list of culprits was long: miscalculations, crop failures, attitudinal deficiencies of participants, lack of government funds, diminishing popular support, IMF's countervailing policies—all these and more were blamed for the failure of the program. By 1980 Phases I and II had placed 37,661 farmers on 74,568 acres, while Phase III was in total disarray.

These factors aside, our analysis points to the class discrepancies generated by these policies as an important source of difficulties. In particular, the politics of national popularism meant that the potentially conflict ing interests of small and large farmers had to be reconciled. While all agrarian classes favored the policies initially, the response varied as certain limits were reached. In the meantime, the export-oriented agricultural fraction of the capitalists had reason to rejoice because its failing enterprises would now be acquired by the state. Added to the purchase of Tate and Lyle's sugar interests and those of the United Fruit (Jamaica) Company in 1971, there was now a coherent policy that did not rule out the acquisition of other agricultural enterprises. These salvage operations, along with the purchase of bauxite lands from the multinationals, could serve as the initial basis for agrarian reform. However, it would only be a matter of time before the issue of privately owned idle lands would be raised.

Other members of the class voiced their concerns, sprinkled with occasionally loud opposition. The Jamaica Agricultural Society, a creation of plantation society, raised the issue of the likely effects of land fragmentation and the supposedly lower levels of productivity of small and medium-sized farmers. The Jamaica Livestock Association required firm assurances that the pens would not be affected—which the government gave, albeit with a marked penchant for inconsistency (*Daily Gleaner*, July 4, 1977, p. 8). In practice, such reactions meant that Project Land Lease would falter to the extent that these interests were threatened. Opponents within the capitalist class would do their best to frustrate the legislation. For instance, they would comply with the letter of the law in ways that presented obstacles to its implementation. A former accountant for a large construction company told us how his former employers went about it:

> They owned this large property. . . . Nothing much on it. They prepared a fairly convincing development plan for the property, took it to the government with a request for financial assistance. The government told them that it had no money. End of the matter; they left the ball in the government's court. For as long as I was there the issue was not raised again by the government.

Whether the problem was insufficient government resources or the government's relative lack of power vis-à-vis the class, the fact was that these regulations were not enforced, thus creating a de facto limit to the land available for distribution under Project Land Lease. In fact, land captures, often under the banner of socialism, occurred rather frequently in the second term of the PNP regime, causing much discomfort for a government bent more and more on compromise.

Phase III of Project Land Lease became the victim of a somewhat different set of circumstances. While its greater costs were an issue, the old problem of patronage proved especially intractable. Leaseholds were generally assigned regardless of political affiliation, but state-affiliated training farm projects became known—whether correctly or not—as PNP bastions, and as such appropriate targets for praedial larceny, vandalism, and attacks. In the end, only 3 percent of the farmers who benefited from Project Land Lease were involved in this program, thus ensuring that the most significant changes in the rural socio-economic structure would not occur.

The failure of Phase III was not a problem, however, given the petty capitalist tendencies of rural subordinate classes. Many middle-sized farmers wanted their own land and were attracted to what amounted to a guaranteed minimum income of $1,500 annually. Research conducted in 1975 revealed that the aspiration of this fraction of the class included increased land ownership and cash income, house ownership, and proprietorship of small businesses.[26] Before these reforms, 37 percent of these farmers had been forced to seek outside employment; with Project Land Lease only 17 percent did so and 67 percent were now hiring labor occasionally (Williams, 1975, p. 34). It should also be noted that the minimum annual income of $1,500 was in excess of the national per capita income of $1,164 for 1975 (*Pocketbook of Statistics, Jamaica 1976*, Table 77, p. 12).

Small farmers now increased production with the knowledge of guaranteed markets and prices through the Agricultural Marketing Corporation. These developments led to a revitalized "higglering" system—a network of petty traders in agricultural products throughout the urban areas. There was ample inducement to participate. In 1975 food imports were $118 million; conservatively, the network could capture most of the 5–6 percent or so that annually went for fruits and vegetables. The cause would also be helped by the pattern of import quotas the government would impose. Thus Phase I of Project Land Lease, which was meant to be temporary was the phase that experienced the most activity and success. Indeed, what likely accounts for much of its relative success is the fact that it built and improved on the land settlement projects of the past.

▶ The Lumpen Proletariat and the Subproletariat

The patterns of mediation we found with regard to the working class were repeated for this constituency. If in the long run it was hoped that improved economic performance would alleviate the circumstances of these groups and the problems they created, in the short run there were distributive programs to provide immediate relief, improved educational opportunities and training schemes, along with more repressive measures demanded by other social classes to reduce crime. True to its national popularist credentials, the PNP was trying to find a middle ground between needed structural change and the requirements of the status quo—a new common good.

The constituency needed immediate attention, if only because of the real and potential threat it represented. Comprising both the un- and underemployed and those with disreputable occupations, it far outstripped the number of members of both political parties. In fact, it constituted the majority for the entire modern period (Gannon, 1976, p. 173). By any account, these largely urban-based elements were a disruptive force, including among them not only common ne'er-do-wells but also a ready army of thugs available to do the bidding of either political party. Further, there had been a notable increase in the rate and violence of crime since the late 1960s. While the size of the grouping was troublesome in itself (we estimate a figure of some 100,000 to 160,000 in 1974), of even greater concern was the high representation of young people among them.

Provisions made for this constituency took two forms. One entailed government allocations for the "crash work programme," which funded projects such as clearing road embankments and gully courses, road and building maintenance, and some construction work. Such practices had been the mainstay of the government's approach to chronic unemployment, but had largely been confined to holiday periods ("Christmas work"). While the PNP continued to view them as a stopgap measure, they greatly expanded the scope of such activities. An allocation of $10.5 million was immediately made, to be shared by six government ministries; the total provision for these elements in the 1976–1977 budget year was $50 million.[27] Examples of the threat prompting this largesse can be drawn from various sections of the country. Between late 1975 and early 1976, members of the group would often "invade the Ministry of Labour"

(*Daily Gleaner*, February 25, 1976). In Montego Bay, the St. James Parish Council was forced to appoint a subcommittee to investigate the "rascality going on": it transpired that payrolls were padded with bogus names with "hundreds of workers collecting paychecks" fraudulently (*Daily News*, July 14, 1976).

Phase III of Project Land Lease, discussed earlier, was the second policy aimed at this group. Urban unemployment and marginality were structurally connected to patterns of dislocation in the agricultural sector. The state farms established in the rural parishes, of which Nyerere Farm was the most noteworthy, were to address the needs of these marginal youths. Here the constituency that naturally gravitated toward the urban areas, Kingston in particular, was to be dispersed to these strategically located farms and carefully tutored to become a functioning part of the farming community. The approach was comprehensive, including the establishment of homes and community centers—in fact, the creation of "villages." The policy did seem to have appeal: by 1977 various supporting organizations, including the National Union of Democratic Teachers (NUDT), put at 23,000 the number of youths eager to take advantage of it and dissatisfied with the slow pace of the program's implementation. However, it appears that such claims were inflated.[28]

These programs, designed to ease the hardships experienced by the lumpen proletariat and the subproletariat, were complemented by stringent law enforcement measures to deter violent crime. Expenditures for law and justice, which had already increased considerably in the 1960s, grew even further: between 1971 and 1978 the budget assigned to the police and the military increased twelvefold! These increases went partly to support the creation of a special tribunal. The famed Gun Court was established in 1974 to deal with the widespread illegal use of firearms. After a period of amnesty during which large quantities of such arms were turned in, the Court was required to give indefinite sentences to all offenders. From most accounts, the Court was an instrument to contain the growing numbers of disgruntled elements dissatisfied with the low levels of patronage. These elements had no interest in democratic socialism, except as a means to an end.

Increased law enforcement efforts and distributive policies were not sufficient to cope with the magnitude of the problem affecting this constituency. In earlier years an accommodation of sorts had been effected through the practice of patronage. This division of the constituency (and distribution of resources) based on party affiliation had carried, however, profoundly negative effects that could be seen in the transformation of urban ghettoes into grounds for "tribal" warfare.

The need for structural change and new mediations was recognized by both political parties and sectors of the capitalist class. As Pearnel Charles (1977, p. 19), a JLP senator and labor organizer, wrote:

> Organized violence caused by the scarcity of economic benefits and the consequential practice of victimization has been the plague of Jamaican politics for several years. A more recent and perhaps more dangerous phenomenon is organized violence as a political strategy. This is itself a product of the hostility which has developed between the political parties.

One of the issues, then, was the perennial call for political intervention. The same politician remarked that political work that used to be done only around election time was becoming a permanent part of the politician's life: "Jamaica used to be divided only at election time. We now have permanent division because there is a permanent election campaign" (Charles, 1977, p. 174). Both political parties were losing control of this potentially disruptive force. Indeed, with democratic socialist slogans being bandied about, it was clear that significant elements from the popular classes had other things in mind. Between June and December 1976, 200 people were killed in politically related violence.

As for the capitalists, calls for law and order would at times be muted by entreaties rooted in the common good, as interests became more enlightened:

> The excess of tribalism . . . threatens to destroy our hopes of achieving national unity. . . . *All* citizens must agree to accept the principle of national unity and concerted effort, even if it means modifying our sectoral interests temporarily while we assure our very survival. (Chamber of Commerce, President's Monthly Statement, January 1978)

Both before the 1972 elections and during its tenure in government the PNP promised a change. The government did attempt to break from the clientelistic cycle and make provisions for the group without political consideration. In response to the call for peace, religious and political leaders joined in several pacts that were, however, of short duration. The political obstacles and economic constraints proved to be insurmountable. In the end, earlier practices continued unabated and little changed for the constituency. Employment in public works projects and the "crash work programme," as well as participation in other government projects (with the important exception of Project Land Lease) would go to PNP supporters.

Acts of buccaneering became more sinister. The 1970s were to witness the institutionalization of the practice whereby gunmen would demand that fictitious workers be placed on the payrolls of private companies. And there were public works projects which, though appearing on the books, could be found nowhere in substance. Bold attempts to end these practices were fatal: the 1977 murder of Edward O'Gilvie, permanent secretary in the Ministry of Works, was such a case. This courageous civil servant threatened to discontinue the practice within his ministry. He was warned that "blood would flow." And it did. Even with its avowed commitment to honesty in government, the PNP was irretrievably caught in the snare. In 1978, some $500,000 from the Pioneer Corps Fund could not be satisfactorily accounted for, along with thirty centers the evidence for whose existence appeared only "on the books."

The criminal element was becoming more effective, less dependent on politics, and better organized. The Gun Court became the regime's answer where redistribution and economic restructuring failed.

In this climate, it was not surprising that the initial goodwill toward these groups quickly evaporated. In unison, all other classes condemned the government initiatives as flawed and their recipients as unworthy of any assistance. Groups of "crash work programme" workers resting on their brooms were the fodder of cartoonists. In the public opinion (to which the conservative daily *The Gleaner* contributed with abandon), the state farms were either training grounds for communist shock troops or living examples of government bungling. There was some truth to the latter assertion; the proposed villages had not materialized, and insufficient provisions had been made for the kind of assistance that would have enabled city dwellers to make the required transition to farming. It was also the case, however, that these projects, known to house PNP supporters, had become the targets of attacks by JLP youths.

Given the volatility of the lumpen proletariat and the subproletariat, PNP policies were doomed merely on account of the sheer size of the constituency relative to the resources that could be allocated to their needs: they simply could not be absorbed by existing networks. In the end, they were left mostly to their own devices and to survival tactics. Mostly denoting marginal occupations, the numbers of the self-employed grew considerably during the 1970s. Their already high numbers in 1972 (30 percent of the labor force) had grown alarmingly: in 1980 they constituted fully 42 percent of the labor force. Meanwhile, political patronage continued along its previous course, generating more tensions than it could mediate, but without any apparent possibility of reversal.

Occasionally, the poor turned to resistance, as when empty government housing units (in Rema and elsewhere) were "captured" by people who overpowered the guards. The government's response was partisan. In the well-publicized "Rema Incident," the Ministry of Housing proceeded against forty "capturers,"—all JLP supporters—ostensibly to collect unpaid rents. A commission of enquiry that followed found that although some 90 percent of the tenants in the area were in substantial arrears, no legal action had been instituted against them (*Weekly Gleaner,* February 15, 1977). Housing in Rema, a PNP stronghold, had been provided by the government. JLP supporters living in the area had been subjected to harassment. The Commission also found that the minister of housing had been reorganizing the community with the assistance of hired gunmen. Similar patterns were repeated in Tivoli, a JLP stronghold.

Political warfare and sheer wantonness seemed to reign. In the heavily lumpen area of West Kingston and Lower St. Andrew, for instance, between 1976 and 1982 4,000 homes and housing units were destroyed and 40 percent of the area's population had to flee due to the siege conditions that prevailed. Indeed, disappearing streets and displaced residents became a way of life: "small farming now prevails on a site of one former ghetto which once had a density of 210 persons per acre." (Eyre, 1983, p. 238). Hustling and political brokerage became ends in themselves. These actions spoke more of exploding need and sheer violence than of more conscious and progressive movements for change. It requires little imagination to appreciate the fear such events evoked among the other social classes, who saw themselves standing next in the line of destruction.

▶ Concluding Note

Our analysis of the impact of PNP policies reveals that class activity centered around naked self-interest. Once resources became an issue, the "natural tendency" of Jamaicans to abide by the tenets of democracy was clearly insufficient to ensure the stability of the class alliances on which the PNP project was based. The PNP had refashioned the notions of the common good developed by Norman Manley with some success. Yet in stressing the theme of a natural harmony among political interests it retained a piece of hallowed Lockean liberalism that was devastatingly out of place in the crisis we have analyzed.

The success of democratic socialism is coextensive with the confluence of appropriate class consciousness and political action. These were hampered by historical and structural factors. As a result, the policies could not realize the regime's expectations; instead, it rushed headlong into the stubborn obstacles erected by dependent capitalism. Trade, international financing, and foreign policy issues unerringly made their way through metropolitan filters, not auguring well for the successful implementation of socialist initiatives.

The crisis revealed less a socialist project that was not allowed to be than a historical conjuncture that was simply not favorable to such a project. While in many instances the PNP regime conceived and articulated policies in the spirit of socialism, their actual meanings and outcomes were dictated by the essentially capitalist character of the economy. The consciousness and political action of the social classes were demonstrably anchored to capitalism. Indeed, to the extent that the class structure bore signs of militancy and restiveness, these are explicable in capitalist and not socialist terms. The basic dissonance and contradictions discussed in this chapter derive largely from the alteration of capitalism to meet the required changes, some of which were embodied in those very class expressions. Instead of signaling a call to socialism, then, the crisis was linked to a new phase in capital accumulation.

State-Directed Change: Some Questions

CHAPTER 11

Our principal aim in writing this book has been to highlight the internal socio-political dynamics that accounted for the emergence of the democratic socialist experiment in Jamaica and, in so doing, to contribute to the understanding of social change in other Third World countries. Our analysis brought into focus certain elements that have frequently been considered secondary in Marxist writings: in particular, we speak of the ideological tendencies embedded in political parties such as the PNP and the importance of the ideology of subproletarian elements like the Rastafarians.

The Third World has an abundance of social phenomena that defy easy classification and fiercely challenge established methodologies. Our analysis makes room for more flexible approaches where the boundaries between structural and so-called "epiphenomenal" elements take the shape of fluid and changing dialectical forms. We have tried to overcome a restricted treatment of these relations and, rather, construe them as "phenomena-in-motion."

One of the questions we asked concerned the meaning of radical ideologies in Third World social change. Historically, regardless of radical-sounding pronouncements, sluggish development and chronic scarcity tend to promote "clambering individualism" and rampant pilfering and corruption among "elites" (Kitching, 1983). If historical records hold, although they may ostensibly be inserted in the state to create the necessary preconditions for socialism, these elites will be self-serving and cling tenaciously to power.

It has not escaped the scholars' notice that the resulting phenomenon is of a different ilk. Raymond S. Franklin (1981, p. 68) reminds us that the "undeveloped socialism" of "radical elites" tends to be "stable and long-lasting." Under Ahmed Sekou Toure, Guinean socialism unwittingly served the interests of a bourgeoisie through the pursuit of policies that followed "bourgeois models" (Riviere, 1977, p. 239). In Ethiopia, a popular Marxist-Leninist revolution quickly turned to repression when the popular forces requested that the military regime surrender to the rule of the majority (Markakis and Ayele, 1986). In these and other instances, undeveloped socialism became a politico-economic stage unto itself. Far from being transitional, this stage appeared to immobilize the transformational process and block any movement toward socialism.

Rather than continuing to analyze such events according to the theoretical framework of socialism, now revamped to account for the obstacles to the actual socialist transformation of many Third World countries, we applied an alternate framework—one in which such events become part and parcel of one of a number of variants of the exceptional peripheral *capitalist* state. It would be a disservice to theory and practice, however, to view these variants as merely variations without a difference. In particular, students of radical social change should find national popularism an interesting, though short-lived, phenomenon, since in many ways it promotes—although it also contains—such change.

Our exploration of the social forces involved in the attempted changes in Jamaica reveals several important factors that one is likely to encounter in the study of other Third World countries engaged in agendas of self-directed change. We limit our discussion here to a few promising areas not highlighted in previous chapters, which, if the Jamaican experience holds, should provide fruitful opportunities for further comparative research.

▶ The Role of Movements and Ideologies from Below

When the main push for change comes from above, in the absence of strong action and organization from the subordinate classes (as was clearly the case in Jamaica), there is a temptation to dismiss radical ideologies as mere instruments of cooptation. We found reality to be far more complex.

Historically and at present, the subordinate classes were lacking in organization and the political process clearly involved cooptation—especially evident in Bustamanteism but present, more subtly, in the PNP's liberal interpretations of impulses from below. Nonetheless, representative political structures also provided new channels for popular expressions. If the paternalism of the planter was replaced by the somewhat more contractual but still manipulative clientelism of the parties, the subordinate classes, in the process, emerged as forces to be taken seriously. To an extent, at least, the political leadership had to learn to "read" them and accede to their wishes. Whether directly or not, they shaped politics and ideology in significant ways.

These factors, we argue, were crucial in pushing the PNP toward national popularism rather than some other agenda. In particular, the option of a more authoritarian or one-party state could not be seriously advanced. Radical movements and ideologies from below, as well as the democratic tradition that took shape after the labor rebellion of 1938, played an important part in this regard.

Our research reveals that these transformative impulses succeeded, to various degrees, in inserting the views and aspirations of subordinate classes into PNP agendas, thus providing them with new inflections. Mainstream ideologies did not have the edge forever. In part, it was the rejection by these ideologies of issues of race and color that prepared the ground for the spread of ideologies from below, in which such themes were never forgotten.

In the period of national popularism, Rastafarian ideology provided an important energizing element for the PNP, contributing significantly to the vision behind the attempt to create a self-reliant and democratized society. The intersection of this ideology with the aspirations of subordinate classes and the dominant ideology of the PNP (Manleyism) produced lasting changes at the level of political organization and consciousness. To this day, even its opponents concede that a crowning accomplishment of the PNP was to raise popular consciousness with respect to self-worth and economic power. This is an example of a policy that successfully combined voluntarism and objective factors (see below): if the policy was right and bespoke goodness, it was also based on a correct reading of the prevailing subordinate class dynamics.

▶ The State and Transformation

The Beginnings: Some Factors Prompting the State to Action

As in other Third World countries, patronage through political institutions such as the trade unions and political parties was pronounced in Jamaica. Patronage served the needs of classes at both ends of the social spectrum. From the perspective of the dominant classes, it aided in containing the threat posed by the growing numbers of the lumpen and subproletariats. For the subordinate classes, it was a means by which resources could be pried competitively from the state—indeed, one of the very few means of redress available to them in a climate of chronic scarcity. It is evident from the records of shifting membership in trade unions and the major political parties, that these classes used patronage well.

The principal agents of patronage are members of the middle class, who play a necessary mediatory role in the class structure. The development of fissures in the structure of patronage that threatened the strategic middle class (and the state) in its mediatory role was thus the principal force precipitating change. Of course, the scope and direction of change was fashioned through the combined workings of the social factors discussed earlier.

Social Change and Voluntarism

In every agenda for social change there is room for the influence of subjective action. People can be persuaded to take one desired course of action rather than another. Effective leadership can make a difference. However, as was evident in the PNP regime in the 1970s, governments with liberal leanings may generously overestimate the ability of voluntarism to change the course of events. Also pushed beyond its reasonable limits was the power of the "Great Man with a Vision." In the heat of the political moment, the leadership yielded too often to the tempting assumption that determination, buttressed by policy pronouncements, could achieve more than was objectively possible. Certain events must occur, it would seem, because of their intrinsic capacity for goodness.

The possible could have been extricated from the desirable through the use of sound theory. The Manley government included its share of knowledgeable advisers from the University. Appropriate analysis should have assisted the government to avoid at least certain mistakes. The failure was as much theoretical as it was political; yet we are not sure that things could have been otherwise. Could the coolness of logic and analysis have stood their ground against the "Great Man's Vision?" This raises profound questions on the connections between theory and political practice-in-formation.

Voluntarism played an important role in the evolution of national popularism in Jamaica. There is no question that the personality and orientation of Michael Manley and others around him were of great importance in shaping the overall direction of the PNP program. The daring action vis-à-vis the bauxite companies, the inclusion of the Rastafarians closer to the mainstream, the stubborn support of social programs over productive projects, the pursuit of democratic socialism itself—all these choices were not predetermined and bear the mark of conscious selection.

In reaching toward an overall assessment of the importance of voluntarism we must, however, clearly differentiate between word and deed. Where did subjective action make a difference and where to lasting effect? If voluntarism succeeded in bringing about changes in ideology and consciousness, the story was quite different when it was applied to the economy. At times, the PNP's approach to economic policies appeared geared to building a new model: development by exhortation.

Voluntarism has within it a mischief factor that is released, often disastrously, when its intrinsic limits are exceeded. In some cases, superficially agreeable policies, introduced with the ultimate blessings of voluntarism (because they are right) went awry because objective factors were ignored or unwisely challenged as, for example, in the reorganization of the mining industry. Here foreign capital, with its protection from the IMF, was attacked in circumstances where inadequate forethought was given to the inevitable consequences. The relative autonomy of the Jamaican state is largely without effect if the economy is dependent on foreign capital and its supportive institutions, such as the IMF.

Voluntarism also attempted to operate in an ideological void, with predictably unfavorable results. Many of the shortcomings resulted from the notion that enlightened leadership must coax or drag the electorate (which is, alternatively, apathetic, unruly or politically ignorant) to the trough of political enlightenment. According to John Hearne (1976, p. 62), Manley had been "exercised by the need to bring the electorate into politics, at a level above their votes every five years." The approach was, however, heavily paternalistic, with the predictable consequences that accompany the failure to acknowledge the class basis of state policies.

This is perhaps the most crucial weakness of national popularism as a model: while the relatively undeveloped state of classes allows the political directorate to rally most segments in the society around a national project, it also creates the illusion that the hardening of economic entities into social classes can be avoided. The Manley regime grossly overestimated the power of voluntarism in emphasizing the state's ability to appeal to those classes' rationality and sense of community. If the state is supposed to be the custodian of the community, as some members of various classes did ask, who will keep tabs on the political leadership and the state? *Qui custodes custodiet?*

The State, the Capitalist Class, and Dominance

An important aspect of the state's agenda for change involved supporting the industrial fraction of capital as it wrestled to escape from the tutelage of the commercial fraction. On the surface, there were substantial points of conflict within the capitalist class, as indicated by the pronouncements of the Chamber of Commerce and the Jamaica Manufacturers' Association. It was also clear, however, that while there were voluntaristic longings for separation, the structural basis for full division of the capitalist class into industrial and commercial fractions was only embryonic. While mistakes could undoubtedly account for some of the state's lack of success in this area, we think that the most important reason these policies failed must be located in the insufficient development of the industrial fraction at the underlying structural level.

While the absolute level of development was an important factor, one also needs to pay attention to the relative strength and weakness of the various social actors. A different problem would have emerged, in fact, had the industrial fraction been more strongly developed. It might, then, have had sufficient strength to resist successfully the state's attempt to subordinate its members. Jamaican industrialists, while giving qualified support to state policies, never ceased to uphold the fundamental principle that the private sector should remain at the helm of the economy.

It is useful, then, to distinguish between the concepts of dominance and hegemony, which seem to constitute an arena for common miscalculations in agendas for change. The capitalist class, for instance, may maintain dominance due to its structural role in a system of capitalist relations, but this role is not sufficient, in itself, to bestow hegemony, especially in circumstances of crisis. The same applies to intermediate classes bent on their own hegemonic projects. Here the visibility and power of their base, the state, may convey the impression of absolute autonomy when, instead, the prevailing pattern may be one of dominance within the circumscribed framework formed by existing political and economic structures. This was clearly the case for the PNP's agenda. The initial success of the class alliance was based on the state's ability to generate and redistribute an excess of resources that could satisfy all participants. As these resources dwindled, however, so did the supposed autonomy of the state.

Obstacles to Institutionalization: The Allure of the Capitalist Economy

It is not sufficient to initiate new patterns; they must also be maintained and nurtured. The context of a mixed economy appears to create special difficulties for this task. As our analysis revealed, the prospect of self-advancement through the private sector provided an allure that was hard to resist. Here the requirements of dutiful public service come face-to-face with expectations bred by social mobility within a tradition of chronic scarcity. Having met the necessary educational requirements, one is simply entitled to financial success. Indeed, there are obligations to others incurred in the course of one's career or as part of one's position, that make a sound financial base a requirement.

As we saw, shedding this cognitive style (not to mention the related social obligations) and embracing socialist ideology did not come easily even for highly placed members of the PNP. Part of the difficulty arose from the fact that the emergent system (that is, socialism) had yet to build the required support at the socio-political and ideological levels. For instance, managers of state enterprises were expected to behave as capitalists for the state but not for themselves, when just outside the gate a capitalist economy beckoned.

The allure of capitalist practices included the "other" economy: for Jamaica, the vastly profitable ganja (marijuana) trade. By the mid-1970s and later, the ganja trade had been transformed from the preserve of the lumpen proletariat and Rastafarians to that of capitalist and middle class elements. By some estimates, the trade between Jamaica and the United States steadily grew to $1 billion per year (Stone, 1986, p. 43; Thomas, 1988). There now exist widespread networks encompassing great numbers from the middle class and subordinate classes. "Ganja men" of lowly social background could (and did) purchase mansions and expensive cars, and move into the legal economy by purchasing choice tourist hotels (as one did in Montego Bay) and participating in other capitalist ventures. All this kept the capitalist promise alive and did much to thwart any radical change.

▶ Epilogue

After two consecutive JLP governments led by Edward Seaga, the PNP, with Michael Manley at its helm, was returned to office in February of 1989. In spite of the marked differences between the two governments (the Seaga years were characterized by renewed dependence on Western formulas and aid), there exist continuities that bear out our findings from the earlier period. What are the major trends? As even the Workers' Party of Jamaica (WPJ) was forced to recognize, the weakness of the working class has become more pronounced. More and more, unemployment has led to "individual forms of survival," of which marginal self-employment has been the most prominent (Munroe, 1987). The trend is more striking among male workers, almost one-half of whom were self-employed by 1982 (Doeringer, 1988, p. 476; Munroe, 1987, p. 41). It is arguable that the upsurge of petty bourgeois consciousness is the mere concretization of a mindset having a less than wholehearted commitment to socialism. The rapid rate at which it emerged makes a persuasive statement.

In spite of the laissez-faire ideology of the 1980s, the state has continued to play a significant role in employment and the economy as a whole. It is true that the growth of the civil service was curtailed, as its ranks dropped from 108,000 in 1980 to 99,800 by October 1983. However, this was due more to the attempt by the JLP government to weed out PNP "clients" than to any significant decline in functions.

This does not mean, however, that the differences between the PNP and the JLP have vanished. The WPJ, for its part, has continued to seek "to extend the possibilities of cooperation with the PNP" and to maintain the posture, adopted in the late 1970s, that the PNP is part of the "democratic forces" (Munroe, 1987, pp. 40–41).

A perhaps more demanding force, rebounding from the frustrations of the 1980s, is represented by the bureaucratic/entrepreneurial middle class. Unquestionably, this emergent fraction continues to make its voice heard. Some black middle class professionals, helped by the permanent vacuum left by the emigrants of the 1970s (only a few of whom returned), moved into high positions in the professions and government, making their forays into entrepreneurial activities from those bases. Others became outright entrepreneurs, taking advantage primarily of opportunities in import substitution. Tangible evidence of the continuing trend is provided by the creation of a black merchant bank, Jamincorp, and the inroads of new entrepreneurs into the stock exchange. Here the prime example is that of Mark Ricketts, a black Jamaican with extensive links to government agencies (including consulting for the IMF), who became the most influential newcomer into the stock exchange.

At the same time that it has benefited from these encouraging trends, the fraction has also been stymied by its old antagonists. Tensions with the dominant capitalist class and international capital reemerged early in the new Manley government, expressed largely through the themes of ideology and race. Immediately after the 1989 elections, the more articulate among the fraction as well as their mentors (university professors, lawyers, businessmen, bankers, and politicians) launched a broadside against the new Manley regime and the capitalists. The commentators were particularly irked by the role that racism had been playing in their economic stagnation. They chastised the new regime for failing to renew its earlier commitment to the fraction.

The racial complexion of the big capitalists and heads of influential parastatals—especially those that generated significant amounts of foreign exchange, like Air Jamaica, was raised as a significant issue. Members of the fraction have continued to complain bitterly about the fact that the parastatals that exercise real economic and financial power are controlled by white or near-white individuals, leaving them lesser positions in the administrative and regulatory apparatuses. Early in 1989 a new daily newspaper, spearheaded by the black fraction, *The Jamaica Record,* carried a series of articles focused on race called "The Black/White Power Structure" (March 19; April 2, 1989). One of the writers, Arnold Bertram (a well-known PNP radical and minister of culture in the 1976–80 Manley government), raised the issue of the dominance of whites and near whites in the economic power structure. As Bertram reports it, a black U.S. Congressional support staff member "on . . . being invited to attend . . . functions for the owners of big business in Jamaica [observed] that in India we see Indians, in China we see Chinese but in Jamaica we see very few Blacks in such places" (*Jamaica Record,* March 19, 1989, p. 5A).

Clearly the patterns that were visible in the late 1970s have continued. As the fraction presses on in its bid for power, the capitalist class continues to counterattack. The capitalists were able to maintain a holding pattern in the 1980s. For instance, Ricketts was led to resign from the Kingston Stock Exchange because of blocks in his path to progress there (*Jamaica Record,* March 19, 1989, p. 1B). Likewise, Jamincorp failed, the oft-expressed view being that it was sabotaged by the capitalists.

These developments are important conceptually because they support our conclusions that the vast majority of this powerful fraction of the middle class was more interested in the pursuit of naked capitalist gain than egalitarian principles. Ricketts's words are a vivid testimony of the trend: "I want to own my own estate, my own plantation, my own power base. I will take the risks to achieve these. I just don't want to be a manager of anybody's plantation irrespective of the rewards" (*Jamaica Record,* March 19, 1989). Democratic socialism was the means to a capitalist end.

Notes, Bibliography, and Index

NOTES

Chapter 1

1. See Avramovic et al. (1964) for a presentation of the savings-gap model and Griffin (1978, chap. 3) for a persuasive discussion of its problems.

2. Dependency theory refers to a body of theorizing about Third World development that combines a critique of classical economics with aspects of neo-Marxist analysis. The main tenet of dependency theory is that economic relations between developed capitalist and Third World countries, far from leading to the development of the latter, produce economic and social distortions and "arrested development." Especially in its early formulations, dependency theory focussed on critiques of foreign capital or transnational corporations.

3. See essays in Jefferson and Girvan (1971). Quashie was the name given to the traditionally servile (but cunning) slave. See Patterson (1967) for an excellent profile.

4. *Abeng,* founded on February 1, 1969, was seen as the mouthpiece of the Black Power Movement. In its heydey, it had a circulation of some 20,000 copies annually, many of which went to the rural areas. Its office and equipment were mysteriously destroyed in a manner suggesting foul play. Although its members were accused of armchair socialism, it was credited with pushing Jamaican politics left of center and even won the acclaim of the PNP's David Coore in 1969 for its role in getting the government to initiate tax reforms. See Hinds (1970) and Nettleford (1970, pp. 131–135) for accounts.

5. Interview, September 24, 1976. We wish to thank Dr. Michael Allen of the University of the West Indies, Mona campus, who conducted the interview, for the use of those materials.

6. The main sources for these agreements are Press Release, Consulate General of Jamaica, New York, November 22, 1974; Securities Exchange Commission, Annual Report (Form 10K), fiscal year ending December 21, 1974; *New York Times,* November 21, 1974; Press release, A. F. Sabo Associates for the Government of Jamaica, December 20, 1974; *Daily Gleaner,* December 21, 1974; *Wall Street Journal,* April 11, 1975.

7. The Committee of Twenty, created in 1972, was charged with proposing a major reform of the international monetary system. It included ten members of long standing from industrial countries (the Group of Ten, including European countries, the United States, and Japan) and ten new representatives from developing countries. The new members pressed for more favorable consideration of these countries' interests.

8. See "Document submitted by Jamaica on behalf of countries in African, Asian and Latin American groups and Romania," (UNCTAD, 1973, pp. 2–5).

Chapter 2

1. Nyerere (1973). Scholars like Chaliand (1977, p. 24) note that this is a mere manipulation of tradition designed to mask social antagonisms.

2. "Root of the Matter," *Public Opinion,* September 5, 1952; address to the NEC of the PNP, September 28, 1975; speech to citizens of Montego Bay, April 7, 1976 (all in Hearne, ed., 1976). The latter is an example of distancing, the act of distinguishing old Manleyism from the new version.

3. The Jamaica-Cuba Cooperation Agreement of 1975 included Cuban assistance with regard to projects in agriculture, fishing, technology, education, and so on. Training for these projects included the exchange of personnel. Young Jamaicans received training in Cuba; Cubans in Jamaica were mainly involved in building schools and small dams and working in hospitals.

4. The other main variants are dependent neo-colonialism (when a weak state and a weak bourgeoisie initiate policies largely for the benefit of foreign capital); national developmentalism (when a strong national bourgeoisie takes on foreign capital mainly in its own interest without any redistribution of the extracted surplus, as in the case of national popularism); and national populism (very much akin to national popularism, except that the subordinate classes tend to be represented by formal political organizations (unions, peasant organizations, etc.). The greater degree of organization in national populism gives these governments a gauge of their support that is lacking in national popularism. These governments' ability to use resources appropriately, in turn, accounts in large measure for their greater stability. See Petras (1975) and Camillieri (1976). We take these largely descriptive categories as our starting point but then develop the concept of national popularism as an analytical tool.

5. These concepts define classes that have either a nonexistent or tenuous connection to the formal capitalist economy. In classical terminology, lumpens (or the lumpen proletariat) comprise the "ne'er-do-wells," the criminal element, those who eke out a living through legal or illegal means on the margins of the capitalist economy. The subproletariat includes those who have occasional, unstable employment in the capitalist economy. The difficulties inherent in applying these concepts in Third World settings are widely recognized, since in such settings those at the lowest ranks frequently shift from one economic activity to another in attempts to accommodate to structural unemployment. The term "marginals" is sometimes applied to these groupings. We prefer the terms lumpen and subproletariat, however. We often employ them together, recognizing the importance of conceptual distinctions but also the difficulty of separating them in practice.

6. Color differentiation in Jamaica has its origins in sociohistorical circumstances. The "free Coloureds" of the slave period (often the offspring of planters or overseers and slaves) were the precursors of the "brown" middle class of this century. The "black" middle class is of more recent vintage and its social origins are more firmly in the lower reaches of society, where African parentage is more marked. Perhaps more important, the terms connote different social status and orientation.

Chapter 3

1. At the time, Jamaica was still a Crown Colony. This type of authoritarian government was imposed after the bloody Morant Bay Rebellion in 1865. Afraid that blacks would control power, the Jamaican Assembly surrendered power to Britain; the plantocracy reasoned that Britain would protect white rule. For good accounts of the rebellion, see Semmel (1962), and Robotham (1981).

2. It goes without saying that the liberal democratic model bestowed upon Jamaica by its colonial masters has been redefined by the realities of local political life. In Jamaica and elsewhere in the Third World these realities—scarcity most prominent among them—have given rise to non-classical, authoritarian forms of democracy (see Therborn, 1979). Clientelism, discussed in later chapters, provides an example of such forms in which, paradoxically, non-democratic patron-client political relations serve to maintain formally democratic institutions.

3. The nucleus consisted of members of the Jamaica Progressive League, formed and located in New York City on September 1, 1936. W. Adolphe Roberts was its president and Jamie O'Meally was one of its two vice-presidents. Local foundation members of the PNP were N. N. Nethersole, a fellow Rhodes Scholar, H. P. Jacobs, and O. T. Fairclough, the first editor of the *Public Opinion,* which was to become a mouthpiece for the PNP.

4. *Daily Gleaner,* August 31, 1938; Bustamante to Norman, July 19, 1938, British Colonial Office document (hereinafter C.O.) 137/835.

5. Failure in electoral politics was guaranteed by the tensions generated by color and the scourge of slavery. Capitalists would also later finance the campaigns of the PNP and exercise power by becoming heads of, or influential forces within, quasi-governmental agencies.

6. Observation made by Major Orde-Browne whose *Report on Labour Conditions in the West Indies* (1939) was one of the bases for Britain's subsequent labor policies toward the region.

7. The *Gleaner* newspaper was founded by powerful capitalist interests in 1834.

8. Petition published in *Plain Talk,* April 30, 1938, p. 7 (cited in Post, 1978, p. 263). The identical posture was struck by the St. Thomas Tax and Ratepayers Association.

9. This policy defined the limits of British radicalism. During the late 1930s and early 1940s, powerful British politicians (Sir Walter Citrine and Sir Stafford Cripps, for example) advocated a strong labor movement and even, in Cripps' case, the installation of socialism. They did not, however, support Marxism-Leninism.

10. This was demanded by striking workers at Islington, St. Ann. They were less concerned with increased wages than with better working conditions and work of longer duration; see Phelps (1960, p. 425).

11. The political role of religion is nothing new. The British labor movement owed much to the agitational politics of the Methodists; see Oxnam (1944, pp. 13–20).

12. Two of McNeil's sons were to become ministers of government: Roy (minister of home affairs under the JLP) and Kenneth (minister of health under the PNP).

13. McLaughlin is remembered for contesting his election loss to George Seymour-Seymour, who was successfully represented by Norman Manley.

14. Buchanan is an interesting case. By 1944 he joined the Democratic Labour Party, the capitalist party, reasoning that, in Marxist terms, its progressive development of the forces of production would promote working-class consciousness.

15. Members of the black petty bourgeoisie also relied on Garveyism to agitate for an industrial bank; they also formed the Jamaica Traders and Consumers Association, which maintained a hostile posture against ascendant Lebanese and Chinese merchants (Post, 1978, pp. 208–209).

16. *Garvey,* 1967, I, p. 44.

17. *Public Opinion,* December 24, 1937, p. 13 (cited in Post, 1978, p. 227).

18. *The Workman,* December 8, 1939, p. 4.

19. Sydney Webb, *British Weekly,* June 29, 1883, p. 146 (cited in Britain, 1982). Douglas Hall (1972a, p. 38) similarly observed: "[Garvey's] great lesson . . . was that the giving and receiving of respect had nothing to do with black, brown, or white, but with character and behavior."

20. Rubin and Zavalloni (1969, pp. 368–369). This was a study of Trinidadian youth conducted in 1957 and 1961.

21. When the Department of Education encouraged schools to plant market gardens in 1937, the resistance movement was led by the Western Federation of Teachers (Carnegie, 1973, p. 31).

Chapter 4

1. See Sir Frank Stockdale, *Development and Welfare in the West Indies, 1943–1944* (Colonial no. 189; London: HMSO, 1945); *Report of Mission of United Kingdom Industrialists on Industrial Development in Jamaica, Trinidad, Barbados and British Guiana* (Colonial no. 294; London: HMSO, 1953).

2. The first concerted attempt resulted in a World Bank study in 1952; see International Bank for Reconstruction and Development, *Report by the Mission on the Economic Development of Jamaica* (Baltimore: Johns Hopkins University Press, 1952).

3. This thesis of cultural destruction (seemingly shared by Post) is suspect. Cultural accommodation is more appropriate, as writers such as Schuler (1980), Brathwaite (1971), and Nettleford (1978) advocate convincingly.

4. Capt. Ernest Platt to Bernard Rickatson-Hatt, April 18, 1941, C.O. 137/850 (cited in Post, 1981, I, p. 183).

5. The 1953 and 1959 reforms made the cabinet the "principal instrument of policy," with overall responsibility to direct and control government. The premier was to be consulted in the appointment of future governors, and ministers were empowered to negotiate international aid and loan agreements. See Barnett (1977, pp. 18–22) for a succinct account of these new powers. Other areas involved successes against the old political order (Legislative Council), negotiating farm workers' contracts with the United States, and providing educational opportunities for members of the middle and subordinate classes; see *Annual Report—Jamaica* (1946, 1948).

6. Members of the JPL, a founding organization of the PNP, and the Rev. Ethelred Brown were favorably received when they took the reform message to the country parishes.

7. "Memorandum on Constitutional Change in Jamaica," February 9, 1939, C.O. 137/834 (cited in Post, 1981, I, p. 53).

8. The words of the voice of Jamaican capitalism of the day—H. G. DeLisser, editor of the *Daily Gleaner*.

9. "Memorandum on the Political and Social Conditions in the West Indies and British Guiana presented by the International African Bureau, the League of Coloured Peoples and the Negro Welfare Association," C.O. 950/30.

10. "Memorandum on Constitutional Change in Jamaica," February 9, 1939, C.O. 137/834, 68714 (cited in Post, 1981, I, p. 49).

11. Under Governor Sir Arthur Richards, political activists were interned on the flimsiest of charges. Bustamante as well as Richard Hart were so interned between 1940 and 1942. As a vilified group, Rastafarians had good reasons to support these initiatives.

12. Kingston and St. Andrew Civic League; see *Daily Gleaner,* January 26, 1937, p. 31 (cited in Post, 1978; p. 191).

13. People's National Party, 1949, "Report of the Second Annual Conference" (Kingston: People's National Party, p. 11).

14. *Ibid.,* pp. 12–13.

15. These associations led rank and file leaders like Hugh Buchanan to divert support from the PNP. In the meantime, Bustamante was demonstrating his independence by flouting his "wealth" to some capitalists: "Today you are almost wealthy—not as wealthy as I am of course" (cited in Post, 1981, 2:318). These displays diminished mass belief in his duplicity.

16. Garvey in *Blackman,* January 1930 (cited in Hart, 1967).

17. A. A. Morris, *The Worker,* December 8, 1938, p. 4 (cited in Post, 1981, I, p. 47).

18. Marx explains the concept in the *Paris Manuscripts;* see also Avineri (1971, chap. 3).

19. These hasten the development of the productive forces, a process that, for Marx, meant a corresponding development of working-class consciousness and the undermining of capitalism.

20. The revisions were merely formal. Indications are that Bustamante never departed from his autocratic approach to leadership (Bradley, 1960, pp. 404–410).

21. People's National Party, *Plan for Progress,* Kingston, 1954, p. 9.

22. C.O. 950/86, 24, cited in Post, 1978, p. 369.

23. PNP, "Report of the Second Annual Conference," p. 11.

24. Cited in Post, 1981, p. 120.

Chapter 5

1. Eyre to Newcastle, January 27, 1864, C.O. 137/378 (cited in Sires, 1955, p. 14). The indications are that the Jews were opportunistic; most of their alliances were based on expediency.

2. C.O. 137/379, February 19, 1864 (cited in Sires, 1955, p. 14).

3. In time, the Exchange became the center of vital information relative to sugar. There are indications that at times valuable information on prices was turned against the interest of the planter.

4. These resulted from mercantilist economics: colonies should produce raw materials, not manufactured goods. Sugar has the potential to develop several industrial link operations, but these were discouraged and even legislated against; see Beckford (1972) for a structural-historical analysis of the process.

5. It is estimated that 2,750 peasants were being forced off the land annually (with an additional annual decline of 2,000 agricultural wage earners) between 1921 and 1943; see Post (1978, p. 132).

6. Kirkwood always saw the merchants as a force that would undermine agriculture as the engine of economic development; personal interview with Gannon, March 4, 1974 (Gannon, 1976, p. 213). Governor Richards even ascribed the 1938 rebellion to their tenacious power (Post, 1981, I, p. 291).

7. In 1938 the merchants had a voice of their own—the *Jamaica Standard*. It clearly challenged the old order and promoted fundamentally commercial interests.

8. By the 1970s, the Matalon interests (the ICD Group) achieved conglomerate status, with manufacturing providing the main base. We will see also that members of the family became advisers and supporters of the PNP.

9. In 1967, the JMA launched its "Buy Jamaican, Build Jamaica" exhibition with a marked emphasis on building consumer confidence in locally manufactured goods. In 1969, the organization's growing importance was dramatized when the prime minister (Hugh Shearer) addressed its annual general meeting. This was the first time any prime minister had ever done so.

10. A parallel problem was excessive price hiking. In 1973, 77 special investigations were conducted; this figure increased to 116 by 1974 (Ministry Paper #30, 1975).

11. The term "comprador" (from the Spanish for buyer) has been used by dependency theorists to denote a group or clan (i.e., the comprador bourgeoisie) that bears a subservient, clientelistic relationship to its metropolitan counterpart. This bias would later help to frustrate certain policies under democratic socialism; see Nelson, (1977, chap. 9, p. 18).

12. During the bauxite levy negotiations, Matalon was held in awe by fellow negotiators. Inside sources indicate that while the others suggested a top levy figure of $80 million, Matalon convinced them to settle for a much higher figure. The final figure was $200 million.

13. Pat Rousseau, a partner of Myers, Fletcher and Gordon, was in charge of assessing the value of the bauxite companies for their restructuring under the PNP.

Chapter 6

1. This was the position of Lord John Russell; see Bell and Morrell, eds. (1968, p. 412).

2. These tabloids, controlled by coloreds, stoutly advocated a political future for Jamaica with blacks and coloreds in the ascendancy. They also stressed the key role of education in the future of these groups.

3. Campbell (1976, p. 239). The coloreds grew in political importance. They really became a critical political mass after the new constitution of 1854. Whites even supported a few black candidates in the hopes of preventing or undermining any alliance between blacks and coloreds.

4. Barkly to Newcastle, March 22, 1854, C.O. 137/322.

5. Samuel W. Blackwell to Henry Taylor, February 5, 1854, C.O. 71/117.

6. Grey to Pakington, June 26, 1853, C.O. 137/313.

7. In the Morant Bay Rebellion (1865), 439 people lost their lives, 600 were flogged, and 1,000 or so homes were destroyed.

8. This attitude was also widespread in Trinidad (Williams, 1962, p. 214) and Guyana (Brereton, 1979, p. 77).

9. Robert Kirkwood, head of the Sugar Manufacturers' Association and a member of the Tate and Lyle family, took these threats from such powerful Jews as Douglas Judah very seriously.

10. Lord Olivier's memorandum to the West India Royal Commission, First Session, September 25, 1938, C.O. 950/28 E. 55.

11. W. Adolphe Roberts, "Self Government for Jamaica," New York: Jamaica Progressive League, 1937 (pamphlet).

12. Jamaica, Report for the Year 1961 (London: HMSO, 1965), p. 21.

13. These mainly middle-class consumers were referred to as the "been to's." The label implies the passion for giving information on the various foreign locations visited and the formal gatherings attended.

14. Partly as a result of emigration, such traditionally "white" law firms as Myers, Fletcher and Gordon and Dunn, Cox and Orrett either merged with black firms or recruited black lawyers.

15. Comments by an entrepreneur in Montego Bay. "Cuffie" is a cornerstone member of the powerful and influential DeLisser family.

16. Anonymous interview, July 15, 1981.

17. These are exclusive, residential sections of Montego Bay. The issue of "ganja money" raises the important question of the impact of drugs on the development of democratic socialist consciousness. By some estimates, between US$500 million and US$1 billion is generated by the underground economy. A sizeable portion remains in the economy and serves to perpetuate capitalist, consumerist practices.

18. Interview with Neville James, Executive Director, PSOJ, June 4, 1983. In time organizations like the JMA and the Jamaica Tourist Board were headed by the prominent black middle-class members.

19. This refers to the 1985 Seaga administration (anonymous source).

20. Speech by Winston Meeks, president of the Jamaican Chamber of Commerce, February 27, 1975 (p. 4 of mimeo, provided by the Chamber).

Chapter 7

1. This characteristic had much to do with the adjunct role trade unions play in political parties and, in turn, the largely capitalist role of these political parties.

2. *Area Handbook for Jamaica* (Washington, D.C.: Government Printing Office, 1976), p. 280.

3. There is a notable exception, Sam Brown. This prominent Rasta philosopher contested a seat in Kingston in 1962. He was one of the first to link the movement with freedom and development at home.

4. This continuing stance is supported by the belief that Rastafarians are the true Israelites, also by the claim that the Ark was taken to Ethiopia (Campbell, 1988, p. 77).

5. This was quite an achievement, as the brethren did not trust Norman Manley's motives. They were troubled by Manley's reluctance to deal squarely with the race issue.

6. Henry held the younger Manley in high esteem. He published a pamphlet in 1959 that designated himself, Manley, and Haile Selassie as the "Trinity of the Godhead," the leadership structure to emancipate black people.

7. As elements within the class became more educated and gained political power, they increased their demands. An excellent critical retrospective is provided by the *Jamaica Record,* March 19 and April 2, 1989; see especially Rex Nettleford's "Black Empowerment—Between Assertion and Delicate Balancing" on p. 3A of April 2nd issue.

8. Even those sympathetic to the Rastafarian cause comment disparagingly on the mindset. For example, Campbell (1988, p. 149) sees the movement's philosophy built on "a form of bourgeois ideology which tolerated the individualism of the western capitalist systems."

9. Historically, this was the time that the title "small man" was replaced by "sufferer," reflecting profound changes in socio-economic perceptions. Manley (1974, p. 47) declared the importance of this new approach: "This is why it is vital for political leadership to 'back its jacket' and get in among the people at the roughest working levels."

10. This was a departure from tradition. Katrin Norris (1969) notes that in 1962 Jamaican men married "to raise the colour." Many well-educated black women were left single or without the opportunity to marry social equals.

Chapter 8

1. Political slogans have always been an important part of Jamaican politics. By this time they began to lose their usual frivolity and theatricality. Instead one found them to be linked to specific and more ideological issues; see Miller (1981).

2. The main detractors were the middle class. We now know that the Rastafarians were engaged in a project of cultural transformation to gel with their basic doctrine of African Redemption. The transformation of language is perhaps the centerpiece of this project; see Pollard (1980); Alleyne (1988); Simpson (1986, pp. 147–149).

3. Agency for Public Information, "Minimum Wage?" information sheet (Kingston: A.P.I., n.d.). A complete breakdown of wages is provided by the *Pocketbook of Statistics, Jamaica* 1977, p. 55.

4. There were isolated pockets of constructive radicalism. One was led between late 1968 and early 1969 by Milton "Scully" Scott, a former NWU organizer. He managed to secure a following of 6,000 workers from the Jamaica Maritime Union as well as many others from among dockworkers loading bananas.

5. Emigration to Britain, which grew dramatically over this period, contributed to this growth. Many from the rural areas sold their possessions to go abroad; others moved to the urban areas, creating a bottleneck that still exists.

6. A good example was the bauxite industry. It was widely known that certain categories of casual workers (sample men, for example) would make far more than a Grade I clerk in the civil service.

7. It is noteworthy that some of the radicals of the 1970s accord Henry the same stature as other Jamaican popular martyred leaders such as Paul Bogle, leader of the Morant Bay Rebellion.

8. These were powerful, charismatic figures who could leave safe constituencies and win seats in the JLP camp. There is also the case of F. L. B. Evans ("Slave Boy"), who in 1951 fought and won against Gladys Longbridge (Lady Bustamante) in a by-election in Westmoreland. This was not Evans' prime constituency. In their fall from grace, none were able to establish independent political bases. After the immensely popular Ken Hill was ousted from the PNP in 1952, he could raise no more than 1.4 percent of his constituency vote for his new National Labour Party.

9. Among them was Richard Hart, a former stalwart of the PNP. His National Freedom Movement was seen as a nuisance and his advocacy of radical unionism as dangerous; his impressive but practically ineffective appearance before the Joint Committee for Jamaica's independence in 1961 made the political mainstream uneasy. The PNP and JLP feared any anti-establishment ideology: a founding member (V. L. Arnett) stressed the PNP's awareness of this, especially since the unemployed, the un-unionized, and the lumpen proletariat outnumbered the unionized, upon which the two-party system is built. The PNP jealously guard the two-party system even in cases, like that of the YSL, where the opposition shares a common ideological bent. Norman Manley referred to them as "intellectual Rastafarians" bent on destroying the PNP (*Daily Gleaner,* February 28, 1966).

10. *Socialism!* 2, no. 6 (June 1975): 17.

11. Up to 1968 Rudie's music was more anti-colonial than class-oriented or racial. Songs like "Sounds and Pressure" and "Keep up the Pressure" were also directed at rival political gangs (Waters, 1985, p. 89). Nettleford (1970, p. 99) reminds us that many songwriters were often not conscious of the lyrics they dispensed.

12. *Abeng* 1, no. 33 (September 13, 1969).

13. This is not to suggest unbroken racial harmony. Ethnic prejudice compounded with such accusations as selfishness, lack of national spirit, clannishness, and so forth have led Jamaicans to attack Chinese and drive fear into the hearts of other minorities; see Lowenthal (1972, pp. 204–207), and Holzberg (1987). The Chinese population in Jamaica declined from 78,000 in 1972 to 20,000 in 1980.

Chapter 9

1. There was a contradiction inherent in their position, in that a considerable part of the economy was foreign-owned. Important examples include utilities (100%) (before government purchase), tourism (55%), and finance (60%). Foreign ownership was generally substantial for export-oriented manufacturing firms, less so for smaller, locally oriented ones (Owen Jefferson, "Is the Jamaican Economy Developing?" *New World Quarterly* 5, no. 4 [1972]:4).

2. Capitalist involvement in state-initiated ventures included such projects as housing estates, agribusiness, and so forth. See Chapter 10 for a discussion of the changing relationship between the PNP and the capitalist class. The class became much more united in its resistance with the downturn of the economy after 1975.

3. Sources informed us that they were encouraged to introduce new lines and goods but feared taking the risks.

4. The debt service ratio increased from 7.3% in 1973 to 15.3% in 1979.

5. Manley (1987, pp. 79–80) can hardly conceal his disappointment in Guinea's chicanery in capturing a bigger market share. Disappointment in the Soviet Union was revealed in an interview with Dr. Carlton Davis, Executive Director, Jamaica Bauxite Institute (August 19, 1977).

6. Agency for Public Information, "Highlights of the Opening Speech of the Budget Debate, May 15th, 1975 by Hon. David Coore, Minister of Finance."

7. These events signaled the beginning of a campaign by the *Gleaner* against the regime. At its height (mid-1977), three prominent columnists (Wilmot Perkins, John Hearne, and David D'Costa) were engaged in an all-out attack. It is estimated that emigrants took $300 million out of the Jamaican economy in 1976 alone.

8. This was a most defiant, anti-imperialist speech. Popular support for the regime was present but not appreciable. The "support" of the lumpen and subproletariats is particularly interesting. Progressively after 1976, rival PNP and JLP gangs attempted to patch up their differences, on the assumption that they were being used destructively by the major political parties. In early 1978, rival gangs for Rema and Concrete Jungle, tough gang areas, called a truce and started to work with the police to set up counseling centers for the youths in the area; see *The Star,* January 12 and 16, 1978. There are some indications that these ventures were destined to function outside of the formal political system.

9. According to anonymous sources, Manley became more and more impatient with his economic advisers. He felt that they were more interested in ousting foreign capital than in finding ways to promote production.

10. A well-placed member of the JLP informed us that the party was confident that Manley would not declare Jamaica a one-party state. He cited the weight of local political traditions. The fear of a one-party state was used by the JLP for propaganda appeal.

11. Duncan was reinstalled as the PNP's General Secretary in 1979 due mainly to pressure from the rank and file. However, the party's equilibrium was already shattered.

12. After 1976, names of prominent "socialists" became linked to real estate deals and business operations that really stretched the meaning of democratic socialism. In one case, the dealings involved a prominent conservative capitalist and supporter of the JLP.

13. The industrialists expressed a concern mainly with the "non-productive" nature of the fraction's input and its role in "overload[ing] Government bureaucracy" (*Daily Gleaner,* July 10, 1976, p. 6).

14. This was the opinion of T. G. Mignott, Chairman of the All Island Cane Farmers' Association. See *Daily News,* July 16, 1977, p. 15.

15. See, for instance, "Intellectuals and Power," and "To Bright Young Men," *Weekly Gleaner,* May 28, 1979, pp. 11, 27.

16. "Worker Participation—B.I.T.U." in Trade Union Center, *Papers on Worker Participation* (Mona, Jamaica: The Centre, n.d.), pp. 11–12.

17. It was widely known, for example, that a prominent civil servant, with access to vital state information, was leaking it to the opposition, yet nothing was done about him. This gentleman is known to maintain his department's filing system in his head.

18. Deduced in an interview with Eddie Hall, Executive Director, JMA (June 27, 1983).

19. The judiciary was also among the recalcitrant state apparatuses. The courts often attacked government policies (labor laws, for example) and in a notable case a High Court judge (Parnell) encouraged law suits against the regime.

Chapter 10

1. From the onset of industrialization in the 1950s, arable land in cultivation declined partly owing to loss of land to bauxite mining and construction. In terms of percentage of the GDP, agriculture plunged from 24 percent in 1950 to 7 percent at the time of democratic socialism. In the 1960s agriculture (especially sugar) had become a reservoir for surplus labor that could not be absorbed by the other economic sectors. The government had systematically held back the mechanization of the sugar industry in order to save jobs. However, in the long run, this policy added to the problems of the industry.

2. Interview, Dr. Reginald Gonzalez, executive director, JMA (August 3, 1977).

3. Jamaica Chamber of Commerce, "Report from Trade (Standing) Committee," July 18, 1977, p. 3. The merchants were quite resourceful. In 1974 they established a liaison committee to hold meetings with the minister of commerce and marketing on matters affecting the distributive trade (*Daily News,* December 13, 1974).

4. "Report from Trade (Standing) Committee," p. 4.

5. During the foreign exchange shortages of 1977, the Chamber tried to make a case for its members getting preference over the manufacturers (Jamaica Chamber of Commerce, "President's Monthly Statement," June 1977).

6. See Agency for Public Information, "Many Manufacturers Are Taking the Bull by the Horns" (Kingston: The Agency, n.d.), mimeo.

7. The *Plan* laid great stress on the role of Community Enterprise Organizations (CEO's) to give "economic power to the people." CEO's were designed to promote worker-peasant economic leadership "once foreign capital is driven out" (Beckford and Witter, 1982, pp. 116–124).

8. Some thirty-two agreements between Jamaica and socialist countries were signed between 1976 and 1980. It seems clear that, while many were not a good fit in terms of Jamaica's long-term economic needs, many others had simply not been taken advantage of.

9. Interview, anonymous JDB official (August 11, 1977).

10. Manley, 1987, pp. 284–285, tables 17–18.

11. The volume index of bauxite production fell from 116 in 1974 to 86 in 1975; the alumina volume index plunged from 138 in 1974 to 108 in 1976 (Boyd, 1988, p. 2).

12. We have personal experience with the automotive parts project. As consultants, we detected great reluctance on the part of investors to undertake risks without some measure of protection from the state.

13. Redistributive policies concerning the middle class were discussed in Chapter 9. Thus the class is not included in the present discussion.

14. "Excerpt from Statement in the House, 27th May, 1975," in Hearne (1976, pp. 242–243). Again, an appeal to the capitalist's sense of fair play is made.

15. This was legislation based on the traditional master-servant relationship rather than on modern concepts of legal rights and rational contract.

16. Statement by Derrick Rochester, parliamentary secretary, Ministry of Labour, *Daily News,* May 5, 1977.

17. Some details can be found in "Worker Participation in the Frome Monymusk Land Company Limited," in *Papers on Worker Participation,* 1975. This is a memorandum submitted to the Committee on Worker Participation. Also see *Daily Gleaner,* September 27, 1975.

18. Jamaica Employers' Federation Memo, p. 1, in *Committee on Worker Participation,* in 1976, p. 25.

19. For the position of the PSOJ, see *Daily Gleaner,* January 28, 1977.

20. "Workers Participation—B.I.T.U," in *Papers on Worker Participation,* 1975, p. 5.

21. The arbitrary power to widen the range of essential services greatly troubled all trade unionists; see "State of the Nation Debate (Senate)," July 22, 1977, reported in *Daily News,* July 23, 1977, p. 2. This was confirmed in an interview with Lascelles Perry, president, United Portworkers' Union (March 22, 1989). Perry held a number of influential positions in the NWU (general secretary and assistant island supervisor).

22. Statement attributed to the Jamaica Exporters' Association. *Daily News,* April 28, 1977.

23. *Project Land Lease,* (Agency for Public Information pamphlet), November 15, 1974, p. 2.

24. *Ibid.,* p. 3.

25. *Ibid.*

26. Carl Stone, "Tenant Farming under State Capitalism," in Stone and Brown, eds. (1976, p. 331).

27. *Jamaica Hansard* 1, no. 1 (Session 1972–73, May 16, 1972), p. 40. The pressure exerted by the lumpen proletariat was dramatically reflected in the budget. The Special Employment Programme (SEP) was started by the JLP government with an allocation of $5.2 million (2.8 percent of total government expenditures. By 1971–72 government expenditures had grown by 38 percent but the SEP allocations had grown by 58 percent, or 3.2 percent of total expenditures (Brown, 1981, p. 12).

28. Nyerere Farm (7,000 acres) was established in 1974. The project was designed to cater to 150 young farmers. Each was to receive five acres individually, with the rest to be farmed communally (*Daily News,* July 17, 1977, p. 4). The project failed from management difficulties, partisan attacks, and a lack of discipline and ideological commitment on the part of the young farmers. As late as April 1977 the regime was still having difficulty getting urban youths to relocate (*Daily Gleaner,* April 3, 1977). There was also substantial support for claims that the farm was used to traffic in ganja, guns, and stolen goods; indeed the democratic socialist consciousness was conspicuously absent (*Daily Gleaner,* June 17, 1976).

BIBLIOGRAPHY

Adams, Nassau. 1969. "An Analysis of Food Consumption and Food Import Trends in Jamaica, 1950–1963." *Social and Economic Studies* 17, no. 1 (March): 1–22.

———. 1971. "Import Structure and Economic Growth in Jamaica, 1954–1967." *Social and Economic Studies* 20, no. 3 (September): 235–266.

Agency for Public Information, Jamaica (API). 1974–1977. (p. 397).

Ahmad, Aijaz. 1978. "Democracy and Dictatorship in Pakistan." *Journal of Contemporary Asia* 8: 477–512.

Alleyne, Mervin C. 1988. *Roots of Jamaican Culture*. London: Pluto.

Althusser, Louis. 1971. *Lenin and Philosophy*. New York: Monthly Review.

Ambursley, F. 1983. "Jamaica: From Michael Manley to Edward Seaga." In *Crisis in the Caribbean,* ed. F. Ambursley and R. Cohen, 72–104. New York: Monthly Review.

Amin, Samir. 1974. *Accumulation on a World Scale*. New York: Monthly Review.

———. 1976. *Unequal Development: An Essay on the Social Formations under Peripheral Capitalism*. New York: Monthly Review.

Amsden, Alice H. 1979. "Taiwan's Economic History: A Case of Etatisme and a Challenge to Dependency Theory." *Modern China* 5, no. 3 (July): 341–380.

Andrade, Jacob A. P. M. 1941. *A Record of the Jews in Jamaica from the English Conquest to the Present Time*. Kingston: Jamaica Times.

Annual Report—Jamaica. 1953–1961. London: HMSO and Kingston: Government Printing Office.

Augier, F. R. 1962. "The Consequences of the Morant Bay Rebellion." *New World Quarterly* 8, no. 3, (September): 21–42.

Augier, F. R., et al. 1960. *The Making of the West Indies*. London: Longmans.

Avineri, Shlomo. 1971. *Social and Political Thought of Karl Marx*. Cambridge, Eng.: Cambridge University Press.

Avramovic, Dragoslav, et al. 1964. *Economic Growth and External Debt.* International Bank for Reconstruction and Development (World Bank), Economic Department. Baltimore: Johns Hopkins University Press.

Ayearst, Morley. 1960. *The British West Indies: The Search for Self-Government.* London: Allen and Unwin.

Ayub, Mahmood Ali. 1981. *Made in Jamaica.* Baltimore: Johns Hopkins University Press.

Balibar, Etienne. 1977. *On the Dictatorship of the Proletariat.* London: New Left.

Balogh, Thomas. 1970. *The Economics of Poverty.* London: Wiedenfeld and Nicholson.

Bamat, Thomas. 1988. "Political Change and the Catholic Church in Brazil and Nicaragua." In *New Perspectives on Social Class and Socioeconomic Development in the Periphery,* ed. Nelson W. Keith and Novella Keith, 119–149. Greenwich Conn.: Greenwood.

Barkan, Joel D. 1975. *An African Dilemma: University Students, Development and Politics in Ghana, Tanzania and Uganda.* New York: Oxford University Press.

Barnett, Lloyd G. 1977. *The Constitutional Law of Jamaica.* London: Oxford University Press.

Barrett, Leonard. 1977. *The Rastafarians: Sounds of Cultural Dissonance.* Boston: Beacon.

Beachey, R. W. 1977. *The British West Indies Sugar Industry in the Late Nineteenth Century.* Oxford: Basil Blackwell.

Beckford, George L. 1967. *The West Indian Banana Industry.* Mona: Institute for Social and Economic Research, University of the West Indies.

———. 1972. *Persistent Poverty.* New York: Oxford University Press.

Beckford, George L., and Michael Witter. 1982. *Small Garden . . . Bitter Weed: Struggle and Change in Jamaica.* London: Zed.

Beckwith, Martha. 1929. *Black Roadways.* Chapel Hill: University of North Carolina Press.

Bell, Kenneth N., and W. P. Morrell, eds. 1968. *Select Documents on British Colonial Policy, 1830–1860.* London: Oxford University Press.

Bell, Wendell. 1964. *Jamaican Leaders.* Berkeley: University of California Press.

Bell, W., and J. W. Gibson. 1978. "Independent Jamaica Faces the Outside World." *International Studies Quarterly* 22, no. 1: 5–47.

Bernal, Richard. 1984. "The IMF and Class Struggle in Jamaica, 1977–1980." *Latin American Perspectives* (Issue 42) 11, no. 3 (Summer): 53–82.

Bienen, Henry. 1977. *Kenya: The Politics of Participation and Control.* Princeton, N.J.: Princeton University Press.

Blanchard, Paul. 1947. *Democracy and Empire in the Caribbean.* New York: Macmillan.

Board of Inquiry into Labour Disputes between Trade Unions. 1950. *Report.* Kingston: Government Printing Office.

Borodulina, T. K. 1972. *K. Marx, F. Engels, V. Lenin: A Collection.* Moscow: Progress Publishers.

Bourne, Compton. 1981. "Government Foreign Borrowing and Economic Growth: The Jamaica Experience." *Social and Economic Studies* 30, no. 4 (December): 52–71.

Boyd, Derick A. C. 1988. *Economic Management, Income Distribution, and Poverty in Jamaica*. New York: Praeger.

Bradley, C. Paul. 1960. "Mass Parties in Jamaica: Structure and Organization." *Social and Economic Studies* 9, no. 4: 375–416.

Brathwaite, Edward. 1971. *The Development of Creole Society in Jamaica, 1770–1820*. London: Oxford University Press.

Brereton, Bridget. 1980. *Race Relations in Colonial Trinidad, 1870–1900*. Cambridge, Eng.: Cambridge University Press.

———. 1981. *A History of Modern Trinidad, 1783–1962*. London: Heinemann.

Brewster, Havelock, and Clive Thomas. 1967. *The Dynamics of West Indian Economic Integration*. Mona: Institute for Social and Economic Research, University of the West Indies.

Britain, Ian. 1982. *Fabianism and Culture*. Cambridge, Eng.: Cambridge University Press.

Brown, Adlith. 1981. "Economic Policy and the IMF in Jamaica." *Social and Economic Studies* 30, no. 4 (December): 1–49.

Brown, E. Ethelred. 1919. "Labor Conditions in Jamaica Prior to 1917." *Journal of Negro History* 4 (October): 349–360.

Camillieri, Joe. 1976. "Dependence and the Politics of Disorder." *Arena*.

Campbell, Horace. 1988. *Rasta and Resistance: From Marcus Garvey to Walter Rodney*. Trenton, N.J.: Africa World.

Campbell, Mavis Christine. 1976. *The Dynamics of Change in a Slave Society: A Sociopolitical History of the Free Coloreds of Jamaica*. Rutherford, N.J.: Fairleigh Dickinson University Press.

Capital Development Fund (CDF). 1975, 1976. *Annual Report*. Kingston: Ministry of Finance.

Carnegie, James. 1973. *Some Aspects of Jamaica's Politics, 1918–1938*. Kingston: Sangsters.

Carnoy, Martin. 1984. *The State and Political Theory*. Princeton, N.J.: Princeton University Press.

Cassidy, Frederic G. 1971. *Jamaica Talk*. London: Macmillan Education.

Central Bureau of Statistics. 1949. Review of Economic Conditions in Jamaica, 1939–1948. Bulletin no. 20. Kingston: The Bureau.

Chaliand, Gerard. 1977. *Revolution in the Third World*. New York: Viking.

Chapman, Esther. 1938. "The Truth about Jamaica," *West Indian Review*, no. 10 (July).

———. 1954. *Development in Jamaica: Year of Progress, 1954*. Kingston: Arawak.

Charles, Pearnel. 1977. *Detained*. Kingston: Kingston Publishers Limited.

Chen-Young, Paul. 1977a. "Commentary." *Economic Report, Jamaica* 2, no. 11 (January): 1–11.

———. 1977b. "Commentary." *Economic Report, Jamaica* 3, no. 2 (April): 1–14.

Chevannes, Barry. 1977. "The Literature of Rastafari." *Social and Economic Studies* 26: 239–262.

———. 1981. "The Rastafari and the Urban Youth." In *Perspectives of Jamaica in the Seventies,* ed. Carl Stone and Aggrey Brown, 392–422. Kingston: Jamaica Publishing House.

Clarke, Colin G. 1975. *Kingston, Jamaica: Urban Development and Social Change, 1692–1962.* Berkeley: University of California Press.

———. 1983. "Dependency and Marginality in Kingston, Jamaica." *Journal of Geography* 82, no. 5: 227–235.

Clarke, Edith. 1969. "The Social Worker." *Public Opinion,* September 6.

Cliffe, Lionel. 1982. "Class Formation as an 'Articulation' Process: East African Cases." In *Introduction to the Sociology of "Developing" Countries,* ed. H. Alavi and T. Shanin, 262–278. London: Macmillan.

Cohen, David, and Jack P. Greene, eds., *Neither Slave nor Free.* Baltimore: Johns Hopkins University Press.

Colburn, Forrest D. 1986. *A Post-Revolutionary Nicaragua.* Berkeley: University of California Press.

Committee on Worker Participation. 1976. Report on Worker Participation in Jamaica. Kingston: The Committee.

Consolidated Development Fund. *See* Capital Development Fund.

Cox, Edward L. 1984. *Free Coloreds in the Slave Societies of St. Kitts and Grenada, 1763–1833.* Knoxville: University of Tennessee Press.

Craton, Michael, and James Walvin. 1970. *A Jamaican Plantation: The History of Worthy Park, 1670–1970.* Toronto: University of Toronto Press.

Cronon, Edmund. 1955. *Black Moses.* Madison: University of Wisconsin Press.

Cross, Malcolm. 1979. *Urbanization and Urban Growth in the Caribbean.* Cambridge, Eng.: Cambridge University Press.

Debray, Regis. 1971. *The Chilean Revolution.* New York: Vintage.

deLisser, H. G. 1944–1945. "Triumphant Squalitone: A Tropical Extravaganza." *Planter's Punch* (Kingston) 1: 5–72.

Delson, Roberta Marx. 1981. *Caribbean History and Economics.* London: Gordon and Breach.

Democratic Socialism in Jamaica: The Government's Policy for National Development. Pamphlet circa 1974, p. 1.

Diaz-Alejandro, Carlos. 1978. "Delinking North and South: Unshackled or Unhinged?" In *Rich and Poor Nations in the World Economy,* ed. Albert Fishlow et al., 105–122. New York: McGraw-Hill.

Doeringer, Peter. 1988. "Market Structure, Jobs and Productivity: Observations from Jamaica." *World Development* 16, no. 4: 465–482.

Dutt, R. Palme. 1935. *Fascism and Social Revolution.* New York: International Publishers.

Eaton, George E. 1975. *Alexander Bustamante and Modern Jamaica.* Kingston: Kingston Publishers Limited.

Economic Report, Jamaica. 1977a. Vol. 2, no. 11 (January).

———. 1977b. Vol. 3, no. 2 (April).

Economic and Social Survey, Jamaica. 1972, 1973, 1976, 1981, 1982. Kingston: National Planning Agency.

Economic Survey of Jamaica. 1957–1971. Kingston: Central Planning Unit.

Edie, Carlene. 1986. "Domestic Policies and External Relations in Jamaica under Michael Manley." *Studies in Comparative International Development* 21, no. 1 (Spring): 71–94.

———. 1989. "From Manley to Seaga: The Persistence of Clientelist Politics in Jamaica." *Social and Economic Studies* 38, no. 1: 1–35.

Eisner, Gisela. 1961. *Jamaica, 1830–1930: A Study in Economic Growth.* Manchester, Eng.: Manchester University Press.

Elkins, W. F. 1978. "Revolt of the British West India Regiment." *Jamaica Journal* 11, nos. 3–4: 73–75.

Erskine, Noel Leo. 1981. *Decolonizing Theory: A Caribbean Perspective.* New York: Orbis.

Essien-Udom, E. U. 1966. *Black Nationalism.* Harmondsworth, Eng.: Penguin.

Evans, Peter. 1979. *Dependent Development: The Alliance of Multinational, State and Local Capital in Brazil.* Princeton, N.J.: Princeton University Press.

Eyre, L. Alan. 1983. "The Ghettoization of an Island Paradise." *Journal of Geography* 82, no. 5 (September–October): 36–39.

Fagen, Richard R. et al., eds. 1986. *Transition and Development: Problems of Third World Socialism.* New York: Monthly Review.

Faristzaddi, Millard. 1982. *Itations of Jamaica and I Rastafari.* New York: Grove.

Floyd, Barry. 1979. *Jamaica: An Island Microcosm.* New York: St. Martin's.

Forsythe, Dennis. 1974a. "West Indian Culture through the Prism of Rastafarianism." *Caribbean Quarterly* 26, no. 4 (April): 62–81.

———. 1974b. "Repression, Radicalism, and Change in the West Indies." *Race* 15, no. 4 (April): 401–429.

Foster-Carter, Aidan. 1978. "The Modes of Production Controversy." *New Left Review* (January–February): 47–77.

Foweraker, Joe. 1981. *The Struggle for Land: A Political Economy of Pioneer Frontier in Brazil from 1930 to the Present Day.* Cambridge, Eng.: Cambridge University Press.

Francis, A. A. 1981. *Taxing the Transnationals in the Struggle over Bauxite.* The Hague: Institute of Social Studies.

Franklin, Raymond S. 1981. "Party, Class, and State: A Leninist and Non-Leninist View." In *Democratic Socialism,* ed. Bogdan Denitch, 61–81. Montclair, N.J.: Allanheld, Osmun.

Gannon, John Charles. 1976. "The Origins and Development of Jamaica's Two-Party System, 1930–1975." Doctoral dissertation, Washington University, St. Louis, Mo.

Garvey, Amy Jacques, ed. 1967. *Philosophy and Opinions of Marcus Garvey.* 2 vols. New York: Atheneum.

Gerth, Hans, and C. W. Mills. 1958. *From Max Weber: Essays in Sociology*. New York: Oxford University Press.

Girvan, Norman. 1967. *The Caribbean Bauxite Industry*. Mona: Institute for Social and Economic Research, University of the West Indies.

———. 1968. "After Rodney: The Politics of Student Protest in Jamaica." *New World Quarterly* 4, no. 3 (High Season): 59–61.

———. 1971. *Foreign Capital and Economic Underdevelopment in Jamaica*. Mona: Institute of Social and Economic Research, University of the West Indies.

Girvan, Norman, and Richard Bernal. 1982. "The IMF and Foreclosure of Development Options: The Case of Jamaica." *Monthly Review* 33, no. 9 (February): 34–60.

Girvan, Norman, Richard Bernal, and W. Hughes. 1980. "The IMF and the Third World: The Case of Jamaica." *Development Dialogue* 2: 113–155.

Glass, Ruth. 1962. "Ashes of Discontent: The Past as Present in Jamaica." *Monthly Review* 14 (May): 14–19.

Gomez, Walter. 1976. "Bolivia: Problems of a Pre- and Post-Revolutionary Export Economy." *Journal of Developing Areas* 10 (July): 461–483.

Gonsalves, Ralph. 1976. "The Trade Union Movement in Jamaica: Its Growth and Some Resultant Problems." In *Essays on Power and Change in Jamaica*, ed. Carl Stone and Aggrey Brown, 227–278. Mona: Department of Government and Extra Mural Centre, University of the West Indies.

———. 1979. "The Rodney Affair and Its Aftermath." *Caribbean Quarterly* 25, no. 3 (September): 1–24.

Gooding, Earl. 1981. *The West Indies at the Crossroads*. Cambridge, Mass.: Schenkman.

Gordon, Derek. 1978. "Working Class Radicalism in Jamaica: An Exploration of the Privileged Worker Thesis." *Social and Economic Studies* 27, no. 3 (September): 313–341.

Gould, David J. 1980. *Bureaucratic Corruption and Underdevelopment in the Third World: The Case of Zaire*. New York: Pergamon.

Green Paper on Agricultural Development Strategy. 1973. Reprinted in the *Daily News*, November 23 to December 4.

Green Paper on Industrial Development Programme: Jamaica, 1975–1980. Publication details not given.

Griffin, Keith. 1978. *International Inequality and National Poverty*. New York: Holmes and Meier.

Haffner, Sebastian. 1972. *Failure of a Revolution: Germany 1918–19*. New York: Library Press.

Hall, Douglas. 1959. *Free Jamaica, 1838–1865*. New Haven, Conn.: Yale University Press.

———. 1972a. "The Ex-Colonial Society in Jamaica." In *Patterns of Foreign Influence in the Caribbean*, ed. Emmanuel deKadt, 32–48. London: Oxford University Press.

———. 1972b. "Jamaica." In *Neither Slave Nor Free,* ed. David W. Cohen and Jack P. Greene, 193–213. Baltimore: Johns Hopkins University Press.

Halliday, Fred, and Maxine Molyneux. 1981. *The Ethiopian Revolution.* London: New Left.

Halpern, Manfred. 1963. *The Politics of Social Change in the Middle East and North Africa.* Princeton, N.J.: Princeton University Press.

———. 1968. "The New Middle Class." In *Politics in Transitional Societies,* ed. Harvey G. Kebschull, 195–202. New York: Appleton-Century-Crofts.

Hamilton, B. St. J. 1964. *Problems of Administration in an Emergent Nation: A Case Study of Jamaica.* New York: Praeger.

Handbook of Jamaica. 1881–1962. Kingston: Government Printing Office.

Harris, Donald J. 1976. "Notes on the Question of a National Minimum Wage for Jamaica." In *Essays on Power and Change in Jamaica,* ed. Carl Stone and Aggrey Brown, 282–305. Mona: Department of Government and Extra Mural Centre, University of the West Indies.

Harrod, Jeffrey, 1972. *Trade Union Foreign Policy: A Study of British and American Trade Union Activities in Jamaica.* New York: Doubleday.

Hart, Richard. 1967. "The Life and Resurrection of Marcus Garvey." *Race* 9, no. 2 (October): 217–237.

———. 1972. "Jamaica and Self-Determination, 1660–1970." *Race* 13, no. 3 (January): 271–297.

Hearne, John. 1976. *A Search for Solutions: Selections from the Speeches and Writings of Michael Manley.* Oshawa, Ont., Can.: Maple House.

Helleiner, G. K. 1974. "The Less-Developed Countries and the International Monetary System." *Journal of Development Studies* 10, nos. 3–4 (April–July): 347–373.

Henriques, Fernando. 1968. *Family and Colour in Jamaica.* London: MacGibbon and Kee.

Heuman, Gad J. 1981. *Between Black and White: Politics and the Free Coloreds in Jamaica, 1792–1865.* Westport, Conn.: Greenwood.

Hinds, Glenville. 1970. "The Theory and Practice of Abeng." Mona: Department of Government, University of the West Indies. Mimeo.

Holzberg, Carol S. 1977. "Race, Ethnicity, and the Political Economy of National Entrepreneurial Elites in Jamaica." Ph.D. dissertation, Boston University.

———. 1987. *Minorities and Power in a Black Society: The Jewish Community in Jamaica.* Boston: North-South.

Horowitz, Irving Louis. 1982. *Beyond Empire and Revolution.* New York: Oxford University Press.

Howe, James. 1975. "Power in the Third World." *Journal of International Affairs* 29, no. 2 (Fall): 113–127.

Hoyos, F. A. 1974. *Grantley Adams and the Social Revolution.* London: Macmillan.

Hurwitz, Samuel, and Edith Hurwitz. 1965. "The New World Sets an Example for the Old: The Jews of Jamaica and Political Rights, 1661–1831." *American Jewish Historical Quarterly* 55, no. 1: 37–56.

———. 1971. *Jamaica: A Historical Portrait*. New York: Praeger.

International Bank for Reconstruction and Development (IBRD). 1952. *Report by the Mission on the Economic Development of Jamaica*. (Baltimore: Johns Hopkins University Press).

———. 1974. *Current Economic Position and Prospects of Jamaica*. Report no. 257 a-JM (February 21).

Jamaica Chamber of Commerce. 1968–1978. *Reports of the Regular Monthly Meeting of Directors*. Kingston: The Chamber.

———. 1977. *Report from the Trade (Standing) Committee*. Kingston: The Chamber.

Jamaica Hansard. 1968. "Proceedings of the House of Representatives." session 1968–69, vol. 1, no. 1, March 28–October 22.

James, C.L.R. 1969. *A History of Pan African Revolt*. Washington, D.C.: Drum and Spear.

———. 1970. "The West Indian Middle Classes." In *Readings in Government and Politics of the West Indies*, Trevor Munroe and Rupert Lewis, 192–197. Mona: Department of Government, University of the West Indies.

Jefferson, Owen. 1972. *The Post-War Economic Development of Jamaica*. Mona: Institute of Social and Economic Research, University of the West Indies.

Jefferson, Owen, and Norman Girvan, eds. 1971. *Readings in the Political Economy of the Caribbean*. Kingston: New World.

Jeffrey, Henry B., and Colin Baber. 1986. *Guyana: Politics, Economics, and Society*. London: Frank Pinter.

Jones, Edwin. 1978. "The Political Uses of Commissions of Enquiry (2): The Post-Colonial Jamaican Context." *Social and Behavioral Studies* 27, no. 3 (September): 285–312.

Karol, K. A. 1970. *Guerillas in Power*. New York: Hill and Wang.

Kaufman, Michael. 1985. *Jamaica under Manley: Dilemmas of Socialism and Democracy*. London: Zed.

———. 1988. "Democracy and Social Transformation in Jamaica." *Social and Economic Studies* 37, no. 3 (September): 45–73.

Keith, Nelson W. 1988. "Class Analysis in the Periphery: A Partial Analysis." In *New Perspectives on Social Class and Socioeconomic Development in the Periphery*, eds. Nelson W. Keith and Novella Keith, 11–51. Greenwich, Conn.: Greenwood.

Keith, Nelson W., and Novella Keith. 1985. "The Rise of the Middle Class in Jamaica." In *Middle Classes in Dependent Countries*, ed. Dale Johnson, 67–106. Beverly Hills, Calif.: Sage.

———. 1988. *New Perspectives on Social Class and Socioeconomic Development in the Periphery*. Greenwich, Conn.: Greenwood.

Kirton, Claremont, and Mark Figueroa. 1981. "State Trading: A Vital Element in Caribbean Economic Development." In *Perspectives on Jamaica in the Seventies*, ed. Carl Stone and Aggrey Brown, 139–172. Kingston: Jamaica Publishing House.

Kitching, Gavin. 1983. *Rethinking Socialism*. London: Methuen.

Knox, Graham. 1965. "Political Change in Jamaica (1866–1906) and Local Reaction to Crown Colony Government." In *The Caribbean in Transition,* ed. F. M. Andic and T. W. Matthews, 144–162. Rio Pedras, Puerto Rico: Institute of Caribbean Studies.

Kuper, Adam. 1976. *Changing Jamaica*. London: Routledge and Kegan Paul.

The Labour Force. 1976. Kingston: Government of Jamaica, Department of Statistics.

Lacey, Terry. 1977. *Violence and Politics in Jamaica, 1960–1970*. Manchester, Eng.: Manchester University Press.

Laclau, Ernesto. 1977. *Politics and Ideology in Marxist Theory*. London: New Left.

La Guerre, John. 1971. "The Moyne Commission and the West Indian Intelligentsia." *Journal of Commonwealth Political Studies* 9, no. 2 (July): 143–157.

Leeson, Phil. 1981. "Capitalism, Statism, and Socialism." In *The Popular and the Political,* ed. Mike Prior, 1–19. London: Routledge and Kegan Paul.

Lenin, V. I. 1972. "The Proletarian Revolution and the Renegade Kautsky." In *K. Marx, F. Engels, V. Lenin: A Collection,* ed. T. K. Borodulina, 603–618. Moscow: Progress.

Lewin, A. 1982. "The Fall of Michael Manley: A Case Study of the Failure of Reform Socialism." *Monthly Review* 3 (February): 49–60.

Lewis, Gordon K. 1968. *The Growth of the Modern West Indies*. London: MacGibbon and Kee.

Lewis, Rupert. 1976. "Black Nationalism in Jamaica." In *Essays on Power and Change in Jamaica,* ed. Carl Stone and Aggrey Brown, 165–181. Mona: Department of Government and Extra Mural Centre, University of the West Indies.

———. 1987a. "Garvey's Forerunners: Love and Bedward." *Race and Class* 28, no. 3 (Winter): 29–40.

———. 1987b. *Marcus Garvey: Anti-Colonial Champion*. London: Karia.

Litvak, Isaiah A., and Christopher J. Maule. 1975. "Nationalisation in the Caribbean Bauxite Industry." *International Affairs* 51, no. 1 (January): 43–50.

Lofchie, Michael F. 1974. "The Political Origins of Uganda Coup." *Journal of African Studies* 1, no. 4 (Winter): 18–37.

Long, Anton. V. 1956. *Jamaica under the New Order, 1827–1847*. Mona: Institute of Social and Economic Research, University College of the West Indies.

Lowenthal, David. 1972. *West Indian Societies*. New York: Oxford University Press.

MacGaffey, Janet. 1983. "How to Survive and Become Rich Amidst Devastation: The Second Economy in Zaire." *African Affairs* 82, no. 328 (July): 351–366.

McLarty, R. W. 1919. *Jamaica: Our Present Condition and Crisis*. Kingston.

Macmillan, W. M. 1938. *Warning from the West Indies: A Tract for the Empire*. Harmondsworth, Eng.: Penguin.

Magdoff, Harry. "The U.S. Dollar, Petrodollars, and U.S. Imperialism." *Monthly Review* 30, no. 8 (January): 1–13.
Mandel, Ernest. 1970. *Europe versus America*. New York: Monthly Review.
Manley, Michael. 1970. "Overcoming Insularity in Jamaica." *Foreign Affairs Quarterly* 49, no. 1 (October): 100–110.
———. 1974. *The Politics of Change*. London: Andre Deutsch.
———. 1976. *Not for Sale*. San Francisco: Editorial Consultants. Year approximate.
———. 1982. *Jamaica: Struggle in the Periphery*. Oxford: Third World Media.
———. 1987. *Up the Down Escalator*. London: Andre Deutsch.
Marcus, Jacob R. 1970. *The Colonial Jew, 1492–1776*, vol. 1. Detroit: Wayne University Press.
Maritain, Jacques. 1951. *Man and State*. Chicago: Chicago University Press.
Markakis, John, and Nega Ayele. 1986. *Class and Revolution in Ethiopia*. Trenton, N.J.: Red Sea.
Markovic, Mihailo. 1982. *Democratic Socialism: Theory and Practice*. New York: St. Martin's.
Marshall, Woodville K. 1968. "Notes on Peasant Development in the West Indies since 1838." *Social and Economic Studies* 17, no. 13 (September): 252–263.
Marx, Karl. 1968. *The Economic and Philosophical Manuscripts of 1844*, ed. David J. Struik, New York: International Publishers.
———. 1975. *Capital*, ed. Frederick Engels. 3 volumes. New York: International Publishers.
Marx, Karl, and Frederick Engels. 1972. *The German Ideology*, ed. C. J. Arthur. New York: International Publishers.
Michels, Robert. 1915. *Political Parties*. Glencoe, Ill.: Free Press.
Miliband, Ralph. 1969. *The State in Capitalist Society*. New York: Basic Books.
———. 1977. *Marxism and Politics*. London: Oxford University Press.
Miller, Claude. 1976. "A Review of Our Political Line." *Socialism!* 3, nos. 8 and 9 (August–September): 2–41.
Miller, J.D.B. 1962. *The Nature of Politics*. London: Duckworth.
Miller, Marion. 1981. "A Comparative Content Analysis of the Issue Content of the 1967 and 1976 General Elections in Jamaica." In *Perspectives on Jamaica in the Seventies*, ed. Carl Stone and Aggrey Brown, 244–256. Kingston: Jamaica.
Millette, James. 1968. "The Caribbean Free Trade Association: The West Indies at the Crossroads." *New World Quarterly* 4, no. 4: 32–40.
Ministry Paper (Jamaica) no. 13. 1977. "Incomes Policy 1977." Circa April 21.
——— no. 28. 1977. "The IMF Agreement." Ministry of Finance, File no. 406/02, circa April 21.
——— no. 30. 1975. "Prices and the Prices Commission." Ministry of Marketing and Commerce, June 16.
——— no. 38. 1976. "The Special Employment Programme 1976–1977." September.

Morrell, W. P. 1966. *The British Colonial Policy in the Age of Peel and Russell.* London: Frank Cass.
Morrish, Ivor. 1982. *Obeah, Christ and Rastaman.* Cambridge, Eng.: James Clarke.
Moyne Commission: Report. 1938. Report of the Commission to Inquire into the Disturbances Which Occurred in Jamaica between 23rd May and the 8th June 1938, Kingston. Submitted memoranda used in the text are Memoranda submitted by Messrs. N. W. Manley, N. N. Nethersole, and Dr. W. E. McCulloch: British Colonial Office document (hereafter C.O.) 950/86; Memorandum on Constitutional Change in Jamaica: C.O. 137/834; and Memorandum on the Political and Social Conditions in the West Indies and Guiana presented by the International African Bureau, the League of Colored Peoples, and the Negro Welfare Association, C.O. 950/30.
Munroe, Trevor. 1966. "The People's National Party: A View of the Early Nationalist Movement in Jamaica." Master's thesis, University of the West Indies.
———. 1972. *The Politics of Constitutional Decolonization.* Mona: Institute of Social and Economic Research, University of the West Indies.
———. 1977. "The Marxist 'Left' in Jamaica, 1940–1950." Working Paper no. 15. Mona: Institute of Social and Economic Research, University of the West Indies.
———. 1981. *Social Classes and National Liberation in Jamaica.* Kingston: Vanguard.
———. 1987. "Jamaica: The WPJ in Struggle for Democracy." *World Marxist Review* 30, no. 2 (February): 34–42.
Munroe, Trevor, and Rupert Lewis, eds. 1971. *Readings in Government and Politics of the West Indies.* Mona: Department of Government, University of the West Indies.
Munroe, Trevor, and Don Robotham. 1977. *Struggles of the Jamaican People.* Kingston: Workers' Liberation League.
Napier, Robert. 1957. "The First Arrest of Bedward." *Jamaica Historical Review Bulletin* 2, no. 1 (March).
National Accounts Income and Expenditure, 1950–1957. 1959. Kingston: Government of Jamaica, Department of Statistics.
National Income and Product, 1976. 1976. Kingston: Government of Jamaica, Department of Statistics.
National Income and Product, 1978. 1978. Kingston: Government of Jamaica, Department of Statistics.
National Income and Product, 1982. 1982. Kingston: Government of Jamaica, Department of Statistics.
Nelson, Cecil. 1974. "The Class Structure of Jamaican Society." *Socialism!* 1, no. 1: 15–33.
———. 1977. "Turn the Setback into an Advance." *Socialism!* 2, no. 2 (June): 2–26.
Nettleford, Rex M. 1970. *Mirror, Mirror.* Kingston: William Collins and Sangster.

———. 1971. *Norman Washington Manley: Manley and the New Jamaica.* New York: Africana.
———. 1978. *Caribbean Cultural Identity.* Los Angeles: Center for Afro-American Studies and UCLA Latin American Center Publications, University of California.
Nicholas, Tracy. 1979. *Rastafari: A Way of Life.* Garden City, N.Y.: Anchor.
Norris, Katrin. 1969. *Jamaica: The Search for Identity.* London: Oxford University Press.
Nyerere, Julius K. *Ujamaa: Essays on Socialism.* London: Oxford University Press.
Ogbuagu, Chibuzo S. A. 1983. "The Nigerian Indigenization Policy: Nationalism or Pragmatism?" *African Affairs* 82, no. 327 (April): 241–266.
Olivier, Lord Sydney. 1936. *Jamaica, the Blessed Island.* London: Faber and Faber.
Orde-Brown, G. St. J. 1939. *Labour Conditions in the West Indies.* London: HMSO.
Ottaway, David, and Marina Ottaway. 1981. *Afrocommunism.* New York: Africana.
Owens, Joseph. 1976. *Dread: The Rastafarians of Jamaica.* London: Heinemann.
Oxnam, G. Bromley. 1944. *Labor and Tomorrow's World.* New York: Abingdon-Cokesbury.
Palmer, Colin. 1989. "Identity, Race, and Black Power in Independent Jamaica." In *The Modern Caribbean,* ed. Franklin W. Knight and Colin Palmer, 111–128. Chapel Hill: University of North Carolina Press.
Palmer, Ransford. 1979. *Caribbean Dependence on the United States Economy.* New York: Praeger.
Papers on Worker Participation. 1975. Mona: Trade Union Education Institute.
Patterson, Orlando. 1967. *The Sociology of Slavery: An Analysis of the Origins, Development, and Structure of Negro Society in Jamaica.* Rutherford, N.J.: Fairleigh Dickinson University Press.
Payne, Anthony, and Paul Sutton. 1984. *Dependency under Challenge: The Political Economy of Commonwealth Caribbean.* Manchester, Eng.: Manchester University Press.
People's National Party (PNP). 1939. *Outline of Policy and Programme.* People's National Party. Kingston: People's National Party.
———. 1949. "Report of the 2nd Annual Conference." Kingston: People's National Party.
———. 1954. *Plan for Progress.* Kingston: People's National Party.
———. 1966. A Plan for Action Today. Kingston: People's National Party.
———. 1974. *Democratic Socialism for Jamaica: The Government Policy for National Development.* Kingston: Agency for Public Information.
———. 1979. "Principles and Objectives." Approved by the 40th Annual Party Conference, September 1978.
Petras, James. 1975. "New Perspectives on Imperialism and Social Classes in the Periphery." *Journal of Contemporary Asia* 5, no. 3: 291–308.
Phelps, O. W. 1960. "Rise of the Labour Movement in Jamaica." *Social and Economic Studies* 9, no. 4 (December): 417–468.

Phillips, Peter. 1976. "Capitalist Elites in Jamaica." In *Essays on Power and Change in Jamaica,* ed. Carl Stone and Aggrey Brown, 93–120. Mona: Department of Government and Extra Mural Centre, University of the West Indies.

———. 1981. "Community Mobilisation in Squatter Communities." In *Perspectives on Jamaica in the Seventies,* ed. Carl Stone and Aggrey Brown, 423–436. Kingston: Jamaica Publishing House.

Pitman, Frank W. 1917. *The Development of the British West Indies, 1700–1763,* New Haven, Conn.: Yale University Press.

Pocketbook of Statistics, Jamaica 1976. 1977. Kingston: Government of Jamaica, Department of Statistics.

Pollard, Velma. 1980. "Dread Talk: The Speech of the Rastafarian in Jamaica." *Caribbean Quarterly* 26: 32–41.

Portes, Alejandro, and John Walton. 1981. *Labor, Class and the International System.* New York: Academic.

Post, Ken. 1969. "The Politics of Protest in Jamaica: Some Problems of Analysis and Conceptualization." *Social and Economic Studies* 18, no. 4 (December): 374–390.

———. 1978. *Arise Ye Starvelings: The Jamaican Labour Rebellion of 1938 and Its Aftermath.* The Hague: Martinus Nijhoff.

———. 1981. *Strike the Iron.* 2 vols. Atlantic Highlands, N.J.: Humanities.

Poulantzas, Nicos. 1974. *Fascism and Dictatorship.* London: New Left.

———. 1975. *Classes in Contemporary Capitalism.* London: New Left.

———. 1980. *State, Power, Socialism.* London: New Left.

Powell, John Duncan. 1971. *Political Mobilization of the Venezuelan Peasant.* Cambridge, Mass.: Harvard University Press.

Private Sector Organisation of Jamaica. n.d. "What's It All About: A Brief."

———. Task Force on Agriculture. 1977. "An Agricultural Programme for Jamaica."

Proctor, Jesse H. 1962. "British West Indian Society and Government in Transition, 1920–1960." *Social and Economic Studies* 11, no. 4 (December): 273–304.

Rance, Sir Hubert. 1950. *Development and Welfare in the West Indies 1947–1949.* (Colonial Report no. 264.) London: HMSO.

Reid, Stanley. 1976. "An Introductory Approach to the Concentration of Power in the Jamaican Corporate Economy and Notes on Its Origin." In *Essays on Power and Change in Jamaica,* ed. Carl Stone and Aggrey Brown, 37–92. Mona: Department of Government and Extra Mural Centre, University of the West Indies.

———. 1977. "Strategy and Policy Issues in Resource Bargaining: Case Study of the Jamaican Bauxite-Alumina Industry since 1974. Mona: Department of Management Studies, University of the West Indies. Unpublished manuscript.

Report of Mission of United Kingdom Industrialists on Industrial Development in Jamaica, Trinidad, Barbados and British Guiana. (Colonial Report no. 294.) 1953. London: HMSO.

Report of the Jamaican Sugar Industry Commission, 1944–45. Kingston: Government Printing Office.

Riviere, Claude. 1977. *Guinea: Mobilization of a People,* trans. Virginia Thompson. Ithaca, N.Y.: Cornell University Press.

Roberts, G. W. 1957. *The Population of Jamaica.* Cambridge, Eng.: Cambridge University Press.

Roberts, W. Adolphe. 1951. *Six Great Jamaicans.* Kingston: Institute of Jamaica.

———. 1955. *Jamaica: The Portrait of an Island.* New York: Coward-McCann.

Robinson, Joan. 1979. *Aspects of Development and Underdevelopment.* Cambridge, Eng.: Cambridge University Press.

Robotham, Don. 1981. "The Notorious Riot: The Socio-Economic and Political Bases of Paul Bogle's Revolt." Working Paper no. 28. Mona: Institute of Social and Economic Research, University of the West Indies.

Rodney, Walter. 1970. *The Groundings with My Brothers.* London: Bogle-L'Ouverture.

Ross, Shala. 1975. "Stage Set for Bourgeois Anti-Worker Law." *Socialism!* 2, no. 3 (March): 8–14.

Rubin, Vera, and Marisa Zavalloni. 1969. *We Wish to Be Looked Upon: A Study of the Aspirations of Youth in a Developing Society.* New York: Teachers' College Press.

Ryan, Selwyn D. 1974. *Race and Nationalism in Trinidad and Tobago: A Study of Decolonization in a Multiracial Society.* Mona: Institute of Social and Economic Research, University of the West Indies.

Saul, John S. 1979. "The Dialectics of Class and Tribe." *Race and Class* 10, no. 4: 347–372.

———. 1986. "The Role of Ideology in the Transition to Socialism." In *Transition and Development: Problems of Third World Socialism,* ed. Richard R. Fagan et al., 212–230. New York: Monthly Review.

Schuler, Monica. 1980. *Alas, Alas, Kongo.* Baltimore, Johns Hopkins University Press.

Schumpeter, Joseph A. 1942. *Capitalism, Socialism and Democracy.* New York: Harper & Brothers.

Schweitzer, Pierre-Paul. 1976. "Political Aspects of Managing the International Monetary System." *International Affairs* 52, no. 2 (April): 208–218.

Scobie, Edward. 1972. *Black Britannia.* Chicago: Johnson.

Semaj, Leahcim. 1980. "Rastafari: From Religion to Social Theory." *Caribbean Quarterly* 26, no. 4 (December): 22–31.

Semmel, Bernard. 1962. *The Governor Eyre Controversy.* London: MacGibbon and Kee.

Senior, Olive. 1972. *The Message Is Change: A Perspective on the 1972 General Elections.* Kingston: Kingston Publishers.

Shivji, Issa G. 1976. *Class Struggles in Tanzania*. New York: Monthly Review.

Sigmund, Paul E., ed. 1967. *Ideologies of Developing Nations*. New York: Praeger.

Simey, T. S. 1946. *Welfare and Planning in the West Indies*. Oxford, Eng.: Clarendon.

Simon, Yves. 1980. *A General Theory of Authority*. Notre Dame, Ind.: University of Notre Dame Press.

Simpson, George Eaton. 1986. "Religion and Justice: Some Reflections on the Ras Tafari Movement." *Journal of Caribbean Studies* 5, no. 3 (Fall): 145–154.

Singham, A. W. 1968. *The Hero and the Crowd in Colonial Polity*. New Haven, Conn.: Yale University Press.

Sires, Ronald V. 1955. "The Experience of Jamaica with Modified Crown Colony Government." *Social and Economic Studies* 4, no. 2: 150–167.

Sklar, Richard. 1979. "Beyond Capitalism and Socialism in Africa." *Journal of Modern African Studies* 17, no. 4: 1–21.

Skocpol, Theda. 1980. "Political Response to Capitalist Crisis: Neo-Marxist Theories of the State and the Case of the New Deal." *Politics and Society* 10, no. 2: 155–201.

Smith, M. G., Roy Augier, and Rex Nettleford. 1960. *The Ras Tafari Movement in Kingston, Jamaica*. Mona: Institute of Social and Economic Research, University of the West Indies.

Smith, Raymond T. 1970. "Social Stratification in the Caribbean." In *Essays in Comparative Social Stratification*, ed. L. Plotnicov, and A. Tuden, 43–76. Pittsburgh: University of Pittsburgh Press.

Smith, Shirley. 1976. "Industrial Growth, Employment Opportunity and Migration within and from Jamaica, 1943–1970." Ph.D. dissertation, University of Pennsylvania.

Smith, W. R., W.O.D. Sealy, and S. Gordon. 1972. *The Creation of New Public Agencies: Case Studies of the Barbados Marketing Corporation (Barbados); The Ministry of Rural Land Development (and Land Authority), Jamaica; The Port Authority (British Virgin Islands)*. Mona: University of the West Indies.

Socialism! Organ of the Workers' Liberation League (Kingston).

Statistical Digest. 1980. Kingston: Bank of Jamaica.

Statistical Yearbook of Jamaica. 1973–1990. Kingston: Department of Statistics, Government of Jamaica.

Stein, Judith. 1986. *The World of Marcus Garvey: Race and Class in Modern Society*. Baton Rouge, La.: Louisiana University Press.

Stephens, Evelyne Huber, and John Stephens. 1986. *Democratic Socialism in Jamaica: The Political Movement and Social Transformation in Dependent Capitalism*. Princeton, N.J.: Princeton University Press.

Stewart, John. 1808. *An Account of Jamaica and Its Inhabitants*. London: Longman, Hurst, Rees, and Orme.

Stewart, Robert. 1984. "The 1872 Diary of James Splaine, S.J., Catholic Missionary in Jamaica: A Documentary Note." *Caribbean Quarterly* 30, nos. 3–4 (September and December): 99–109.

Stockdale, Sir Frank. 1945. *Development and Welfare in the West Indies.* (Colonial Report no. 189.) London: HMSO.

Stone, Carl. 1973. *Class, Race and Political Behaviour in Urban Jamaica.* Mona: Institute for Social and Economic Research, University of the West Indies.

———. 1974. *Electoral Behaviour and Public Opinion in Jamaica.* Mona: Institute of Social and Economic Research, University of the West Indies.

———. 1976a. "Bauxite and National Development in Jamaica." In *Essays on Power and Change in Jamaica,* ed. Carl Stone and Aggrey Brown, 343–358. Mona: Department of Government and Extra Mural Centre, University of the West Indies.

———. 1976b. "Tenant Farming under State Capitalism." In *Essays on Power and Change in Jamaica,* ed. Carl Stone and Aggrey Brown, 307–342. Mona: Department of Government and Extra Mural Centre, University of the West Indies.

———. 1976c. "Worker Participation in Industry: A Survey of Workers' Opinions." In Committee on Worker Participation. *Report on Worker Participation in Jamaica,* 77–108. (No publication details given.)

———. 1980. *Democracy and Clientelism in Jamaica.* New Brunswick, N.J.: Transaction.

———. 1986. *Class, State, and Democracy in Jamaica.* New York: Praeger.

Stone, Carl, and Aggrey Brown. 1976. "Prices and Markups." In *Essays on Power and Change in Jamaica,* ed. Carl Stone and Aggrey Brown, 359–366. Mona: Department of Government and Extra Mural Centre, University of the West Indies.

———. 1981. *Perspectives on Jamaica in the Seventies.* Kingston: Jamaica.

Stone, Carl, and Aggrey Brown, eds. 1976. *Essays on Power and Change in Jamaica.* Mona: Department of Government and Extra Mural Centre, University of the West Indies.

Sugar Industry Enquiry Commission. 1966. *Report.* Kingston: Government Printing Office.

Sweezy, Paul M. 1968. *The Theory of Capitalist Development.* New York: Monthly Review.

Sweezy, Paul M., and Charles Bettelheim. 1971. *On the Transition to Socialism.* New York: Monthly Review.

Tafari, I. Jubulani. 1980. "The Rastafari: Successors of Marcus Garvey." *Caribbean Quarterly* 26, no. 4 (December): 1–12.

Taussig, M. 1978. "Peasant Economics and the Development of Capitalist Agriculture in Conca Valley, Colombia." *Latin American Perspectives* 5, no. 3: 62–91.

Tawney, R. H. 1926. *Religion and the Rise of Capitalism.* New York: Harcourt, Brace.

Taylor, John. 1979. *From Modernization to Modes of Production.* Atlantic Highlands, N.J.: Humanities.

Therborn, Goran. 1983. "Why Some Classes are More Successful Than Others." *New Left Review*, no. 138 (March–April): 37–56.

Thomas, Clive Y. 1984. *The Rise of the Authoritarian State*. New York: Monthly Review.

———. 1988. *The Poor and the Powerless*. New York: Monthly Review.

Tocqueville, Alexis de. 1945. *Democracy in America*. New York: Alfred A. Knopf.

Turner, Mary. 1982. *Slaves and Missionaries: The Disintegration of Jamaica Slave Society, 1787–1834*. Urbana: University of Illinois Press.

UNCTAD, Trade and Development Board. 1973. *Official Records*. Fifth Special Session. Supplement no. 1, Annex A. New York: United Nations.

UNIDO. 1975. *Declaration and Plan of Action of the General Conference of the United Nations Industrial Development Organisation*. Lima, Peru: UNIDO.

Vergopoulos, Kostas. 1983. "L'Etat dans le capitalisme peripherique." *Tiers Monde* 24: 35–52.

Verity, D. J. 1938. "The Sugar Industry and the Present Situation." *West Indian Review* 4, no. 12 (August): 6–12.

Vincent, Theodore G. 1976. *Black Power and the Garvey Movement*. Palo Alto, Calif.: Ramparts.

Waters, Anita M. 1985. *Race, Class, and Political Symbols: Rastafari and Reggae in Jamaican Politics*. New Brunswick, N.J.: Transaction.

Weber, Max. 1964. *The Sociology of Religion*. Boston: Beacon.

White, Garth. 1967. "Rudie, Oh Rudie." *Caribbean Quarterly* 13, no. 3: 39–44.

Will, H. A. 1970. *Constitutional Change in the British West Indies, 1880–1903*. Oxford, Eng.: Clarendon.

Williams, Eric. 1962. *History of the People of Trinidad and Tobago*. Port of Spain, Trinidad: People's National Movement (PNM).

Williams, James. 1975. "The Essence of Land Lease." *Socialism!* 2, no. 3 (March): 34–38.

Workers' Party of Jamaica (WPJ). 1981. Report of the Central Committee to the Second Congress of the Workers' Party of Jamaica. Kingston: Vanguard Press.

Wright, Philip. 1966. *Lady Nugent's Journal*. Kingston, Jamaica: Institute of Jamaica.

INDEX

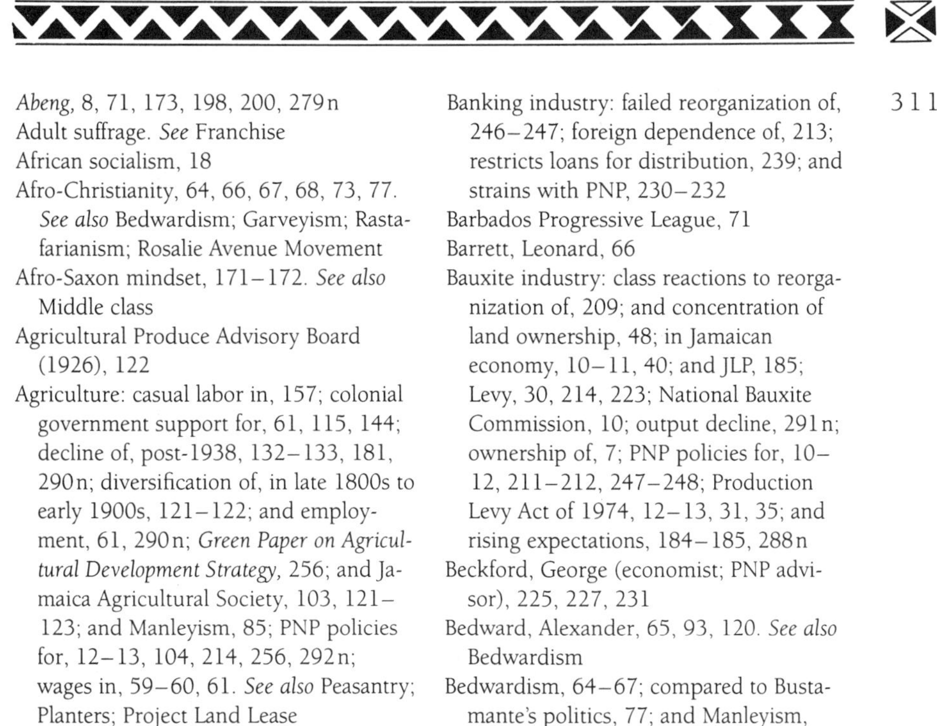

Abeng, 8, 71, 173, 198, 200, 279n
Adult suffrage. *See* Franchise
African socialism, 18
Afro-Christianity, 64, 66, 67, 68, 73, 77. *See also* Bedwardism; Garveyism; Rastafarianism; Rosalie Avenue Movement
Afro-Saxon mindset, 171–172. *See also* Middle class
Agricultural Produce Advisory Board (1926), 122
Agriculture: casual labor in, 157; colonial government support for, 61, 115, 144; decline of, post-1938, 132–133, 181, 290n; diversification of, in late 1800s to early 1900s, 121–122; and employment, 61, 290n; *Green Paper on Agricultural Development Strategy,* 256; and Jamaica Agricultural Society, 103, 121–123; and Manleyism, 85; PNP policies for, 12–13, 104, 214, 256, 292n; wages in, 59–60, 61. *See also* Peasantry; Planters; Project Land Lease
Anderson, Beverley, 173
Armed forces, and anti-government plot, 233–234

Banana industry: as alternative to sugar for planters, 117; challenges monopoly, 122; growth of, 115; ownership of by merchants, 116, 122–123
Banking industry: failed reorganization of, 246–247; foreign dependence of, 213; restricts loans for distribution, 239; and strains with PNP, 230–232
Barbados Progressive League, 71
Barrett, Leonard, 66
Bauxite industry: class reactions to reorganization of, 209; and concentration of land ownership, 48; in Jamaican economy, 10–11, 40; and JLP, 185; Levy, 30, 214, 223; National Bauxite Commission, 10; output decline, 291n; ownership of, 7; PNP policies for, 10–12, 211–212, 247–248; Production Levy Act of 1974, 12–13, 31, 35; and rising expectations, 184–185, 288n
Beckford, George (economist; PNP advisor), 225, 227, 231
Bedward, Alexander, 65, 93, 120. *See also* Bedwardism
Bedwardism, 64–67; compared to Bustamante's politics, 77; and Manleyism, 75–76
Bell, Eric (PNP), 214, 221, 232
Bernal, Richard (advisor to Michael Manley), 13
Bertram, Arnold (PNP), 199, 200, 214, 275
Black Power Movement, 67, 161, 184, 186, 191, 203. *See also* Bedwardism; Rastafarianism; Rodney, Walter; Rudie boys

Blake, Vivian (PNP), 214
Bogle, Paul, 66, 199–200
Brewster, Havelock, 8
Brigade League ("Brigadistas"), 22, 233–234
British West India Sugar Association, 117
Buchanan, Hugh C. (labor organizer), 59, 71, 282n, 284n
Bureaucratic/entrepreneurial fraction (of middle class): ideology of, 167–174; in 1980s, 274–275; increased power of, 223–225; origins of, 148–150; links between political and economic activities of, 29–30, 35, 139, 150–153. *See also* National popularism
Burnham, Forbes, 28
Bustamante, Alexander, 99, 100, 283n; and Bedwardism, 77; and capitalist class, 113; compared to Norman Manley, 89, 90, 92; and constitutional reforms, 90–91; criticizes early PNP socialist policies, 104; and elections of 1949, 103; founds JLP, 54; leadership style of, 55–57, 284n; and race and color issues, 72, 93–94; and self-government, 87, 104, 108. *See also* Bustamante Industrial Trades Union; Jamaica Labour Party (JLP)
Bustamante Industrial Trades Union (BITU), 39, 54, 59, 99, 102

Campbell, A. S. (planter; politician), 130
Capitalism: and democratic socialism, 37; factors favoring restructuring of, xx–xxi, 28–29, 209–210; and intraclass competition, 137, 238; and national popularism, 26–27; peripheral and advanced, 39, 40; role of state in, 31, 39, 249; support for, by JLP, 54, 56, 104; support for, by PNP, 104–105. *See also* Social classes; *names of individual classes*
Capitalist class: anti-union sentiments of, 251–252; barriers to entry into, 78; and capital accumulation, 237–238; evolution of, 1939–1980, 42–43; intraclass tensions, 114, 137; power of "21 Families" in, 49, 113, 120, 125, 136; support for PNP in 1972, 160; systemic weakness of, 114, 137–138; unity, after 1938 rebellion, 113. *See also* Industrial fraction (of capitalist class); Merchant/commercial fraction (of capitalist class); Planters
Castro, Fidel, 187
Charles, Pearnel (JLP), 170, 262
Chile, 210, 228
Chin, Noel (director of JDB), 152
Clarke, Fred (planter), 114
Class fractions, defined, xxiii. *See also names of individual classes and class fractions*
Clientelism. *See* Patronage
Coke, B. B. (PNP), 172
Colonial government (in Jamaica), 61, 90, 100, 238, 283n; opposes industrialization, 61, 85, 144, 284n; and relationship to "coloreds," 140, 142; and support for planters, 114–115, 117, 144. *See also* Crown Colony government
Coloreds: anti-planter posture of, 141–142; and attitudes to ex-slaves, 142; defined, 281n; and education, 143, 144, 285n; and entrepreneurial activities, 143–144; and franchise in Crown Colony, 87; as instruments of interclass mediation, 139, 140, 143; post-abolition ascendance, 140–142. *See also* Middle class
Common good, 78; avoidance of class issues in, 95; avoidance of race and color in, 93–95, 162; capitalists' call for, 262; as defined in Manleyism, 84, 87; and liberal democracy, 86–91; limitations of PNP's version of, 91–98; in PNP after 1944 election, 103; and Rastafarianism, 166; redefined by 1960s radicals, 200–203; two versions of, 83–84. *See also* Manley, Norman; Manleyism; People's National Party (PNP)
Commonwealth Sugar Agreement, 117
Constitutional reforms. *See* Liberal democracy
Cooke, Howard F. (educator; PNP), 172, 214
Coombs, A.G.S. (labor organizer), 88
Coore, David (PNP), 214, 231–232, 279n

"Crash work programme," 260, 262
Crown Colony government, 79, 87, 116, 146, 281n; and coloreds, 143–146; opposition to industrialization, 116–117; opposition to merchants, 119; support for planters, 116, 144. *See also* Colonial government (in Jamaica); Morant Bay Rebellion
Cuba: Jamaican migrant workers in, 59; Michael Manley's visit to, 215–216; military training in, 234; and PNP foreign policy, 15–16, 77, 280n; and Rosalie Avenue Movement, 187. *See also* Brigade League ("Brigadistas")

DaCosta, Sir Alfred, 124, 127
Daily Gleaner. See Gleaner
deLisser, H. G. (*Gleaner* editor), 114, 120
Democratic socialism: and class relations, 36; compared to national popularism, 28, 32–34, 248, 265; conceptual criticisms of, 24–26; definitions of, 17–26; indicators of, 37–38; introduction of, in 1974, xx, 3; and Jamaican crisis, 3–5; and Manleyism, 18; and Marxism-Leninism, 22–23; reasons for decline of, 16; and Third Worldism, 9. *See also* Marxism; National popularism; People's National Party (PNP); Radicalism
Denham, Sir Edward (governor), 55
Dependency theory, 168; as critique of foreign capital, 6–8; defined, 279n. *See also* New World Group; People's National Party (PNP)
Domingo, W. A., 75
Duncan, D. K. (PNP), 21, 22, 23, 214, 219, 221, 290n

Economic nationalism, 6, 178–179, 185
Economic policy: and economic delinking, 210–214, 235; and "industrialization by invitation," 4. *See also* Dependency theory; Industrialization
Economy: and class disparities, 6; control of, 113, 120, 125, 136; decline of, in late 1970s, 50; and foreign exchange flow, 240; foreign ownership in, 5–6, 289n; and ganja trade, 273, 286n; importance of bauxite industry in, 10–11; after independence, 48–49; indicators of performance, 40–41, 125–126, 132–133, 217, 230, 244; informal sector in, 183, 222, 288n; state intervention in, 10, 35, 49–50, 217; by "21 Families," 49. *See also* Agriculture; Industrialization; *names of individual classes and class fractions*
Education: Alexander Bustamante's attitude toward, 56; and Garveyism, 68, 77; as mimesis, 79–80; and Norman Manley, 77, 185; and paternalism, 86; and PNP policies, 180, 214; and radical intellectuals, 198–203; and social mobility, 79, 144–145, 185–186, 189; and subordinate classes, 78–81. *See also* Middle class
Employment: creation, between 1973 and 1976, 243; and economic restructuring, 61–62; Employment Termination and Redundancy Payments, 255; seasonal, in Cuba, 59; in state sector, 147, 185; among urban youth, 193. *See also* Agriculture; Industrialization
Ethiopia, 204, 268
Ethiopian Federation Movement, 196
Evans, Peter, 249

Fabianism, 95
Fascism: and national popularism, 27. *See also* Rastafarianism
Fletcher, Richard (adviser to Michael Manley), 232
Foreign capital; anger toward, 6–9; and democratic socialism, 30; in Jamaican economy, 4–6; joint ventures with, 246. *See also* Bauxite industry; Dependency theory; Industrialization
Foreign lending, 214, 215. *See also* International Monetary Fund (IMF)
Foreign policy, Third World orientation of, in 1970s, 14–16. *See also* Manley, Michael; National popularism
Franchise: JLP failure to update electoral lists (1967), 201; between 1901 and 1935, 86; in 1944, 62, 91; post-

Franchise (*continued*)
abolition, 142; push for adult suffrage, 90. *See also* Common good

Garvey, Marcus, 59, 120; as advocate of self-government, 76; and education, 77; imprisonment of, 69–79; and People's Political Party (Garvey), 68–69, 70, 74. *See also* Garveyism; Rastafarianism
Garveyism, 67–72; and Bedwardism, 68, 69, 70; class support for, 71; and labor organizing in BITU, 72; and Manleyism, 69, 72, 76–77; and Rastafarianism, 72, 73. *See also* People's Political Party (Garvey)
Girvan, Norman, 8, 199
Glasspole, Sir Florizel (PNP; governor general), 172, 214, 219, 222
Gleaner, 8, 55, 65, 72, 75, 100, 102; anti-PNP stance of, in 1970s, 289n; and capitalist interests, 120, 282n; as conservative voice, 59, 114.
Government. *See* State
Gramsci, Antonio, 188
Grenada, 195–196
Guinea, 214, 268
Gun Court, 263
Guyana, 28
Gyles, J. P. (planter; JLP), 130

Hart, Richard (PNP left), 74, 100–101, 283n
Henriques, Owen K. (merchant-industrialist), 56, 124, 127, 131
Henry, the Reverend Claudius (164, 187, 189–190, 287n. *See also* Rosalie Avenue Movement
Henry, Ronald, 187–188, 189, 199
Hill, Ken (PNP left), 101, 103
Hill, Wilton (JLP), 173
Hinds, Robert (Rastafarian leader), 67
Howell, Leonard (Rastafarian leader), 67, 75
Hungary, 243

Ideology: contribution of Rastafarianism to, 161; and interpellations, xxi, 166–167; and national popularism, 166–175, 269; and relative autonomy of middle class, 198. *See also* Manleyism; National popularism; People's National Party (PNP); Rastafarianism
Independent Trade Union Action Council (ITAC), 182
Industrial Disputes Tribunal (IDT), 251, 254, 255
Industrial fraction (of capitalist class): and merchant/commercial fraction, 241–244; origins of, 131–135; preferential treatment for, under national popularism, 241. *See also* Industrialization
Industrialization: and colonial government's opposition to, 61, 85, 116–117, 144, 284n; and employment, 5–6, 181; and foreign capital, 6, 85; *Green Paper on Industrial Development Programme,* 13, 249; high import content of, 5, 213, 240; and import substitution, 132, 241; indicators of, 117, 131; and intra-capitalist competition, 130; Jamaica Industrial Development Corporation (JIDC), 48, 105, 185, 246; merchant/commercial influences in, 133–135; and 1938 labor rebellion, 126, 130; obstacles to, 272–273; and PNP support of, 13, 105–106, 239, 241; push toward, 62, 129–133; supportive legislation, 105–106, 116–117, 131, 132. *See also* Economy; Industrial fraction (of capitalist class)
Intermediate classes, evolution of, 1939–1980, 44–45. *See also* Middle class
International Bauxite Association, 11–12
International Monetary Fund (IMF), 14, 217, 218–220, 225, 226, 227, 231, 232, 251
International Sugar Agreement (1937), 117
Iran, 210
Isaacs, Allan (PNP), 23, 214
Isaacs, William (PNP), 220
Isaacs, Wills O. (PNP), 102, 214

Jamaica Agricultural Development Corporation (JADC), 185

Jamaica Agricultural Society (JAS), 103, 121
Jamaica Broadcasting Corporation (JBC), 232
Jamaica Chamber of Commerce, 124, 224, 225, 242, 249, 253, 262, 272, 291n
Jamaica Democratic Party, 56, 113, 241
Jamaica Development Bank (JDB), 134–135, 151–152, 245–246
Jamaica Imperial Association (JIA), 81, 97, 114, 117, 121, 127
Jamaica Industrial Development Corporation (JIDC), 48, 105, 185, 246
Jamaica Labour Party (JLP), 38–39; attacks PNP socialism, 218; as capitalist party, 56; and capitalists' anti-union pressure, 251; constitution of 1951, 104, 108; and economic nationalism, 185; formation of, 54; interprets workers' demands in 1938 rebellion, 60–61, 62; leftist tendency in, 178; and 1944 elections, 56, 62, 94; and 1949 elections, 103; and PNP, 108, 222; and radicalism, 191–192; sticks to tradition, 201–203. *See also* Bustamante, Alexander
Jamaica Labour Weekly (working class newspaper), 59, 74
Jamaica Manufacturers' Association (JMA), 134, 241–242, 243, 272, 285n
Jamaican Council for Civil Liberties, 94
Jamaican Employers' Federation (JEF), 251–252, 253
Jamaica Progressive League (JPL), 57, 93, 97, 146, 281n
Jamaica Union of Teachers, 97
Jamaica Welfare Limited, 84–85, 102. *See also* Manleyism
Jamaica Workers' and Tradesmen's Union, 62
James, C.L.R., 153
Jewish population, 116; appointed to PNP administrative posts, 136–137; in 1850s Assembly, 141; and *Gleaner* newspaper, 120; representation among merchants, 119; and "21 Families," 136.
JLP. *See* Jamaica Labour Party (JLP)
Johnson, Millard (founder of PPP), 188; and JLP, 190
Kenya, 28
Kingston Federation of Citizens' Associations (KFCA), 97
Kingston St. Andrew Corporation (KSAC), 101
Kirkwood, Sir Robert, 8, 55, 129, 285n, 286n
Kitching, Gavin, 267
Korea, 40

Labor rebellion (1938), 21, 53, 81, 126; as beginnings of modern period, 40; and consciousness of subordinate classes, 58–63; and policies of 1940's and 1950s, 89. *See also* Colonial government (in Jamaica); Subordinate classes
Labor Relations and Industrial Disputes Act (LRIDA), 250–251. *See also* Working class
Labor unions: and links to political parties, 39; push for democratization of, 182–183, 288n; worker disaffection with, 252; and worker participation policies, 254–255, 292n; and working class opportunism, 182. *See also* Bustamante Industrial Trades Union (BITU); National Workers' Union; Trades Union Congress (TUC); University and Allied Workers' Union (UAWU)
Lenin, V. I., 100
Lewis, Rupert, 188
Liberal democracy, 281n; and franchise, 86, 107–108; and Garveyism, 68; in Marxism and Manleyism, 98, 101; and PNP philosophy, 87; and post-1938 constitutional reforms, 90, 106–107; and Rastafarianism, 170, 174; and socialism, xxii–xxiii; and subordinate class consciousness, 87–89, 169–170; unfulfilled promises of, 170–171, 202. *See also* Common good; Manleyism
Liberalism, 99; and voluntarism, 270. *See also* Common good; Manleyism, People's National Party (PNP)
Lightbourne, Robert (JLP), 134, 185
Lumpen proletariat: activities of, 158, 184, 260–261, 263–264; consciousness of, 183–184, 289n; defined, 281n; growth

Lumpen proletariat (*continued*) of, 46–47, 159, 260, 263; and PNP 1970s policies, 260–264, 292n; and political parties, 159, 160, 262; and street crime, 159; and threat to stability, 30, 48, 159. *See also* Rastafarianism; Subordinate classes

McLarty, R. W., 97
McNeil, Eustace (Garveyite), 68, 282n
McNeil, Kenneth (PNP), 151, 221
McNeil, Roy A., (planter; JLP), 130
Manley, Michael, 15, 256; blames external forces for failure, 211; and economic nationalism, 7; and PNP left, 214, 215, 217; and *The Politics of Change,* 230, 234; and Rastafarian symbols, 164, 172; and redefinition of common good, 203; trip to Cuba, 215–216; voluntaristic appeals by, 234, 271. *See also* Democratic socialism; People's National Party (PNP)
Manley, Norman: and Alexander Bustamante, 89, 90, 92, 104; and capitalist class, 113; and economic nationalism, 178–179; as educated model for subordinate classes, 80; farewell speech of, 178, 203; and founding of PNP, 15, 54, 58; and Jamaica Welfare Limited, 84, 85; and philosophy of common good, 83–84; and PNP Marxists, 101–102; testimony to Moyne Commission, 105. *See also* Manleyism; People's National Party (PNP)
Manleyism, 57–58; and Bedwardism, 67, 75–76; continuing appeal of, 189; defined, 18; and Garveyism, 72, 76–77; ideological tensions in, 105; and nationalism, 95; and Rastafarianism, 94–95; rationalist tendencies of, 98–99, 108; and social classes, 19, 99; and subordinate classes, 89. *See also* Common good; People's National Party (PNP); Race
Marley, Bob (reggae singer), 196, 197
Marx, Karl, 100, 128, 130, 198
Marxism, 18, 168, 184, 186, 267; in Cuba, 59; and democratic socialism, 20–22; in early PNP, 98–102; and radical intellectuals, 199–200. *See also* Radicalism; Workers' Liberation League (WLL); Workers' Party of Jamaica (WPJ)
Matalon, Aaron (industrialist), 134
Matalon, Eli (PNP), 137
Matalon, Mayer (industrialist, bauxite negotiator), 137, 285n
Meeks, Winston (Chamber of Commerce president), 223–224
Merchant/commercial fraction (of capitalist class): between abolition and Crown Colony government, 116, 119; and anti-merchant bias, 118–119; ascendance of, 123–125; continued dominance of, in national popularism, 241, 243–244; during Crown Colony government, 116–126; excluded from national popularist agenda, 238–241; high profit margins of, 135; and industrial fraction, 129–134, 241–244; Jewish representation in, 116; in 1940s, 108; during slavery, 115; structural factors in, 126, 128
Mexico, 212
Middle class: 22, 48, 64, 153; Afro-Saxon traits of, 172–174; attitudes toward subordinate classes, 97; black and brown divisions in, 93; consumerism of, 147–148, 223; and educated blacks, 146; growth of, in modern period, 147; and nationalism, 146; as political functionaries, 146; and pre-1970s conditions, 184–186; radical intellectual youth in, 198–203; and Rastafarians, 164; in self-government movement, 97, 146; in state apparatuses, 147; and support for democratic socialism, 154, 216, 223–226; takes flight, 216. *See also* Bureaucratic/entrepreneurial fraction (of middle class); Coloreds; People's National Party (PNP)
Migration, 256: external, during 1970s, 216, 226, 286n; internal, 41, 157–158, 284n
Miliband, Ralph, 226, 227
Morant Bay Rebellion, 66, 92; and Crown Colony government, 142–43
Moyne Commission, 53, 56, 86, 90, 93,

145; Norman Manley's testimony to, 105; report of, 59, 61
Munn, Keble (PNP), 233
Munroe, Trevor, 58, 148, 200; as critic of self-government movement, 107; in politics, 198
Myers, Fred L. (merchant), 114, 121, 124

National Housing Trust, 214
National income: and farmers, 125–126, 259; in 1832 and 1850, 116; in 1930, 126
National Planning Agency, 218, 224
National popularism: and capitalist ideology, 272–273; compared to democratic socialism, xx, 28, 31–33, 36, 248; defined, xx, 29, 36–37; economic factors, in rise of, 179–186; external antagonism to, 211, 218–219, 226; and fascism, 27–28; indicators of, 37–38; and movements from below, 268–269; political factors in rise of, 186–187, 204–205; and Rastafarian ideology, 166–174; and resource distribution, 243, 248, 250, 272; and restructuring of capitalism, 177, 179, 203–205, 210, 244–249, 265; and social classes, 29–36; and the state, 35, 204; as Third World phenomenon, 267–268; variants of, 280n; and voluntarism, 270–271. *See also* People's National Party (PNP); *names of individual classes and class fractions*
National Reform Association (NRA), 97
National Workers' Union (NWU), 39, 99; and PNP, 101
Native Defenders' Committee, 72
Native Industries Bill (1930), 115, 131
Negro Workers' Educational League, 62, 93
Nettleford, Rex, 186, 196
New Negro Voice (UNIA newspaper), 75, 93
New World Group: and critique of foreign capital, 7–9; in PNP government, 224; radical intellectuals in, 198. *See also* Black Power Movement; Marxism; Radicalism
New World Quarterly, 198
Nyerere, Julius, 13, 18

O'Gilvie, Edward (PNP), 263
Organization of Petroleum Exporting Countries (OPEC), 9, 12, 15

Paternalism, xxi; political implications of, 86–87
Patronage, 281n; and class interests, 269–270; and national popularism, 21, 203, 262–263; in organized labor, 156; and People's Political Party (Johnson), 191; and political parties, 159–160, 184, 268–269, 274; in rural areas, 157
Patterson, Orlando, 88
Patterson, P. J. (PNP), 170, 214
Peasantry: displaced by sugar industry, 125, 129; evolution of, 1939–1980, 46–47; income of, 125–126, 259; and Manleyism, 85; mixture of class tendencies in, 157; and PNP 1970s agrarian policies, 257–258, 259; and post-Emancipation franchise, 142. *See also* Agriculture; Sugar industry
People's National Party (PNP), 38–39, 54, 69; and armed forces, 233–234; and capitalist class, 216, 217, 232; conservatives and moderates, 214, 216, 218, 219, 222, 231; and dependency theory, 6; and economic decline, 227; and economic nationalism, 178–189; elections (1949), 101, 102–103; elections (1955, 1959), 106; elections (1972), 177, 179–180, 186–187, 223; elections (1976), 20, 21, 216, 223, 250; elections (1986), 273; *Emergency Production Plan* (1977), 242–243, 255; ideology of, in early period, 37–38, 57–58, 77; and IMF, 217, 218–220, 227; intra-party conflicts, in 1970s, 214–223, 231; leftists (1940s), 96, 98–102; leftists (1970s), 214, 215, 217, 219, 221, 228, 234, 267, 288n; *People's Plan* (1976), 217, 242, 291n; *Plan for Progress* (1949), 105; policies in 1940s, 90, 104; policies in 1950s, 106; policies in

People's National Party (PNP) (*continued*) 1970s, 10–13, 22, 104, 166, 214; *Principles and Objectives* (1979), xxiii, 23, 215; *Programme for Action Now* (1945), 103; and race and color, 93–95, 96, 121, 216; and radicalism, xxii, 20–24, 104–106, 108–109, 191, 200–201; and socialist change, 226–227; *Statement of Policy* (1940), 84, 100, 103, 108; unable to control state organs, 228–234; Youth Organization (PNPYO), 22, 218. *See also* Agriculture; Bauxite industry; Common good; Democratic socialism; Economic policy; Industrialization; Manleyism; Marxism; Race; Rastafarianism; Worker participation policy (PNP)
People's Political Party (Garvey), 68–69, 70, 72, 74
People's Political Party (Millard Johnson), 188; failure of, 190–191. *See also* Garveyism; Race
Peronism, 27
Phillips, Peter, 133
Pioneer Industries (Encouragement) Act, 41, 72, 105, 132
Plain Talk, 59
Planters: between abolition and Crown Colony government, 115, 141; anti-merchant and anti-semitic bias of, 118; and colonial state, 118; during Crown Colony government, 114–115, 116–118, 119–123; decline of, after 1938, 41, 108, 115, 117, 127–130, 136; Planters' and Farmers' Party, 129–130; political power of, 117–118; and resistance to industrialization, 115; support organizations, 117, 121, 127
PNP. *See* People's National Party
Political parties: and labor unions, 38–39; structure of, 220. *See also* Jamaica Labour Party (JLP); People's National Party (PNP); Workers' Party of Jamaica (WPJ)
Poor Man's Improvement Land Settlement and Labour Association, 60
Post, Ken, 53, 71, 74, 122, 127
Poulantzas, Nicos, 121
Private Sector Organization of Jamaica (PSOJ), 232, 241, 253; Task Force on Agriculture of, 257
Project Land Lease, 13, 214, 257–260
Public Opinion, 74
Puerto Rico, 106

Race: and bureaucratic/entrepreneurial fraction, 170; consciousness, 87, 203, 204–205, 281n; in 1940s and 1970s, 93; and political parties, 92–94, 190, 192, 195–196, 216; and social hierarchy, 115. *See also* Bedwardism; Coloreds; Garveyism; Marxism; Middle class; Rastafarianism; Subordinate classes
Radicalism: convergence toward, 197; expressed through music, 201, 288n; and Rastafarianism, 196–197; subordinate class expressions of, 187–188, 193–194; in Third World change, 267; and young intellectuals, 198–203, 204. *See also* Black Power Movement; Marxism; People's National Party (PNP); leftists (1940s); People's National Party (PNP): leftists (1970s); People's National Party (PNP): Youth Organization (PNPYO); Rastafarianism
Rastafarianism, 18, 30; and Afro-Christianity, 73; attitudes toward, 94–95, 162, 164; and Black Power Movement, 195–196; Coral Gardens incident, 163; and Ethiopianism, 74, 161, 196; gains recognition by state, 196; in Grenada, 195–196; growth of, 160–161, 164–165; and Haile Selassie, 72, 73, 163, 164; ideological impact of, 269; and JLP, 163; and Marcus Garvey, 72–73; and Marxism, 195; middle class members of, 186; music of, 165, 186; and national popularist ideology, 166–175; and PNP, 74–75, 91, 94, 163–164; and radicalism, 196–197; as redefined common good, 166; and Rosalie Avenue Movement, 187–190; and subordinate class consciousness, 88, 160–162, 186; and traditional religion, 196–197;

Twelve Tribes, 161; University report on, 163; and young intellectuals, 200
Reggae music, 164, 165, 186, 193, 196; and political slogans, 172, 179–180. *See also* Marley, Bob (reggae singer)
Reid, Stanley, 247
Religion: and established churches, 63, 196; importance of, in Jamaican life, 66; paternalistic tendencies of, 86; use of, by political parties, 103. *See also* Afro-Christianity; Bedwardism; Garveyism; Rastafarianism; Rosalie Avenue Movement
Rema, 264
Revolution, 192; and the state, 226. *See also* Radicalism
Richards, Sir Arthur (governor), 61, 101, 283n
Ricketts, Mark, 274, 275
Rodney, Walter, 191–192, 197, 198, 201; and Rodney Affair, 201–202
Rosalie Avenue Movement, 187–189
Rose, Dexter, 151
Royal Empire Society (1938), 117
Rudie boys, 193–195, 201, 288n; and middle class youth, 194
Rumble, Robert, 60

Sandinista revolution, 26, 36
Seaga, Edward (JLP), 159–160, 163, 178, 273
Selassie, Haile (emperor of Ethiopia), 72–73, 74, 163, 164. *See also* Rastafarianism
Senghor, Leopold, 24
Seymour-Seymour, George (politician/merchant), 123–124, 282n
Sharpe, Sam, 66
Shearer, Hugh (JLP), 101, 164, 185, 191–192, 210; launches anti-communist policy, 202
Small, Hugh, 214, 222
Smith, J.A.G., 80, 90, 145–146
Social classes: and concepts of dominance and hegemony, 272; evolution of, 1939–1980, 42–47; and interclass tensions, 118–121; in Jamaican class structure (1973), 49; and 1976 election, 21–22; in PNP definition of democratic socialism, 19–20; and PNP redistributionist policies, 50; and the state, 29, 38–50. *See also names of individual classes*
Socialism, 25–26, 234; many paths to, 26; PNP's position on, 95–96; and the state, 25–26; "undeveloped," 268; and weakness of socialist solidarity, 213. *See also* Democratic socialism; Marxism
Soviet Union, 214
Spaulding, Anthony (PNP), 173, 214, 222, 233
Special Employment Programmes, 214
Squatters Laws, 157
Standard Fruit Company, 97
State: as agent of revolution, 226; as capitalist entity, 189–190; central role of, xxi–xxii, 38–39; intervention in economy, 30, 35, 240, 244–249, 274; and maintenance of status quo, 193; and mediation of class conflict, 29, 31, 204, 250; peripheral capitalist, 39–40, 268; and socialism, 25–26; and state power, 228–235, 243, 271–272; and support for planters, 116. *See also* Crown Colony government; Democratic socialism; National popularism.
State Trading Corporation (STC), 151, 232, 240, 241, 244
Stephens, Evelyne Huber, 219, 227
Stephens, John, 219, 227
Stock exchange (Jamaica), 136
Stone, Carl, 192, 205, 209, 227
Strategic fraction (of middle class); aligns economic power to political power, 167–168. *See also* Bureaucratic/entrepreneurial fraction (of middle class)
Subordinate classes: consciousness of, 58–63, 189; criminal activities among, 159–160; dislocation of, 129; effect of IMF's policies on, 222; evolution of, 1939–1980, 46–47; and ideological convergence in 1970s, 169–170; importance of education for, 48, 78–81, 173; lack of organization of, 268–269; and Manleyism, 92, 96; and Marxism-

Subordinate classes (*continued*)
Leninism, 22–23; and PNP policies, in 1970s, 249–250; and pre-1970s economic conditions, 180–184; and pre-1970s political conditions, 187–197; and race and color, 92–93, 96–97, 168; and rural-to-urban migration, 41, 48, 157–158; and self-government movement, 107–108; weakness of, 39. *See also* Bedwardism; Lumpen proletariat; Peasantry; Rastafarianism; Working class
Sub-proletariat. *See* Lumpen proletariat
Sugar industry: in decline, 117, 128; employment in, 290n; International Sugar Agreement of 1937, 117; merchant ownership of, 120; structural requirements of, 125; support for, in 1920s, 121–122; and worker cooperatives, 249, 251, 252, 253. *See also* Planters
Sugar Manufacturers' Association (SMA), 8, 121

Taiwan, 40
Tate and Lyle, 8, 41, 55, 117
Third Worldism, 9–10, 14–15, 26, 168
Thomas, Clive, 8, 235
Trades Union Congress (TUC), 39, 99, 101, 102
Trade unions. *See* Labor unions

Uganda, 228
United Fruit Company, 41, 122
United States, 25; attitudes of Jamaican capitalists toward, 209; and economic delinking, 210–211; political and economic agenda of, 38; as threat to political autonomy, xxi, 202
Universal Negro Improvement Association (UNIA), 67–68, 91, 93. *See also* Garveyism
University and Allied Workers' Union (UAWU), 39, 183, 198
University of West Indies (UWI), 163, 198, 217; radicalism in, 186, 192–193; as University College of the West Indies, 189
Urban Development Corporation (UDC), 246

Weber, Max, 66
West India Committee, 117
West India Sugar Company, 55
Williams, R. Danny (PNP), 220
Witter, Michael, 227, 231
Worker and Peasant, 93
Worker participation policy (PNP), 214, 250–256
Workers' Liberation League (WLL). *See* Workers' Party of Jamaica (WPJ)
Workers' Party of Jamaica (WPJ), 22, 39, 102, 198; analysis of working class consciousness, 156, 157, 182, 234, 273; and democratic socialism, 21
Working class: antagonism to bureaucratic-entrepreneurial fraction, 225; casual labor in, 157; consciousness of, 155–157, 181–182; and dissatisfaction with labor unions, 251–252; evolution of, 1939–1980, 46–47; Labour Relations and Industrial Disputes Act (1975), 251, 252, 254–255; PNP policies in 1970s for, 249–250; political apathy of, 255–256; relations with capitalist class, 251–253, 255; strike activity of, 182–183; and worker participation, 214, 250–251, 252, 253–254
Workman, The, 98
World Bank, 153, 229, 232; developmental formulas of, 4–5; development committee of, 14; 1952 recommendations, 130
Worthy Park Sugar Estates, 55, 114